Genetic Resources and Adaptive Management of Conifers in a Changing World

Genetic Resources and Adaptive Management of Conifers in a Changing World

Editor

Csaba Mátyás

MDPI • Basel • Beijing • Wuhan • Barcelona • Belgrade • Manchester • Tokyo • Cluj • Tianjin

Editor
Csaba Mátyás
University of Sopron
Hungary

Editorial Office
MDPI
St. Alban-Anlage 66
4052 Basel, Switzerland

This is a reprint of articles from the Special Issue published online in the open access journal *Forests* (ISSN 1999-4907) (available at: https://www.mdpi.com/journal/forests/special_issues/genetic_resources_conifer).

For citation purposes, cite each article independently as indicated on the article page online and as indicated below:

LastName, A.A.; LastName, B.B.; LastName, C.C. Article Title. *Journal Name* **Year**, *Volume Number*, Page Range.

ISBN 978-3-0365-2281-4 (Hbk)
ISBN 978-3-0365-2282-1 (PDF)

Contents

About the Editor

Csaba Mátyás Forest Engineer, graduate of University of Technology, Dresden, Germany, 1967. Researcher of the Forest Research Institute, Hungary until 1985. Professor at the University of Sopron, Hungary, from 1986. Visiting professor, Toronto University, Canada. Fulbright scholar, Institute of Forest Genetics, California. Fellow, Hungarian Academy of Sciences since 2004. Coordinator, forest genetic research for IUFRO. Organised 4th Consultation of IUFRO on Forest Genetic Resources in Beijing, China, 1998. Co-founder, EUFORGEN network, 2000. Founder, NEESPI Regional Focus Research Center for Nonboreal Eastern Europe, 2008. Main extension project: "Agrárklíma" Decision Support System for Forestry and Agriculture, 2018. Received IUFRO's medal for Distinguished Services. Professor emeritus since 2013.

forests

MDPI

Editorial

Genetic Resources and Adaptive Management of Conifers in a Changing World

Csaba Mátyás

NEESPI Regional Focus Research Center for Nonboreal Eastern Europe, Institute of Environmental and Geosciences, Faculty of Forestry, University of Sopron, P.O. Box 132, H-9401 Sopron, Hungary; matyas.csaba@uni-sopron.hu

Human activities have widely exploited and transformed the resources of coniferous species and ecosystems. The global threat of climate change further amplifies their vulnerability. There is an urgent need to support their resilience and to conserve genetic resources in order to maintain their adaptive potential.

Climatic change causes a mismatch between tree populations in the sites they occupy in the future and the climate to which they have adapted in the past. The maintenance of productivity, carbon sequestration, ecological and societal services require stable populations and ecosystems. The response of conifer species to worsening climatic scenarios is still insufficiently investigated, particularly at the vulnerable trailing (xeric) range limits. Management strategies require more information on their adaptive and evolutionary capacity. Provenance tests are obvious sources, providing valuable details on the performance of populations planted in diverse environments. Unsurprisingly, the majority of papers in this Special Issue utilize their results, pondering the odds of matching seed sources with projected climates, to increase resilience and prevent extinction.

The papers have disparate geographical and climatic backgrounds, from Central and Northwestern Europe, the Mediterranean, Russia, China, and North and Central America. They represent very diverse approaches, which was the aim of the guest editor.

Selective effects of extreme habitats have also maintained high genetic and phenotypic diversity among isolated, peripheral populations [1]. Due to the high survival risks, extreme conditions require careful management planning [2]. Detailed studies clearly indicate that differences in stability are genetically controlled, but local site interactions may be unpredictable [3]. Soil conditions can mask climate effects, and a separated analysis is proposed to better interpret their interaction [4].

The unique potential of provenance tests is illustrated by the results of trial networks; response projections are less dramatic than those provided by inventory data analyses [5]. In mountainous terrain, shifting trees upwards in altitude to compensate for climatic warming appears to be a feasible management action [6]. When selecting the proper provenance for assisted migration, the choice of a single population should be preferred over admixtures and composite methods [7]. Seed weight increases with the warming climate, improving seedling survival. This trait should also be considered, especially on extreme sites [8]. Some further queries for assisted migration are the transfer of populations outside their current natural distribution, selective breeding for disease-resistant trees, differentiation of measures according to the position in the range of the species, or the invasive potential of transferred species, etc. Research into these issues requires multiple field trials in disparate climates that contain populations from a representative range of habitats [9,10].

Funding: This research received no external funding.

Informed Consent Statement: Not applicable.

Conflicts of Interest: The author declares no conflict of interest.

Citation: Mátyás, C. Genetic Resources and Adaptive Management of Conifers in a Changing World. *Forests* **2021**, *12*, 1213. https://doi.org/10.3390/f12091213

Received: 31 August 2021
Accepted: 2 September 2021
Published: 6 September 2021

References

1. Palla, B.; Ladányi, M.; Cseke, K.; Buczkó, K.; Höhn, M. Wood anatomical traits reveal different structure of peat bog and lowland populations of *Pinus sylvestris* L. in the Carpathian region. *Forests* **2021**, *12*, 494. [CrossRef]
2. Jandl, R.; Kindermann, G.; Foldal, C.; Schüler, S.; Bouissou, C. Early Performance of Tree Species in a Mountain Reforestation Experiment. *Forests* **2021**, *12*, 256. [CrossRef]
3. Skrøppa, T.; Steffenrem, A. Performance and Phenotypic Stability of Norway Spruce Provenances, Families and Clones Growing Under Diverse Climatic Conditions in Four Nordic Countries. *Forests* **2021**, *12*, 230. [CrossRef]
4. Feng, L.; Sun, J.; Shi, Y.; Wang, G.; Wang, T. Predicting Suitable Habitats of *Camptotheca acuminata* Considering Both Climatic and Soil Variables. *Forests* **2021**, *11*, 891. [CrossRef]
5. Mátyás, C.; Beran, F.; Dostál, J.; Čáp, J.; Fulín, M.; Vejpustková, M.; Božič, G.; Balázs, P.; Frýdl, J. Surprising Drought Tolerance of Fir *(Abies)* Species Between Past Climatic Adaptation and Future Projections Reveals New Chances for Adaptive Forest Management. *Forests* **2021**, *12*, 821. [CrossRef]
6. Cruzado-Vargas, A.L.; Blanco-García, A.; Lindig-Cisneros, R.; Gómez-Romero, M.; Lopez-Toledo, L.; de la Barrera, E.; Sáenz-Romero, C. Reciprocal Common Garden Altitudinal Transplants Reveal Potential Negative Impacts of Climate Change on *Abies religiosa* Populations in the Monarch Butterfly Biosphere Reserve Overwintering Sites. *Forests* **2021**, *12*, 69. [CrossRef]
7. Notivol, E.; Santos del Blanco, L.; Chambel, R.; Climent, J.; Alía, R. Seed Sourcing Strategies Considering Climate Change Forecasts: A Practical Test in Scots Pine. *Forests* **2021**, *11*, 1222. [CrossRef]
8. Parfenova, E.I.; Kuzmina, N.A.; Kuzmin, S.R.; Tchebakova, N.M. Climate Warming Impacts on Distributions of Scots Pine (*Pinus sylvestris* L.) Seed Zones and Seed Mass Across Russia in the 21st Century. *Forests* **2021**, *12*, 1097. [CrossRef]
9. Sáenz-Romero, C.; O'Neill, G.; Aitken, S.N.; Lindig-Cisneros, R. Assisted Migration Field Tests in Canada and Mexico: Lessons, Limitations, and Challenges. *Forests* **2021**, *12*, 9. [CrossRef]
10. Rehfeldt, G.E.; Warwell, M.V.; Monserud, R.A. Species, Climatypes, Climate Change, and Forest Health: A Conversion of Science to Practice for Inland Northwest (USA) Forests. *Forests* **2021**, *11*, 1237. [CrossRef]

forests

Article

Wood Anatomical Traits Reveal Different Structure of Peat Bog and Lowland Populations of *Pinus sylvestris* L. in the Carpathian Region

Balázs Palla [1], Márta Ladányi [2], Klára Cseke [3], Krisztina Buczkó [4] and Mária Höhn [1,*]

1 Department of Botany, Institute of Agronomy, Hungarian University of Agriculture and Life Sciences, H-1118 Budapest, Hungary; palla.balazs@hallgato.uni-szie.hu
2 Department of Applied Statistics, Institute of Mathematics and Basic Science, Hungarian University of Agriculture and Life Sciences, H-1118 Budapest, Hungary; Ladanyi.Marta@uni-mate.hu
3 Department of Tree Breeding, Forest Research Institute, H-9600 Sárvár, Hungary; erti.dnalab@gmail.com
4 Centre for Ecological Research, Danube Research Institute, H-1113 Budapest, Hungary; buczko.krisztina@ecolres.hu
* Correspondence: Hohn.Maria@uni-mate.hu

Abstract: Ecologically extreme habitats at a species' distribution edges bear significance for biota under adverse climatic conditions and climate change. Range-edge populations adjust their functional traits to the special local ecological conditions, leading to increased intraspecific variability in their morpho-anatomical structure and, consequently, favor population survival in the absence of competitors. On the basis of wood anatomical traits, such as tracheid lumen area (CA), cell wall thickness (CWTrad), cell diameter-to-radial cell wall thickness ratio (CD/CWT), and the number of tracheids in the radial tracheid files (TNo), we investigated the xylem adjustment of *Pinus sylvestris* L. populations from six ecologically extreme habitats from the Eastern Carpathians and the Pannonian Basin. Results indicated significant differences among all studied sites in case of all wood anatomical traits as signs of the local adaptation of trees. Peat bog populations adapted their wood anatomical traits to the generally hydric, cool and anaerobic conditions of the peat bogs, exhibiting smaller CA and proportionally thick CWTrad to ensure the hydraulic safety of the stem, whereas, on the lowland site, trees were characterized by a more effective water-conducting system, developing larger CA with relatively thin CWTrad with lower carbon-per-conduit-costs at the expense of higher vulnerability to cavitation. Radial tree ring growth and TNo also differed markedly among sites, following the temperature and groundwater constraints of the habitats. Wood anatomical variability among tree rings and the corresponding short-term climate response of populations differed from the adaptive responses of the trees to the ecological characteristics of the habitat. In addition to the different phylogeographic origin evidenced in former studies, phenotypic differentiation by the habitat type of the studied populations linked to the variance in morpho-anatomical traits have contributed to the survival of the peripheral Scots pine populations at the species' range margins.

Keywords: Scots pine; adaptation; climate change; wood anatomy; tracheidogram; traits

Citation: Palla, B.; Ladányi, M.; Cseke, K.; Buczkó, K.; Höhn, M. Wood Anatomical Traits Reveal Different Structure of Peat Bog and Lowland Populations of *Pinus sylvestris* L. in the Carpathian Region. *Forests* **2021**, *12*, 494. https://doi.org/10.3390/f12040494

Academic Editor: Csaba Mátyás

Received: 17 March 2021
Accepted: 11 April 2021
Published: 16 April 2021

1. Introduction

Ongoing climate change and the resulting interspecific competition lead to shifts in the geographic distribution of natural forest communities [1]. Cold tolerant species are predicted to expand their area of distribution towards the cooler regions of the poleward latitudes or to upslope elevations, causing local extinctions and population contractions at the southern periphery. Mostly, at the warmer range margins, populations of the cold tolerant species are expected to survive in small remnant habitats called refugia, where local conditions favor their persistence [2–4]. The term "refugia" originally referred to restricted full-glacial locations of temperate and high-latitude taxa [5]. Species' survival in refuge habitats are receiving increasing attention today, as they serve as safe havens to which biota can retreat and persist under adverse climatic conditions [6]. However, sites with special local ecological characteristics where retreating populations are able to survive are also considered as being cryptic refugia [5], and such habitats are now particularly important for being identified and conserved, as they mitigate global biodiversity loss [6,7]. Species' migration and the survival of populations due to in situ adaptive differentiation and selection within refugia strongly determine the properties of remnant communities [8,9]. When populations are at a species' range edge, adaptive processes strongly depend on genetically coded functional traits, affecting fitness, growth, reproduction and survival [10,11]. Adaptation of species and populations to climate constraints may lead to increased intraspecific variability imprinted in the morpho-anatomical traits of populations. Wood anatomical characteristics of tree species are among the most important indicators, as trees have to adjust their xylem to the ecological conditions of the habitats, including water regime and seasonal and interannual climate variations [12].

Pinus sylvestris L. is one of the most important forest tree species of the Eurasian taiga. Its distribution spans more than 14,000 km, extending from the Iberian Peninsula towards the Siberian plain, reaching its easternmost distribution edge at the Okhotski Sea [13,14]. At the southern limit of its distribution range where, during the Pleistocene, it was more widespread, it is represented only by fragmented, island-like populations. In these regions following the Last Glacial Maximum, the warming of the climate and autogenic succession favored broadleaved competitors, forcing the Scots pine southern limit populations to fragmentation, long-term isolation and, finally, to genetic differentiation [15–17]. Isolation in several southern European refugia is probably responsible for the genetic diversity of modern-day populations, and for the high ecological plasticity, with which the species was able to survive in a multitude of environmental conditions, plant communities and diverse habitat types [18,19]. Such fragmented, peripheral populations are found also in East-Central Europe, in the Carpathians, growing in peatbogs with elevated groundwater table, or sustaining on sunny rocky outcrops with shallow soil layer. Small populations have also survived on the lowland area of the Pannonian Plain, where populations persist on calcareous sandy substrates with low nutrient content [20–22]. Despite the ongoing fragmentation and isolation, molecular genetic studies of these peripheral populations have revealed high population genetic diversity. Moreover, former molecular studies could detect two main genetic lineages that originate from two divergent glacial refugia; one extending from the Western Carpathians towards the Eastern Alps and another in the Eastern Carpathians [16,23]. Aside from their peculiar spatial genetic structure and diversity, Carpathian populations exhibit considerable phenotypic variation in both morphological and anatomical traits, most probably as signs of local adaptation to different habitat types [24–26].

Tree rings are among the most important traits affecting individual tree fitness, as they are the main contributors to the radial growth performance of trees. Tree ring width and characteristics depend on the site conditions and interannual climate variations that are discussed in many dendrochronological and densitometry studies [27–29]. Furthermore, the fundament of wood increment, wood anatomy, may offer insights into ecological adaptation through diverse anatomical parameters and fine-scale time resolutions. Studies on central temperate and northern boreal Scots pine populations have revealed long-term wood cell structure chronologies influenced by inter- and intra-annual climatic variation [30–32]. Traits such as cell area (CA), cell diameter (CD), cell wall thickness (CWT), and cell number (Tno) have shown correlations with the current year climate and also depend on the climate variables of the preceding year. The above-listed parameters are also strongly affected by the site conditions [33,34]. Scots pine tracheids respond to interannual and seasonal droughts in the southern range periphery [35,36], to fluctuations in temperature and precipitation evidenced in the Mediterranean region [37,38], or to continental climate variations [39] at different altitudes [40,41] and soil moisture conditions [42]. These plastic responses are either indicators of direct wood structure adaptation to extreme climatic events (e.g., [42]), or they could be related to the optimized carbon allocation and hydraulic conditions of the trees at such circumstances (e.g., [37]). Beside altitude [41], latitudinal position [43], climatic zone [44] on such rear edge locations, Scots pine may adjust its anatomical features to special local site conditions, such as aridity [42,45], constant drought, or cessation of former irrigation [46,47]; these are usually intertwined with processes perceived in interannual climate variations, as mentioned before. Moreover, predicted global warming is expected to have further strong selective pressure, especially on these rear edge populations [48].

Study results referring to wood anatomical characteristics have been reported from a wide range of locations and habitat types at the southern periphery of Scots pine area, namely the Iberian Peninsula, the Swiss Alps, the Vitosha, and South Siberia. However, the specific migration history of the Carpathian and Pannonian populations, their expansion and retreat to different refugia, beside the "geographic and genetic" marks, has left imprints in the wood of anatomical traits of populations strongly affected by natural selection acting over extended periods of time.

In this study, we aimed to (i) reveal wood anatomical trait variation of refugial Scots pine populations from the edaphically extreme habitats of the Carpathians and the Pannonian Basin; (ii) characterize the level of adaptive differentiation in these fragmented, peripheral populations; and (iii) outline seasonal/interannual climate variations reflected by the measured anatomical parameters. The observed phenotypic differentiation sheds light on the adaptive strategy of this long-living woody tree species to extreme habitats facing climate change.

2. Materials and Methods

2.1. Study Sites, Sampling and Tree Ring Analyses

Six natural populations of *Pinus sylvestris* L. were sampled from the Carpathians and the Pannonian Basin in 2019 and 2020 (Romania: Suhardul Mic (RCO), Tinovul Luci (RLU), Băile Tusnad (RMO), Poiana Stampei (RPO), Saru Dornai (RSD), Hungary: Fenyőfő (HFE); Table 1, Figure 1). These fragmented populations in the southern margin of the natural range of the species grow in specific habitat types, such as peat bogs, dry rocky substrates or mixed forests sustaining on sandy soils and calcareous bedrock [1,20,21]. Populations were selected based on their nature conservation importance, as they are considered natural populations, sustaining on undisturbed remnant habitats. The edaphic and climatic properties of these habitats and the plant community characteristics are given in Table S1 and Figure 1, derived from the SoilGrids [49] and CARPATCLIM [50] databases, with 0.25×0.25 km and 10×10 km spatial resolution, respectively.

In total, 57 dominant, undamaged, mature trees were randomly sampled (about 10 trees per site, except RCO–Table 1). Most of the trees were of old age (Table 1). In the cases of RLU, RMO and RSD, considering their shorter generation times [51] and that trees are characterized by restricted apical growth, the sampled trees also appeared to be the oldest from these sites, representing the same mature ontogenetic phase related to the cessation of intense height growth rates [52,53]. Cores were taken with an increment borer at breast height, c. 1.3 m. Standard techniques in dendrochronology [54] were used to collect, transport and prepare the cores for tree ring width (TRW) measurements. After being glued to wooden mounts, cores were sanded and then scanned with a high resolution scanner at 2400 dpi resolution. For each core, we semiautomatically measured TRW series with the Image J 1.52 software [55], using the Object J plugin at a resolution of 0.01 mm. Visual cross-dating was verified using COFECHA [56].

Table 1. Main features of the studied peripheral *Pinus sylvestris* L. populations from the Carpathian Basin and the Eastern Carpathians.

Code	Country	Residential Area	Habitat Type	Community Type	Latitude, Longitude	Altitude (m)	Slope Aspect	Area (km^2)	Sampled Trees	Selected Trees for Wood Anatomy	Mean Age (Year) (SD)	DBH (cm) (SD)
HFE	HU	Fenyőfő	Mixed forest	Festuco vaginatae-Pinetum sylvestris	47°21′26.0″ N 17°46′08.4″ E	252	SE	4.49	10	8	115 (47.0)	55.1 (8.9)
RCO	RO	Suhardul Mic	Rocky substrate	Seslerio-rigidae-Pinetum sylvestris	46°48′06.0″ N 25°47′31.4″ E	981	NW	0.04	3	3	101 (49.0)	19.2 (3.3)
RLU	RO	Tinovul Luci	Peat bog	Sphagnion magellanici, Vaccinio uliginosi-Pinetea	46°17′51.6″ N 25°43′39.2″ E	1076	-	2.73	12	7	77 (6.6)	17.4 (2.8)
RMO	RO	Băile Tusnad	Peat bog	Sphagnion magellanici, Vaccinio uliginosi-Pinetea	46°08′07.9″ N 25°54′11.0″ E	1052	-	0.58	9	8	59 (12.0)	12.5 (3.3)
RPO	RO	Poiana Stampei	Peat bog	Sphagnion magellanici, Vaccinio uliginosi-Pinetea	47°17′56.5″ N 25°06′59.8″ E	878	-	1.43	11	5	95 (37.7)	13.6 (2.8)
RSD	RO	Saru Dornei	Peat bog	Sphagnion magellanici, Vaccinio uliginosi-Pinetea	47°15′28.8″ N 25°21′24.3″ E	913	-	0.41	12	6	61 (17.1)	16.2 (6.1)

DBH—Diameter at breast height.

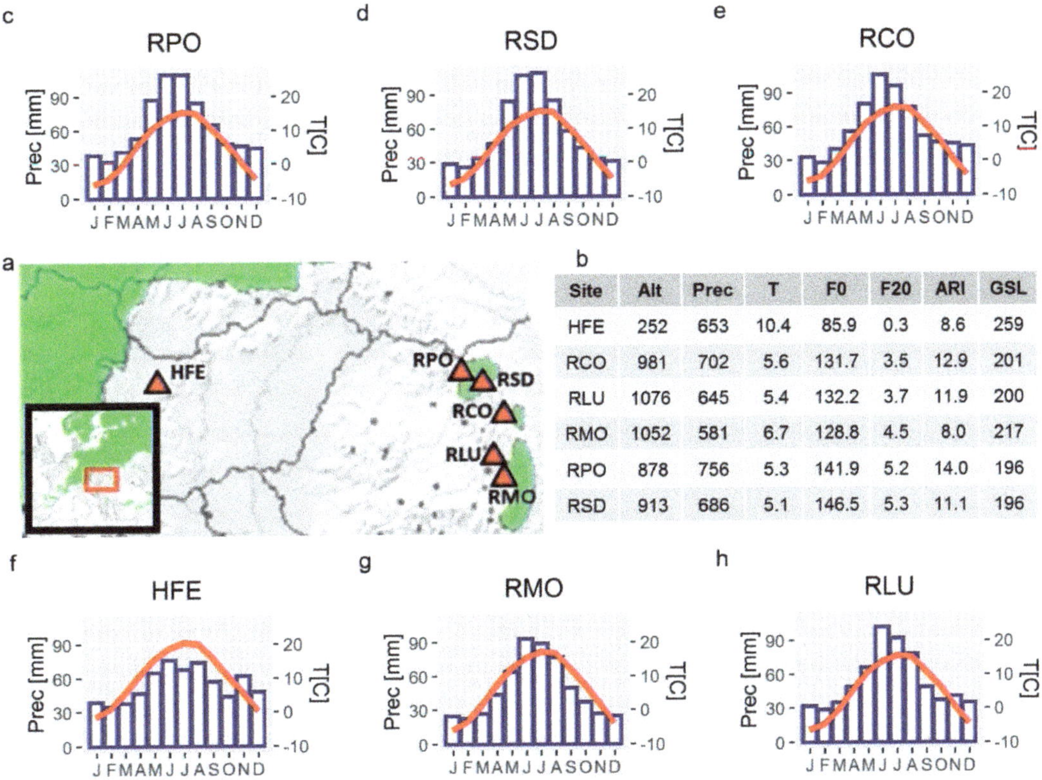

Site	Alt	Prec	T	F0	F20	ARI	GSL
HFE	252	653	10.4	85.9	0.3	8.6	259
RCO	981	702	5.6	131.7	3.5	12.9	201
RLU	1076	645	5.4	132.2	3.7	11.9	200
RMO	1052	581	6.7	128.8	4.5	8.0	217
RPO	878	756	5.3	141.9	5.2	14.0	196
RSD	913	686	5.1	146.5	5.3	11.1	196

Figure 1. Natural distribution of *Pinus sylvestris* in Europe (subfigure green area) and in the Carpathian region (green area), together with the geographic position of sampled *Pinus sylvestris* populations (**a**) with the altitude (Alt), total yearly precipitations (Prec), average yearly temperatures (T), number of frost days with minimum temperatures below 0 °C (F0), number of frost days with minimum temperatures below −20 °C (F20), global aridity indexes (ARI), and average growing season length (GSL) of the sites (**b**), and the corresponding climate diagrams of the sites (**c**–**h**). Blue bars represent monthly precipitation sums, and red lines represent monthly mean temperatures in the climate diagrams, calculated over the period 1961–2010 and derived from the CARPATCLIM database. Site abbreviations: Suhardul Mic (RCO), Tinovul Luci (RLU), Băile Tusnad (RMO), Poiana Stampei (RPO), Saru Dornei (RSD), Fenyőfő (HFE).

2.2. Wood Anatomical Measurements

At least five tree cores from each site (except RCO) were selected for anatomical measurements from the cross-dated samples (Table 1). We preferred mature trees whose individual indexed TRW series highly correlated with the local mean chronology, in order to represent the common growth of the stands better. Five specific annual rings (1990, 1991, 1992, 1993, 2010) were selected for anatomical measurements, based on their climatic characteristics derived from the CARPATCLIM database. The period of 1990–1992 was relatively dry, based on the average yearly aridity indexes of the sites (ARI—as the ratio of mean precipitation and mean evapotranspiration; 7.55, 5.90 and 6.83 respectively) and precipitation sums (PPT; 514, 648 and 584 mm respectively). Both 1993 and 2010 were humid (ARI: 16.56 and 16.63 respectively), with the highest precipitation in the studied years (695 and 1017 mm respectively), from which the latter was one of the most humid and wet year in the last decades in the Carpathian region. In total, 185 annual rings were selected, from which 19 annual rings were excluded from the analysis due to missing rings or low quality samples, with at least 5 replicates of each annual ring per site (except RCO) remaining in the following analyses.

From the cores with the selected annual rings, 15 μm thin cross sections were made with a sliding microtome (HN40, Jung, Heidelberg, Germany) and were stained with safranin and astra blue, according to [57]. Microscopic images of the cross sections were captured with a digital camera (AxioCam HRc, Zeiss, Göttingen, Germany) mounted on a transmitted light microscope (Zeiss Axio Imager A2, Zeiss, Göttingen, Germany), with a 10× objective (image resolution: 0.9699 pixel/μm). In case the annual ring width could not be entirely covered in the field of view of the objective, consecutive images were taken and then merged together using the program PTgui 10.0.11 (New House Internet Services BV, Rotterdam, The Netherlands).

The resulting images were processed with a semiautomatic method, using the image analysis software ROXAS 3.0.1 [58]. For each analyzed ring, after the exclusion of excessive cells, we manually fine-edited the measurements of five radial tracheid files. Tracheid files showing the tangentially largest cells were selected in order to consider only those tracheids that were cut close to the middle of their longitudinal length. From the resulting measurements, the tracheid lumen area (CA), the mean radial cell wall thickness of tracheids (CWTrad), the ratio of the cell diameter and radial cell wall thickness (CD/CWT), and the number of tracheids in the radial tracheid files (TNo) were selected for further analysis. Since trees were not of the same age (Table 1), the increasing power-like ontogenetic trend of the measured CA variable in and across the sites became evident (Figure 2). This trend is most apparent in the early stages of tree life, while it ceases when trees reach their mature height [53]. To account for this trend, we separated and excluded the CA measurements taken from the juvenile wood, following the recommendations of Yang et al. [59]. The intersection of two regression lines, obtained from the mean CA-cambial age coordinate space of all measurements, was treated as a boundary point between the juvenile and mature part of the trees. According to the intersection point, the youth trend ceased at the cambial age of 18 years (Figure 2), which, in its magnitude, corresponds well to observations on *Pinus sylvestris* by Pritzkow et al. [31]. After the exclusion, 155 annual rings, with at least 5 replicates of each annual ring per site (except RCO), remained for the final anatomical evaluation. We found no evident ontogenetic trends in case of the CWTrad measurements, and thus no exclusions or detrending were applied. On the basis of its high correlations with TRW chronologies [30,45] and visual examination, TNo measurements were standardized with a negative exponential function [54] fitted to all pooled measurements in earlywood ($y = 50.36*\exp(-0.05*x) + 11.12$; all p values of parameters <0.001) and latewood ($y = 9.66*\exp(-0.02*x) + 5.75$, p values of parameters <0.001, <0.1 and <0.001, respectively) to control for the decreasing age related trend of the tracheid numbers.

The resulting ROXAS outputs were re-arranged according to the order of tracheids in each tracheid file in the pith-bark direction. From the reordered CA, CWTrad and CD/CWT measurements, initial tracheidograms were produced, following the methods of Vaganov [60]. Using a moving average normalization procedure, this method produces standard curves of cell parameter variations in the radial files as functions of cell position within an annual ring, in order to make cells comparable between growth rings, trees, and sites formed in the same period of the growing season. Based on the average cell number of all measured files, standardized tracheidograms of CA, CWTrad and CD/CWT were obtained based on a normalized number of 24 cells (16 earlywood and 8 latewood) per ring. Latewood tracheids were identified according to the Mork index [61]. Standard tracheidograms were computed with the R 3.6.3 (R foundation) package 'tgram' (available from CRAN; http://cran.r-project.org, accessed on 24 November 2017.).

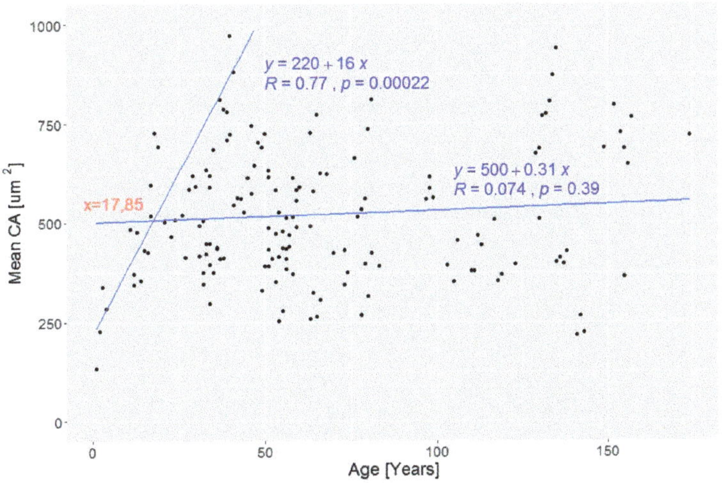

Figure 2. Linear regressions (blue) for the mean annual tracheid area (CA) parameter for all sites as a function of cambial age. The intersection of the two regression lines (red) represents the boundary point between the juvenile and mature wood as it was suggested by Yang et al. [59].

2.3. Statistical Analysis

Linear mixed effects models (LMM) were used to test the effects of sites, years, and monthly weather on the wood anatomical variables. This method allows the modelling of non-independent, hierarchical observations as random factors (such as the normalized tracheid position inside the tree rings or the trees within the sites), aside from the analysis of independent variables (i.e., site, the year of tracheid formation and monthly weather data; [62]). Thus, models included site (S) as a fixed factor, while the normalized tracheid position (NTNo.) and the trees (T) inside the sites were treated as random factors. Moreover, to account for the effect of yearly, seasonal and monthly weather conditions, the year of tracheid formation (Y), the monthly precipitation sums (PPI) and aridity indexes (ARI) of this current year (from April to October; PPI_04–PPI_10 and ARI_04–ARI_10), and the 4-month average of the monthly precipitation sums (pPPI) and aridity indexes (pARI) of the preceding year (from January to December; pPPI_0104–pPPI_0912 and pARI_0104–pARI_0912) were also included as independent variables. The linear combination of five or four (depending on the number of significant variables) monthly and seasonal weather variables (PPI, ARI and pPPI and pARI), with the two other independent (S, Y) and two random variables (NTNo., T) were tested for nested models by Likelihood Ratio Tests, in case of all the possible combinations of the monthly and seasonal weather variables. Every variable of the tested linear combinations (full models) was considered as significant if the likelihood ratio of the reduced model (null model) not containing that certain variable was significantly lower, i.e., more, than 15,000 full models containing five monthly weather variables were tested against more than 77,000 reduced models containing four weather variables, in both earlywood and latewood subsets of three dependent variables—CA, CWTrad and TNo. In the case of the CD/CWT variable, full models included only four weather variables, and they were tested against reduced models with three weather variables. Based on the lowest p values (or the highest χ^2 values), we selected those full linear combinations as final models, where the likelihood ratio tests of all their reduced (null) models were worse. Pooled over the years, statistical differences among the sites were determined by Tukey's tests (homogeneity of variances was accepted by variance ratio test). LMM's and Tukey's tests were computed with the R 3.6.3 (R foundation) package 'lme4' and 'multcomp' (available from CRAN; http://cran.r-project.org, accessed on 1 December 2020 and 08 February 2021.).

3. Results

3.1. Tree Ring Width Chronologies

The mean age at all sites ranged from 59 to 115 years (Table 1). Without standardization, tree ring widths of the HFE site were the largest in the common period 1938–2020, while RSD, RLU, RMO, and RCO sites had smaller increments, with RPO being the smallest (Figure 3). Series intercorrelation values (i.e., the average correlation of each series with a master chronology derived from all other series in a site, denoted by r_{bt}) are also given for all the sites, together with the average mean sensitivity values (a measure of the relative change in ring-width from one year to the next in a given series, m_{sx}) under Figure 3.

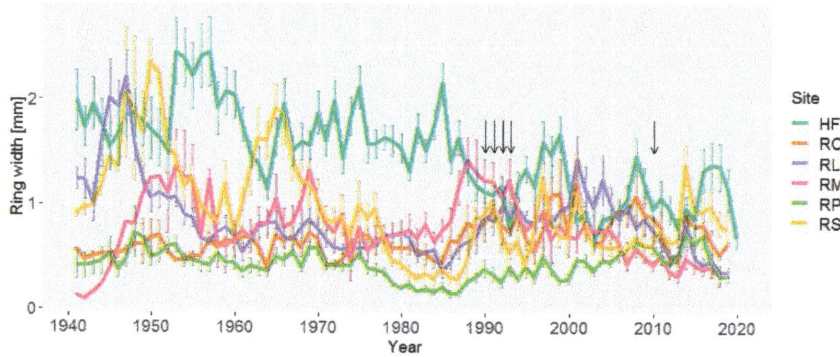

Site	HFE	RCO	RLU	RMO	RPO	RSD
Mean(SD) tree ring width (mm)	1.51 (0.45)	0.67 (0.17)	0.86 (0.36)	0.75 (0.32)	0.42 (0.13)	0.95 (0.46)
Series intercorrelation (r_{bt})	0.473	0.374	0.414	0.498	0.428	0.511
Average mean sensitivity (m_{sx})	0.295	0.344	0.331	0.266	0.320	0.379

Figure 3. Raw ring width chronologies in the common interval of the sites in the period 1940–2020. Lines indicate mean ring widths of all individuals sampled in the sites, vertical bars indicate the standard errors, together with their values indicated in the table below, and completed with the series intercorrelation values (r_{bt}) and the average mean sensibility values (m_{sx}). Site abbreviations: Suhardul Mic (RCO), Tinovul Luci (RLU), Băile Tusnad (RMO), Poiana Stampei (RPO), Saru Dornei (RSD), Fenyőfő (HFE). Annual rings selected for wood anatomical analyses are indicated by arrows.

3.2. Wood Anatomical Differentiation by Habitat Types

In the earlywood, CA was significantly different between the sites ($p < 0.001$, Table 2), and most sites formed separate groups, where HFE had the largest and RPO had the smallest lumen areas: means varied from 914.9 to 1038.2 μm^2 for HFE and from 414.0 to 554.0 μm^2 for RPO in all measured years (Figure 4a). LMM coefficients reflected the same relations between the sites: the intercept (1045.59) representing HFE displayed the largest cells, while the relative coefficient of RPO (−712.27) showed the smallest cells (Table 2). CWTrad of earlywood cells were also statistically different between the sites ($p < 0.001$), with most sites forming separate groups. Earlywood CWTrad means varied from 3.35 to 3.58 μm for HFE, and from 3.37 to 3.90 μm for RLU, having the largest CWTrad (LMM intercept and coefficient were 4.23 and −0.47, respectively); while RPO had the thinnest earlywood cell walls (means 2.59 to 3.06 μm, Figure 4b, Table 2). The CD/CWT ratio in the earlywood was also significantly different between the sites ($p < 0.001$); HFE had the largest (means between 42.01 and 51.07; LMM intercept: 41.01), while RPO had the smallest CD/CWT values (means between 22.48 and 31.49; Figure 4c; Table 2). Earlywood TNo was significantly affected by sites as well ($p < 0.001$), where HFE had the largest mean number of tracheids per radial file (ranging from 1.24 to 1.60, LMM intercept 1.40), and RPO had the smallest (means ranged from 0.44 to 0.86, Figure 4d, Table 2).

Table 2. Linear mixed effect models (LMMs) and likelihood ratio tests conducted on the fixed factors of site, year, and climate factors to test their effects on the wood's anatomical features. Heat maps correspond to coefficient (Coef.) relations between the climate covariates (PPT, ARI) and among factor levels of the site (S) and year (Y) variables. Color hues are from the minimum to the maximum, from red–orange–yellow and light to dark green, proportionally. Reference site is HFE, reference year is 1990, both referred by Int (intercept).

Dependent Variable			Site						Year				Climate Variables				
			Int	**RCO**	**RLU**	**RMO**	**RPO**	**RSD**	**1991**	**1992**	**1993**	**2010**	**PPT 06**	**ARI 04**	**ARI 06**	**ARI 07**	**pPPT_0508**
CA	Ew	Coef	1045.59	−606.72	−505.64	−291.49	−712.27	−412.49	1.48	62.75	32.61	95.25	−5.54	57.07	619.59	62.40	−2.54
		χ^2				2575.65 ***				87.13 ***			40.03 ***	69.73 ***	39.47 ***	35.21 ***	32.48 ***
			Int	RCO	RLU	RMO	RPO	RSD	1991	1992	1993	2010	PPT 05	PPT 10	ARI 04	ARI 05	pARI_0508
	Lw	Coef	337.20	−71.17	−37.42	−10.74	−126.87	−81.41	−99.81	65.23	−44.78	6.45	−1.80	−0.76	14.21	142.02	−175.64
		χ^2				649.76 ***				340.49 ***			53.13 ***	60.79 ***	9.17 **	47.81 ***	196.55 ***
			Int	RCO	RLU	RMO	RPO	RSD	1991	1992	1993	2010	PPT 04	PPT 06	ARI 06	ARI 08	pARI_0508
CWTrad	Ew	Coef	4.23	−0.442	−0.467	−0.648	−1.493	−0.929	−0.486	0.037	−0.838	−0.610	−0.003	−0.029	2.773	0.971	−0.302
		χ^2				1776.65 ***				784.54 ***			36.54 ***	84.26 ***	67.15 ***	212.96 ***	6.37 *
			Int	RCO	RLU	RMO	RPO	RSD	1991	1992	1993	2010	PPT 07	PPT 08	ARI 06	ARI 07	pARI_0508
	Lw	Coef	3.25	−0.12	0.28	−0.43	−1.93	−0.46	−0.06	−0.78	−0.55	−1.59	0.04	0.01	−0.28	−5.31	1.69
		χ^2				1376.55 ***				372.15 ***			38.52 ***	120.78 ***	6.74 **	39.48 ***	62.72 ***
			Int	RCO	RLU	RMO	RPO	RSD	1991	1992	1993	2010	PPT 05	ARI 04	ARI 07	- PPT_0508	
CD/CWT ratio	Ew	Coef	41.01	−26.45	−22.17	−9.42	−24.55	−13.90	3.12	3.35	6.06	8.70	−0.018	3.12	4.21	−0.03	
		χ^2				2147.10 ***				447.95 ***			15.42 ***	63.80 ***	57.11 ***	5.18 *	
			Int	RCO	RLU	RMO	RPO	RSD	1991	1992	1993	2010	PPT 08	ARI 05	ARI 07	ARI 10	
	Lw	Coef	5.67	−3.47	−1.43	−0.02	−2.82	−2.61	−0.97	0.01	0.48	0.85	−0.02	0.48	0.71	0.53	
		χ^2				401.48 ***				78.73 ***			31.40 ***	11.40 ***	13.60 **	33.19 ***	

Table 2. *Cont.*

Dependent Variable				Site					Year				Climate Variables				
			Int	**RCO**	**RLU**	**RMO**	**RPO**	**RSD**	**1991**	**1992**	**1993**	**2010**	**PPT 06**	**ARI 04**	**ARI 06**	**ARI 07**	**pPPT_0508**
TNo indices	**Ew**		Int	RCO	RLU	RMO	RPO	RSD	1991	1992	1993	2010	PPT 05	PPT 08	ARI 04	ARI 07	ARI_08
		Coef	1.40	−1.203	−0.937	−0.541	−1.340	−1.042	−0.339	−0.055	−0.410	−0.428	−0.001	−0.014	0.122	0.317	1.882
		χ^2				273.81 ***				53.08 ***			3.91 *	8.41 **	7.85 **	21.53 ***	13.04 ***
	Lw		Int	RCO	RLU	RMO	RPO	RSD	1991	1992	1993	2010	PPT 05	PPT 08	ARI 05	ARI 09	pARI_0912
		Coef	2.01	−0.05	−0.31	−0.99	−0.90	−0.73	0.35	−0.04	−0.26	−0.41	0.02	0.005	−1.13	−0.45	−0.03
		χ^2				156.40 ***				12.70 *			15.62 ***	5.62 *	10.36 **	6.12 *	13.43 ***

HFE Fenyőfő, RCO Suhardul Mic, RLU Tinovul Luci, RMO Băile Tusnad, RPO Poliana Stampei, RSD Saru Dornei. CA: cell area; CWTrad: radial cell wall thickness; TNo indices: standardized tracheid number within a tracheid file; Ew: earlywood; Lw: latewood. PPT and ARI are monthly precipitation sums and monthly mean aridity indexes, where the number index corresponds to the number of the month within the current year; climatic variables names starting with "p" refer to the previous year with 3-month-means of the monthly values indicated in the indexes. Coef coefficients of the LMM variables and their factor levels. χ^2 (Chi-square values) of the likelihood ratio tests of the LMM variables, with their corresponding p values indicated by asterisks (*, $p < 0.05$; **, $p < 0.01$; ***, $p < 0.001$)). df(Site) = 5; df(Year) = 4; df(climatic variables) = 1.

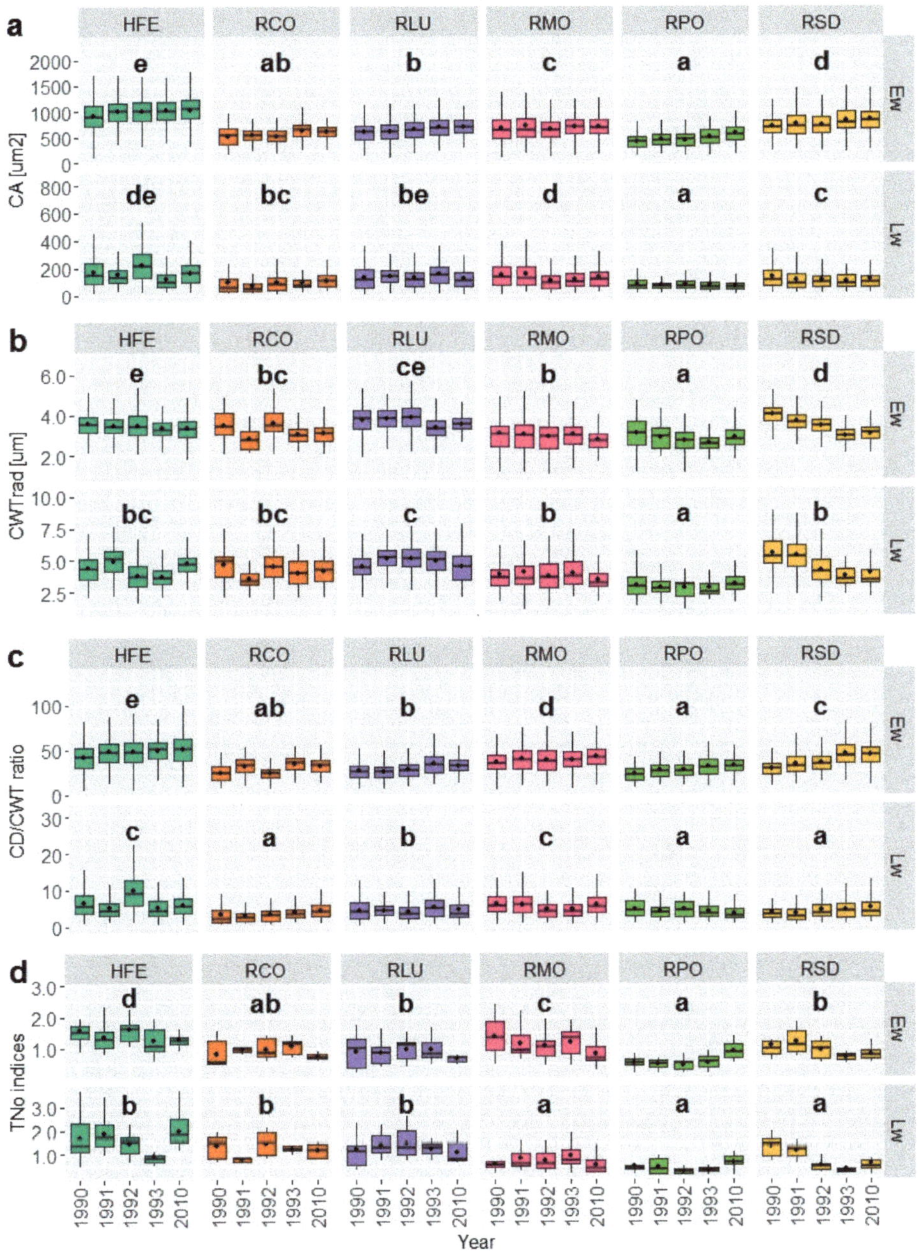

Figure 4. Boxplots (displaying minimum, first quartile, median, third quartile, maximum and mean) of the four measured wood anatomical features of trunk cross sections in earlywood (Ew) and latewood (Lw) from the sampled sites (HFE-RSD) for each studied year (1990–2010): CA cell area (**a**); CWTrad radial cell wall thickness (**b**); CD/CWT ratio of the cell diameter and radial cell wall thickness (**c**); TNo indices standardized tracheid number within a tracheid file (**d**). Tukey's HSD significant differences between sites are indicated with letters, pooled over the studied years.

In the latewood, sites had significant effects on the four measured parameters of CA, CWTrad, CD/CWT and TNo (all with $p < 0.001$; Table 2). RPO displayed significantly smaller CA (means from 66.9 to 79.6 μm^2), thinner CWTrad (2.75–3.04 μm), lower CD/CWT (3.51–4.68), and lower TNo (0.25–0.64) in the latewood compared to HFE (mean CA: 121.2–174.3 μm^2, mean CWTrad: 3.71–4.92 μm, mean CD/CWT: 5.17–9.95, mean TNo: 0.91–1.97, respectively). RPO was also different from the rest of the sites, except in the case of two variables CD/CWT and TNo: RPO was grouped together with RCO and RSD considering CD/CWT, while with RMO and RSD for TNo (Figure 4, Table 2).

Along the normalized tracheidograms in all sites and years, CA decreased in average from mean (SD) = 739.33(156.03) μm^2 in the first tracheid position of the earlywood to mean (SD) = 62.29(18.75) μm^2 in the last tracheid position of the latewood (Figure 5a,d,g,j,m). In earlywood, CWTrad increased slightly from mean (SD) = 3.16(0.32) μm to mean (SD) = 3.67(0.41) μm until the 16th tracheid position, then the latewood wall-thickening accelerated and peaked in the 19th tracheid position at mean (SD) = 4.48(0.64) μm, then, finally, decreased to mean (SD) = 3.15(0.40) μm (Figure 5b,e,h,k,n). In general, the CD/CWT ratio increased from mean (SD) = 37.71(2.23) to mean (SD) = 41.42(2.06) in the 6th tracheid position of earlywood, then it decreased to mean (SD) = 3.25(0.29) in the last tracheid position of the latewood (Figure 5c,f,i,l,o). In the first half of the earlywood, CA tracheidograms (at tracheid positions 1–5), the difference between the means of HFE with mean (SD) = 1151.19(124.03), and the rest of the sites with mean (SD) = 651.78(87.85) were conspicuous. The peat bog sites (RLU, RMO, RPO and RSD) differed from each other, as well in the whole earlywood, with RSD having the largest mean CA, then followed RMO, RLU and RPO (Table 3). This trend is also reflected in the CD/CWT tracheidograms, though the relations between the peat bog sites in the earlywood differed from that in the CA tracheidograms, as RMO was characterized by the second largest average CD/CWT values, then came RSD, RLU and RPO. In 1992, HFE in the latewood showed a deviation from the general trend that is characteristic of intra-annual density fluctuations (IADF) in the latewood (L-type IADF), with a peak in CA (Figure 5g) and CD/CWT (Figure 5i) tracheidograms at the 19th and 21st tracheid positions, with mean (SD) = 278.35(66.34) μm^2 and 12.05(4.50), respectively [63]. In general, the inter- and intra-annual variability of the sites were more pronounced in the parameter CWTrad, especially in the latewood, where it varied greatly between sites, years and tracheid positions.

Table 3. Mean and standard error (SE) of the wood anatomical features of earlywood and latewood for each studied site. CA: cell area; CWTrad: radial cell wall thickness; CD/CWT: ratio of the cell diameter and radial cell wall thickness. Site abbreviations: Suhardul Mic (RCO), Tinovul Luci (RLU), Băile Tusnad (RMO), Poiana Stampei (RPO), Saru Dornei (RSD), Fenyőfő (HFE).

	Parameter	HFE	RCO	RLU	RMO	RPO	RSD
Earlywood	mean(SE) CA (μm^2)	999.27 (5.47)	570.06 (6.45)	645.86 (4.54)	688.37 (5.13)	470.44 (3.76)	758.53 (5.40)
	mean(SE) CWTrad (μm)	3.45 (0.01)	3.25 (0.02)	3.70 (0.01)	2.95 (0.02)	2.81 (0.01)	3.38 (0.01)
	mean(SE) CD/CWT	47.60 (0.30)	29.01 (0.38)	28.61 (0.24)	38.58 (0.27)	27.36 (0.23)	37.04 (0.31)
Latewood	mean(SE) CA (μm^2)	168.30 (2.66)	95.92 (3.78)	135.67 (2.67)	129.18 (2.53)	71.54 (1.34)	111.07 (2.61)
	mean(SE) CWTrad (μm)	4.33 (0.03)	4.21 (0.06)	4.84 (0.04)	3.78 (0.04)	2.87 (0.02)	4.46 (0.05)
	mean(SE) CD/CWT	6.51 (0.11)	3.62 (0.13)	4.48 (0.09)	5.41 (0.08)	4.17 (0.08)	4.16 (0.10)

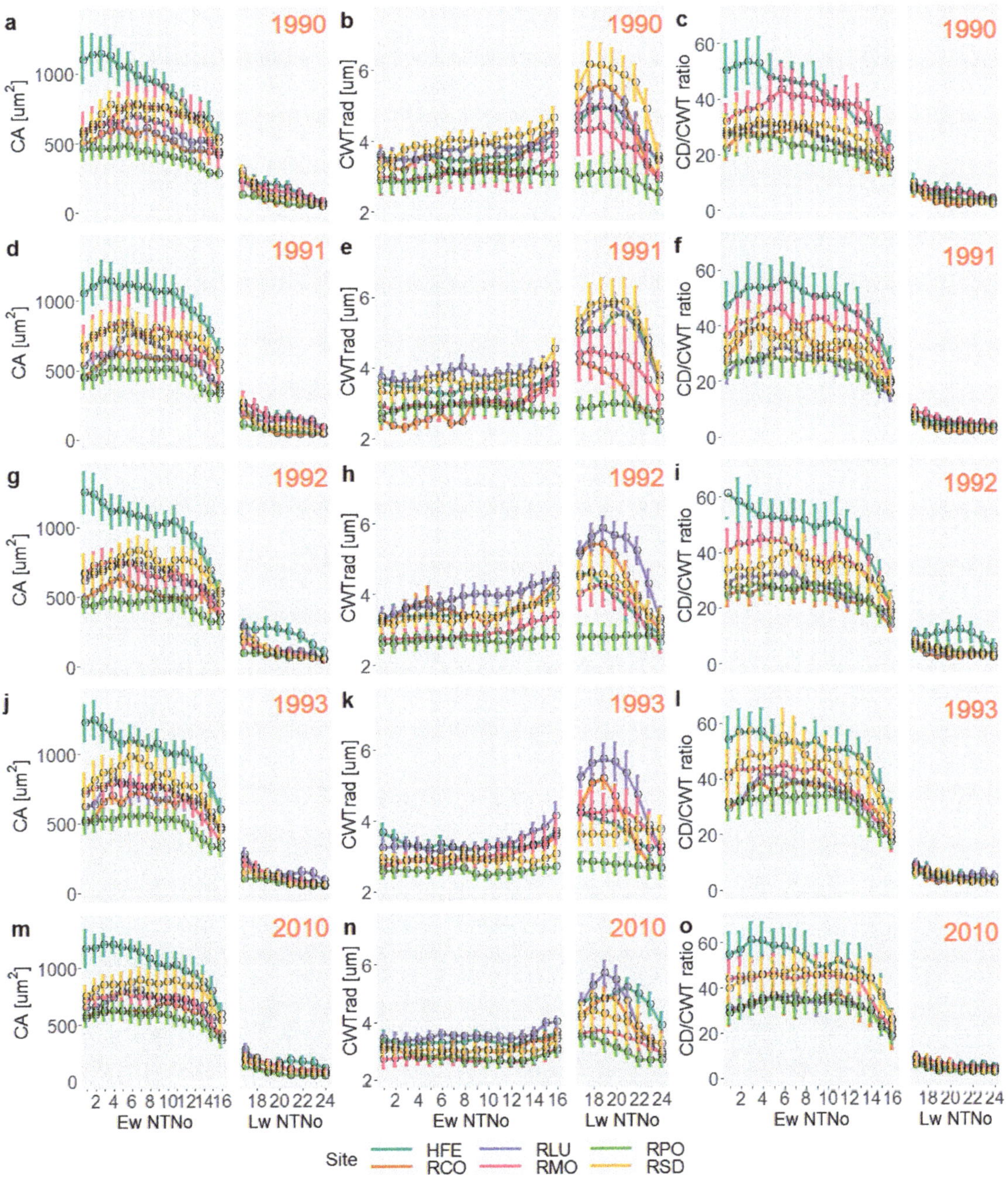

Figure 5. Average tracheidograms (tracheid parameter variations as functions of their normalized positions in radial files–NTNo.) for each studied year (1990–2010) at the studied sites of three measured wood anatomical features in earlywood (Ew) and latewood (Lw): CA cell area (**a**,**d**,**g**,**j**,**m**); CWTrad radial cell wall thickness (**b**,**e**,**h**,**k**,**n**); CD/CWT ratio of the cell diameter and radial cell wall thickness (**c**,**f**,**i**,**l**,**o**). Tracheidograms were obtained based on a normalized number of 24 cells (16 earlywood and 8 latewood). Lines with different colors indicate different sites, error bars indicate standard errors. Site abbreviations: Suhardul Mic (RCO), Tinovul Luci (RLU), Băile Tusnad (RMO), Poiana Stampei (RPO), Saru Dornei (RSD), Fenyőfő (HFE).

3.3. *Effect of Yearly and Monthly Climate on Wood Anatomy*

Earlywood CA, CWTrad, CD/CWT and TNo were significantly affected by the year of cell formation (all with $p < 0.001$, Table 2). The sites produced the smallest earlywood CA and CD/CWT in 1990 (LMM intercepts: 1045.59 and 41.01 respectively), while the largest cells with high CD/CWT were produced in 2010 (Figure 4a,c, Table 2). Earlywood CWTrad and TNo were the highest in 1990 (LMM intercepts are 4.23 and 1.40, respectively) and in 1992, while these parameters displayed the lowest values in 1993 and 2010 (Figure 4b,c, Table 2). Years also had significant effects on the latewood CA, CWTrad and CD/CWT (all with $p < 0.001$), as well as on TNo ($p < 0.05$). Similarly to the relations seen in the earlywood, latewood CWTrad and TNo values were the lowest in 2010 (CWTrad coefficient: −1.59, TNo coefficient: −0.41) and CD/CWT were highest in this same year, while CA were only slightly larger in 2010, compared to 1990 (intercept: 337.30, Figure 4, Table 2). For the means and standard errors of the four parameters in the sites and years, see Table S2. The main monthly climatic variables affecting wood parameters were dependent on the time of formation (earlywood–latewood position) of the measured tracheids. For CWTrad and TNo, earlywood tracheids correlated negatively with precipitation and positively with the current year aridity indexes (in case of all monthly variables at $p < 0.05$ or lower), while latewood tracheids were positively correlated with precipitation and negatively correlated with the current year aridity indexes ($p < 0.05$ or lower). For CA and CD/CWT, earlywood and latewood tracheids show negative correlation with the precipitation and positive correlation with the aridity indexes, at $p < 0.05$ or lower (Table 2).

4. Discussion

In this study, we investigated how *Pinus sylvestris* trees adjust their xylem to the ecological and climatic characteristics of different rear-edge habitats in the Carpathians and in the Pannonian Lowland. Through the examination of tracheid parameters lumen area (CA), radial cell wall thickness (CWTrad), ratio of radial lumen diameter to radial cell wall thickness (CD/CWT), and the number of tracheids in the radial files (TNo), we compared populations inhabiting four different peat bogs, a mountainside habitat with rocky outcrops, and a sandy, calcareous lowland site. To disentangle the influence of yearly and seasonal climate from the effect of sites, the statistical modelling included the year of tracheid formation, monthly climate data (i.e., precipitation and aridity), and the corresponding tracheid positions inside the tree rings, taken from tracheidograms.

Our results revealed significant differences among all studied sites in the case of almost all wood anatomical parameters (Table 2, Figure 4). Even populations growing on the same type of habitat with similar ecological background exhibited anatomical differences.

As expected, the most distinct population proved to be HFE population from the Pannonian Lowland, formerly reported to represent a different genetic lineage from those of the Eastern Carpathians [16]. Ecological conditions of this site were also dissimilar to other populations. Located in Northwest Hungary, at the intersection of the northern slopes of the Bakony Mountains and the Little Hungarian Plain (Figure 1a), the population is situated on less humid, calcareous sandy soil [20,22]. The thick sand layer is coupled with intensive summer evaporation and, consequently, a low groundwater table [28]. According to these circumstances, this site proved to be the most arid habitat among all investigated in this study (Figure 1b). Such long-term drought conditions could lead to enlarged lumen diameters, thinner cell walls, and an adaptive reduction of the thickness-to-span ratio [47]. This mechanism offers a more effective water-conducting system, which requires lower carbon investments in tissue production [64], at the cost of higher vulnerability of the water conducting cells exposed to cavitation failures [47,65]. Compared to larger angiosperm vessels, the narrow and short conduits of conifers—even if relatively enlarged—could be more easily refilled once cavitated, as the gas emboli in these conduits have larger surface-to-volume ratios [64], and conifer tracheids are also less exposed to eventual implosion [66]. Moreover, extreme xylem cavitation can be prevented if the increased specific hydraulic conductivity of wider tracheids in drier sites is associated with a lower leaf/sapwood area ratio [67]. This strategy is more evident if the CD/CWT ratio is taken into account (Figure 4c), and it is compared to the global aridity indexes of the sites. This optimized carbon-per-conduit-cost strategy is reflected in the wood anatomical traits of HFE: the large tracheid lumen areas (Figures 4a and 5a,d,g,j,m) are combined, though with larger, but proportionally thinner, cell walls (Figures 4b and 5b,e,h,k,n), resulting in the highest CD/CWT ratio Figures 4c and 5c,f,i,l,o) among the studied sites. Larger tracheid lumens could also be formed due to site-specific soil conditions: though the soil is basically sandy, a thin layer of brown cambisol soil is present above the calcareous sandy substrate [68]; thus, this site is relatively less nutrient-deprived compared to the peat bog sites. This could lead to the formation of wider tracheids, as well as to more tracheids per radial files and the delayed cessation of tracheid formation [69]. The high number of TNo (Figure 4d) and the large radial increments (Figure 3) could also be attributed to (1) the high mean temperature and long growing seasons (Figure 1b) favoring physiological processes in the direction of increased radial growth performance [70,71]; and (2) the absence of the limiting effect of the over-saturated groundwater (as evidenced in peat bogs), enabling and forcing trees to produce expanded, deep-penetrating rooting systems, enhanced water uptake and promoted radial growth [72,73]. As a result of the high water permeability and low capacity of the sandy soil, the irrigating effect of precipitation is reduced, and the year-to-year growth of the tree rings is more balanced, resulting in relatively low TRW mean sensitivity (0.295; [74]).

Altogether, four peat bog populations and one from a rocky habitat were studied from the Carpathians. By comparing wood anatomical parameters, RPO was the most distinct among the peat bogs. Though the RPO site is characterized by the highest average yearly and summer precipitation and the most humid conditions (Figure 1b,c), the trees displayed the smallest ring widths (Figure 3), lowest TNo (Figure 4d; Table S2), smallest CA (Figures 4a and 5a,d,g,j,m; Table S2), thinnest CWT (Figures 4b and 5b,e,h,k,n; Table S2), and lowest CD/CWT ratio (Figures 4c and 5c,f,i,l,o, Table S2) in both earlywood and latewood. On the other hand, RPO is one of the coldest sites, with a relatively short growing season (Figure 1b,c) compared to HFE and most of the studied peat bogs, as well as RLU and RMO. Lower yearly and seasonal temperatures generally de-escalate the physiological processes related to the growth performance of trees [34,75], with shortened periods of cambial activity [40] and shorter growing seasons resulting in the decrease of tree basal area and tree-ring growth [70,71]. Trees at RPO seem to have reduced their yearly tracheid production and radial growth to adapt to the cooler, shorter, vegetation period, and to avoid possible frost deformations of the wood caused by relatively frequent frost events (Figure 1a, [41]). The high water table at this site could also contribute to smaller increments, as a result of restricted rooting depth, decreased amount of fine root biomass and, thus, a limited nutrient uptake from the flooded soils [73]. The RPO population itself is also surrounded by a tall forest community dominated by *Picea abies*, inducing the more vigorous apical growth of local *P. sylvestris* trees in order to compete for light, resulting in thinner and narrower trunks (Table 1) and restricted radial growth. RPO trees produced the smallest lumen areas with the lowest CD/CWT ratios, probably as a result of two main reasons: (1) the higher number of extreme frost days (Figure 1b) led to the adaptation of tracheids against possible embolism-formations, related to freeze-thaw cycles [76–78], in parallel with the interrelation of higher altitudes and reduced tracheid lengths [33]; or (2) RPO lies in the western part of the Dornelor basin, as a part of a vast wet site complex, between the Bârgău and Călimani Mountains. As an oligotrophic peat bog, its water accumulation is mostly determined by precipitation [79], which is complemented by the irrigating effect of the rivers, Dorna and Dornişoara, in the area [80]. As this site is characterized by the highest precipitation and humidity (Figure 1b), the constantly high water table and anaerobic, hydric conditions led to reduced root and mycorrhizae growth and to shallow root systems [72,73], resulting in small, thick-walled tracheids in the trunk, probably as a consequence of hormonal adjustments [81–84].

The RSD site also resides in the Dornelor basin, between the Bârgău, Călimani and Bistrița Mountains, east of the RPO site and located in the same wet site complex [79]. This is an oligotrophic peat bog, but probably with a different soil water regime compared to RPO, due to its more centralized position in the eastern part of the basin. This fact, and the lesser amount of precipitation (Figure 1b), prevent the constant presence of a high water table and advanced anaerobic conditions, as seen in case of the RPO site. Thus, trees at the RSD site can better utilize available moisture [42,84], producing larger earlywood and latewood tracheids (Figure 4a). Low mean temperatures and a shorter growing season at this site (Figure 1b) result in fewer yearly formed earlywood tracheids (Figure 4d; Table S2), in parallel with the temperature and growing season length dependence of wood growth, as discussed above. In the latewood as well, RSD trees (as peat bog sites RPO and RMO) developed fewer cells (Figure 4d; Table S2). It seems that the higher numbers of frost days, with minimum temperatures below −20 °C at these peat bogs (Figure 1b), is the main driving force behind the reduced number of latewood tracheids. Trees at these sites possibly try to avoid severe frost damage in autumn through the early cessation of cambial activity, with reduced cell production rates and radial growth [34,85]. Shortened root growth periods with slower growth and activity could also contribute to this phenomenon [73]. The high earlywood-to-latewood TNo ratios at these three sites could also improve hydraulic efficiency in order to utilize the higher water supply of the RPO and RSD peat bogs (Figure 1b, [86]).

The RLU site has developed in one old volcano crater within the Harghita Mountains, exhibiting extreme acidic soil, about 11–12 °C summer soil temperature and summer water table at 40–50 cm depth [87]. Both earlywood and latewood CWTrad were the highest at this site, compared to other peat bogs (Figure 4b), with pronounced inter- and intra-annual variation (Figure 5b,e,h,k,n; Table 2); still, there seems to be no clear evidence of connection between the observed high CWTrad values and the long-term climate of the site (Figure 1b,h). Rosner et al. [88] suggested, that climate has a poor effect on CWT, but the conduit wall reinforcement of the tracheids $(t/b)^2$ are responsible for preparing trees for extreme climate events. Trees with lower $(t/b)^2$ are hydraulically more efficient in the long term, but are more prone to extreme sudden drought and eventual dieback through the higher probability of hydraulic failure. Though Borbély and Indreica [87] described *P. sylvestris* at the RLU site with vigorous, competitive growth, this population—at least in some parts—seemed to degrade: snags, standing, dead trees and decaying logs were commonly present in the site, accompanied by low population density. The sampled, healthy trees were possibly more prepared for the drought events of the last 25 years [87] than were dead ones, through their enlarged CWT and higher wall reinforcement [88]. As in the case of RPO and RSD, RLU trees formed fewer earlywood tracheids (Figure 4d; Table S2) due to low mean temperatures and shorter growing seasons, while, in the latewood, the smaller number of extreme cold events (T20, Figure 1b) induced higher latewood TNo compared to other peat bogs [34,85].

The peatland basin of RMO is located within the volcanic crater of the former Ciomadu volcano, situated in the Calimani-Gurghiu-Harghita range. It is an ombrotrophic peat bog, with 15 °C summer temperatures and a deep peat layer that reaches ca. 10 m, with its central part dominated by the 2–3 m tall *P. sylvestris* trees [89]. As pine peat bogs are already relatively more acidic and nutrient-deprived among the peat forest vegetations of Central-East Europe [1], the thick layer of peat is possibly coupled with pronounced N and P deficiency in the soil (Table S1), resulting in the restricted growth and height of *P. sylvestris* trees at this site. The lack of tall trees could also be connected to the cold winter air, which could be trapped in the depression of the volcanic crater of the Ciomadu, resulting in temperature inversions [1]. Despite trees at RMO having the smallest breast height diameters (12.5 ± 3.3; Table 1), the slenderness ratio of the trees—the proportion of tree height and diameter—is most probably the smallest at this site due to their extremely limited height. This is shown to be accompanied by reduced cell wall thickness (Figures 4b and 5b,e,h,k,n), suggesting the adaptation of the internal regulation of the mechanical properties of the stem in the face of constrained dynamic pressure on the smaller trees caused by less windy days [90]. RMO is also characterized by the lowest amount of precipitation, and trees produced relatively large tracheids at this site (Figures 4a and 5a,d,g,j,m). Since 1950, after the earlier wet hydroclimatic periods, the peat bog has been drained [89], thus, the water table became less saturated. If the over-saturated water table was a constraining factor for the anatomical development of the trees, this condition ceased, and trees adapted their tracheid lumens to the changed moisture conditions [84]. Characterized by the higher mean and more equilibrated temperature and humid conditions among the peat bogs (Figure 1b), RMO trees are represented with higher CD/CWT ratios. The RMO site produced higher number of earlywood and a smaller number of latewood cells (Figure 4d; Table S2) in parallel with the temperature and growing season lengths, as discussed above [34,85].

The rocky RCO population from the Bicaz Valley (southern Giurgeu Mountains) sustains on the southern facing cliffs above Lacul Roșu lake. Leptosol, episkeletic podzol and dystric cambisol are the dominant soil types, laying on rocky substrate with outcrops, which separates this site from the peat bogs and the lowland HFE [91]. Average yearly precipitation exceeds 700 mm, coupled with moderate temperatures and relatively high humidity (Figure 1b). Trees are characterized by small average TRW (0.67 ± 0.17, Figure 3), low earlywood TNo and CA (Figure 4a,d), and low earlywood and latewood CD/CWT ratios (Figure 4c), while latewood TNo, CA, and both earlywood and latewood CWT are around the global mean of the sites (Figure 4, Table S2). As trees growing on rocky soils are characterized by shallow root systems and reduced water accessibility, tree ring and wood formation cease earlier, accompanied by inhibited cell expansion and reduced radial earlywood cell widths [92]. On the other hand, the abovementioned carbon-per-conduit-cost strategy, in the face of the relatively high humidity of the site (Figure 1b), could have led to the reduced tracheid lumens and to the proportionally thicker cell walls [47]. Due to the low sample number at this site (Table 1), results should be treated with caution and further measurements are needed.

The patterns of site effects were influenced by local yearly and seasonal climate. The CD/CWT ratio parameter, both in earlywood and latewood, especially corresponded to the current year's humidity conditions and precipitation: the relatively arid 1990–1992 years (lower ARI) induced smaller CD/CWT ratios as a result of smaller lumen areas and thicker cell walls, while in 1993, and especially in 2010, the increased amount of precipitation and wet, humid conditions (higher ARI) led to increased CD/CWT ratios (Table 2, Figure 4c). This trend is reflected on the monthly level as well, as current year spring and mid-summer ARI was positively correlated with the earlywood and latewood CD/CWT ratio and CA (Table 2). However, monthly precipitation showed negative correlations with these parameters, in parallel with the findings of Fernández-de-Uña et al. [38]; thus, the positive relationship between ARI and these cell parameters could be attributed to the indirect effect of corresponding temperature fluctuations [42] affecting PET, probably both on the inter- and intra-annual scale. The seasonal climatic imprint on CWT differs between earlywood and latewood, suggesting more complex effects of climate on the intra-ring scale of this parameter [39]. Regarding tracheid numbers (TNo), there seems to be a negative connection between yearly tracheid production and humidity conditions: trees in general have reduced their TNo in the humid and wet 1993 and 2010 years (Table 2), but site variability differed in this parameter (Figure 4d). While latewood TNo was affected in the same way by late spring and early autumn humidity, tracheid production in the earlywood benefited from moist summer conditions (Table 2, [37,38]). It seems that, rather, the combined effect of current year precipitation and evapotranspiration—through the indirect effect of temperature—is responsible for the short-term wood anatomical responses of the trees at these peripheral communities, as it was shown in other studies [36,37]. Furthermore, as the growing season proceeds, the within-year climate variations affect the earlywood and latewood portions of the trees differently [35,37]. The occurrence of the L-type IADFs in the latewoods of the dry HFE site in 1992 (Figure 5g,i) is probably a consequence of the late summer drought (August PPT: 2.77 mm, August ARI: 0.02) and heat (August T: 24.62 °C) in 1992, followed by a relatively wet and mild autumn (October PPT: 97.49 mm, October T: 9.78 °C) as it was shown in other cases [93–95]; though possible delays between the controlling climatic factor and the occurrence of the IADF makes the determination of the time for triggering difficult to assess [63,74]. It seems that the short-term plastic responses of these populations to yearly and seasonal climate fluctuations differ from the long-term adaptive responses of these trees to the ecological characteristics of their habitats [12].

5. Conclusions

Altogether, *P. sylvestris* in the studied peripheral peat bog populations (RPO, RSD, RLU, RMO) adapted their wood anatomical traits to ensure the hydraulic safety of the stem (smaller tracheids, with proportionally thick walls), whereas, on the sandy, lowland site (HFE), trees are building a more effective water-conducting system (larger tracheids with relatively thin walls) at the cost of higher vulnerability to cavitation. Radial tree ring growth and yearly tracheid production also differed among the habitat types. Wood structure reflected the generally hydric, cool and anaerobic conditions of the peat bogs [82–84], while trees at the lowland site optimized their carbon-per-conduit-cost in the face of the dry, sandy soil, coupled with the arid and long growing seasons [47]. At the same time, further studies are needed to reveal the exact small scale microclimatic effects on wood anatomy at these sites due to a lack of precise local meteorological measurements. Individually, all populations are unique, as they exhibited specificity to site conditions, suggesting the local anatomical adaptation of Scots pine trees at these peripheral habitats. This by-habitat type phenotypic differentiation is reinforced by the adaptation of the cone morphology and needle anatomy of these populations [24], though the phylogeographic history and genetic differentiation of these populations could have also contributed to the variance of the wood's anatomical traits [16]. These phenotypic and genotypic adjustments allow *P. sylvestris* to prosper in these peripheral habitats in the face of climate change, as possible competitors do not pose a risk of being outcompeted through their lack of such variability and smaller ecological tolerance [1].

Supplementary Materials: The following are available online at https://www.mdpi.com/article/10.3390/f12040494/s1, Table S1: Edaphic properties of the studied sites derived from the SoilGrids database. Table S2: Mean and standard error (SE) of the wood anatomical features of earlywood and latewood for each studied site and year.

Author Contributions: Conceptualization, B.P. and M.H.; methodology, B.P.; software, B.P.; formal analysis, M.L. and B.P.; investigation, B.P.; resources, M.H. and K.C.; data curation, B.P. and K.C.; writing—original draft preparation, B.P. and M.L.; writing—review and editing, M.H., B.P., M.L. and K.B.; visualization, B.P.; supervision, M.H.; project administration, M.H. and K.B.; funding acquisition, K.B. All authors have read and agreed to the published version of the manuscript.

Funding: This research was funded by the National Research, Development and Innovation Office, Hungary, with a grant provided by the Hungarian Scientific Research Fund, grant number NKFI-EPR 119208. The authors were invited by the editor of the Special Issue.

Institutional Review Board Statement: Not applicable.

Informed Consent Statement: Not applicable.

Data Availability Statement: The data presented in this study are available in Supplementary Materials Tables S1 and S2.

Acknowledgments: The authors wish to thank the following people for providing planning, sampling, logistic support, and field and laboratory assistance: Károly Bóka, Marco Carrer, Máté Farkas, Katalin Jäger, Zoltán Kern, János Korponai, Zsófia Kovács, László Lőkös, Enikő Major, Roberto Menardi, Dávid Misi, Dániel Pifkó, Peter Prislan, and Tibor Wollner.

Conflicts of Interest: The authors declare no conflict of interest.

References

1. Leuschner, C.; Ellenberg, H. *Ecology of Central European Forests—Vegetation Ecology of Central Europe*; Springer International Publishing: Cham, Switzerland, 2017; Volume I, ISBN 978-3-319-43040-9.
2. Parmesan, C. Ecological and Evolutionary Responses to Recent Climate Change. *Annu. Rev. Ecol. Evol. Syst.* **2006**, *37*, 637–669. [CrossRef]
3. Sheth, S.N.; Angert, A.L. Demographic compensation does not rescue populations at a trailing range edge. *Proc. Natl. Acad. Sci. USA* **2018**, *115*, 201715899. [CrossRef] [PubMed]
4. Fadrique, B.; Báez, S.; Duque, Á.; Malizia, A.; Blundo, C.; Carilla, J.; Osinaga-Acosta, O.; Malizia, L.; Silman, M.; Farfán-Ríos, W.; et al. Widespread but heterogeneous responses of Andean forests to climate change. *Nature* **2018**, *564*, 207–212. [CrossRef] [PubMed]
5. Bennett, K.; Provan, J. What do we mean by 'refugia'? *Quat. Sci. Rev.* **2008**, *27*, 2449–2455. [CrossRef]
6. Keppel, G.; Van Niel, K.P.; Wardell-Johnson, G.W.; Yates, C.J.; Byrne, M.; Mucina, L.; Schut, A.G.T.; Hopper, S.D.; Franklin, S.E. Refugia: Identifying and understanding safe havens for biodiversity under climate change. *Glob. Ecol. Biogeogr.* **2012**, *21*, 393–404. [CrossRef]
7. Rull, V. Microrefugia. *J. Biogeogr.* **2009**, *36*, 481–484. [CrossRef]
8. Losos, J.B. Phylogenetic Perspectives on Community Ecology. *Ecology* **1996**, *77*, 1344–1354. [CrossRef]
9. Davis, M.B.; Shaw, R.G. Range Shifts and Adaptive Responses to Quaternary Climate Change. *Science* **2001**, *292*, 673–679. [CrossRef]
10. Violle, C.; Navas, M.-L.; Vile, D.; Kazakou, E.; Fortunel, C.; Hummel, I.; Garnier, E. Let the concept of trait be functional! *Oikos* **2007**, *116*, 882–892. [CrossRef]
11. Anderson, J.T.; Song, B. Plant adaptation to climate change—Where are we? *J. Syst. Evol.* **2020**, *58*, 533–545. [CrossRef]
12. Fonti, P.; Von Arx, G.; García-González, I.; Eilmann, B.; Sass-Klaassen, U.; Gärtner, H.; Eckstein, D. Studying global change through investigation of the plastic responses of xylem anatomy in tree rings. *New Phytol.* **2010**, *185*, 42–53. [CrossRef] [PubMed]
13. Labra, M.; Grassi, F.; Sgorbati, S.; Ferrari, C. Distribution of genetic variability in southern populations of Scots pine (*Pinus sylvestris* L.) from the Alps to the Apennines. *Flora Morphol. Distrib. Funct. Ecol. Plants* **2006**, *201*, 468–476. [CrossRef]
14. Debreczy, Z.; Rácz, I.; Musial, K. *Conifers around the World: Conifers of the Temperate Zones and Adjacent Regions*; DendroPress: Budapest, Hungary, 2011; ISBN 978-9-632-19061-7.
15. Tóth, E.G.; Köbölkuti, Z.A.; Pedryc, A.; Höhn, M. Evolutionary history and phylogeography of Scots pine (*Pinus sylvestris* L.) in Europe based on molecular markers. *J. For. Res.* **2017**, *28*, 637–651. [CrossRef]
16. Tóth, E.G.; Vendramin, G.G.; Bagnoli, F.; Cseke, K.; Höhn, M. High genetic diversity and distinct origin of recently fragmented Scots pine (*Pinus sylvestris* L.) populations along the Carpathians and the Pannonian Basin. *Tree Genet. Genomes* **2017**, *13*, 47. [CrossRef]
17. Rubiales, J.M.; García-Amorena, I.; Álvarez, S.G.; Manzaneque, F.G. The Late Holocene extinction of *Pinus sylvestris* in the western Cantabrian Range (Spain). *J. Biogeogr.* **2008**, *35*, 1840–1850. [CrossRef]
18. Cheddadi, R.; Vendramin, G.G.; Litt, T.; Francois, L.; Kageyama, M.; Lorentz, S.; Laurent, J.-M.; de Beaulieu, J.-L.; Sadori, L.; Jost, A.; et al. Imprints of glacial refugia in the modern genetic diversity of *Pinus sylvestris*. *Glob. Ecol. Biogeogr.* **2006**, *15*, 271–282. [CrossRef]
19. Prus-Głowacki, W.; Urbaniak, L.; Bujas, E.; Curtu, A.L. Genetic variation of isolated and peripheral populations of *Pinus sylvestris* (L.) from glacial refugia. *Flora Morphol. Distrib. Funct. Ecol. Plants* **2012**, *207*, 150–158. [CrossRef]
20. Borhidi, A. *Magyarország Növénytársulásai*; Akadémiai Kiadó: Budapest, Hungary, 2003; ISBN 963-05-7983-9.
21. Donita, N.; Popescu, A.; Pauca-Comanescu, M.; Mihailescu, S.; Biris, A. *Habitatele din România*; Editura Tehnică Silvică: Bukarest, Romania, 2005; ISBN 973-96001-4-X.
22. Bölöni, J.; Molnár, Z.; Biró, M.; Horváth, F. Distribution of the (semi-)natural habitats in Hungary II. Woodlands and shrublands. *Acta Bot. Hung.* **2008**, *50*, 107–148. [CrossRef]
23. Șofletea, N.; Mihai, G.; Ciocîrlan, E.; Curtu, A.L. Genetic Diversity and Spatial Genetic Structure in Isolated Scots Pine (*Pinus sylvestris* L.) Populations Native to Eastern and Southern Carpathians. *Forests* **2020**, *11*, 1047. [CrossRef]
24. Köbölkuti, Z.A.; Tóth, E.G.; Ladányi, M.; Höhn, M. Morphological and anatomical differentiation in peripheral *Pinus sylvestris* L. populations from the Carpathian region. *Dendrobiology* **2017**, *77*, 105–117. [CrossRef]
25. Köbölkuti, Z.A.; Höhn, M. Habitat type differentiation in peripheral *Pinus sylvestris* populations based on seed traits and germination data. *Stud. Bot. Hung.* **2018**, *49*, 97–119. [CrossRef]
26. Ambroży, S.; Kapsa, M. Regeneration of Scots pine *Pinus sylvestris* L. in the Carpathians depends on site fertility. *For. Res. Pap.* **2019**, *80*, 189–200. [CrossRef]
27. Bouriaud, O.; Popa, I. Comparative dendroclimatic study of Scots pine, Norway spruce, and silver fir in the Vrancea Range, Eastern Carpathian Mountains. *Trees* **2009**, *23*, 95–106. [CrossRef]
28. Misi, D.; Náfrádi, K. Late Winter—Early Spring Thermal Conditions and Their Long-Term Effect on Tree-Ring Growth in Hungary. *Balt. For.* **2016**, *22*, 203–211.
29. Nagavciuc, V.; Roibu, C.-C.; Ionita, M.; Mursa, A.; Cotos, M.-G.; Popa, I. Different climate response of three tree ring proxies of *Pinus sylvestris* from the Eastern Carpathians, Romania. *Dendrochronologia* **2019**, *54*, 56–63. [CrossRef]

30. Liang, W.; Heinrich, I.; Simard, S.; Helle, G.; Linan, I.D.; Heinken, T. Climate signals derived from cell anatomy of Scots pine in NE Germany. *Tree Physiol.* **2013**, *33*, 833–844. [CrossRef]
31. Pritzkow, C.; Heinrich, I.; Grudd, H.; Helle, G. Relationship between wood anatomy, tree-ring widths and wood density of *Pinus sylvestris* L. and climate at high latitudes in northern Sweden. *Dendrochronologia* **2014**, *32*, 295–302. [CrossRef]
32. Björklund, J.; Seftigen, K.; Fonti, P.; Nievergelt, D.; von Arx, G. Dendroclimatic potential of dendroanatomy in temperature-sensitive *Pinus sylvestris*. *Dendrochronologia* **2020**, *60*, 125673. [CrossRef]
33. Ståhl, E.G. Transfer effects and variations in basic density and tracheid length of *Pinus sylvestris* L. populations. *Stud. For. Suec.* **1988**, *180*, 1–15.
34. Schmitt, U.; Jalkanen, R.; Eckstein, D. Cambium dynamics of *Pinus sylvestris* and *Betula* spp. in the northern boreal forest in Finland. *Silva Fenn.* **2004**, *38*, 167–178. [CrossRef]
35. Eilmann, B.; Weber, P.; Rigling, A.; Eckstein, D. Growth reactions of *Pinus sylvestris* L. and *Quercus pubescens* Willd. to drought years at a xeric site in Valais, Switzerland. *Dendrochronologia* **2006**, *23*, 121–132. [CrossRef]
36. Martin-Benito, D.; Anchukaitis, K.; Evans, M.; del Río, M.; Beeckman, H.; Cañellas, I. Effects of Drought on Xylem Anatomy and Water-Use Efficiency of Two Co-Occurring Pine Species. *Forests* **2017**, *8*, 332. [CrossRef]
37. Martin-Benito, D.; Beeckman, H.; Cañellas, I. Influence of drought on tree rings and tracheid features of *Pinus nigra* and *Pinus sylvestris* in a mesic Mediterranean forest. *Eur. J. For. Res.* **2013**, *132*, 33–45. [CrossRef]
38. Fernández-de-Uña, L.; Rossi, S.; Aranda, I.; Fonti, P.; González-González, B.D.; Cañellas, I.; Gea-Izquierdo, G. Xylem and Leaf Functional Adjustments to Drought in *Pinus sylvestris* and *Quercus pyrenaica* at Their Elevational Boundary. *Front. Plant Sci.* **2017**, *8*, 1200. [CrossRef]
39. Belokopytova, L.V.; Babushkina, E.A.; Zhirnova, D.F.; Panyushkina, I.P.; Vaganov, E.A. Pine and larch tracheids capture seasonal variations of climatic signal at moisture-limited sites. *Trees* **2019**, *33*, 227–242. [CrossRef]
40. Swidrak, I.; Gruber, A.; Oberhuber, W. Cambial activity and xylem cell development in *Pinus cembra* and *Pinus sylvestris* at their climatic limits in the Eastern Alps in 2007. *Phyton* **2011**, *51*, 299–313.
41. Panayotov, M.P.; Zafirov, N.; Cherubini, P. Fingerprints of extreme climate events in *Pinus sylvestris* tree rings from Bulgaria. *Trees* **2013**, *27*, 211–227. [CrossRef]
42. Babushkina, E.A.; Belokopytova, L.V.; Kostyakova, T.V.; Kokova, V.I. Earlywood and Latewood Features of *Pinus sylvestris* in Semiarid Natural Zones of South Siberia. *Russ. J. Ecol.* **2018**, *49*, 209–217. [CrossRef]
43. Del Castillo, E.M.; Longares, L.A.; Gričar, J.; Prislan, P.; Gil-Pelegrín, E.; Čufar, K.; de Luis, M. Living on the Edge: Contrasted Wood-Formation Dynamics in *Fagus sylvatica* and *Pinus sylvestris* under Mediterranean Conditions. *Front. Plant Sci.* **2016**, *7*, 370. [CrossRef] [PubMed]
44. Prislan, P.; Gričar, J.; de Luis, M.; Novak, K.; Martinez del Castillo, E.; Schmitt, U.; Koch, G.; Štrus, J.; Mrak, P.; Žnidarič, M.T.; et al. Annual Cambial Rhythm in *Pinus halepensis* and *Pinus sylvestris* as Indicator for Climate Adaptation. *Front. Plant Sci.* **2016**, *7*, 1923. [CrossRef]
45. Hereş, A.-M.; Camarero, J.J.; López, B.C.; Martínez-Vilalta, J. Declining hydraulic performances and low carbon investments in tree rings predate Scots pine drought-induced mortality. *Trees* **2014**, *28*, 1737–1750. [CrossRef]
46. Eilmann, B.; Zweifel, R.; Buchmann, N.; Graf Pannatier, E.; Rigling, A. Drought alters timing, quantity, and quality of wood formation in Scots pine. *J. Exp. Bot.* **2011**, *62*, 2763–2771. [CrossRef] [PubMed]
47. Eilmann, B.; Zweifel, R.; Buchmann, N.; Fonti, P.; Rigling, A. Drought-induced adaptation of the xylem in Scots pine and pubescent oak. *Tree Physiol.* **2009**, *29*, 1011–1020. [CrossRef] [PubMed]
48. Hampe, A.; Petit, R.J. Conserving biodiversity under climate change: The rear edge matters. *Ecol. Lett.* **2005**, *8*, 461–467. [CrossRef] [PubMed]
49. Hengl, T.; Mendes de Jesus, J.; Heuvelink, G.B.M.; Ruiperez Gonzalez, M.; Kilibarda, M.; Blagotić, A.; Shangguan, W.; Wright, M.N.; Geng, X.; Bauer-Marschallinger, B.; et al. SoilGrids250m: Global gridded soil information based on machine learning. *PLoS ONE* **2017**, *12*, e0169748. [CrossRef] [PubMed]
50. Szalai, S.; Auer, I.; Hiebl, J.; Milkovich, J.; Radim, T.; Stepanek, P.; Zahradnicek, P.; Bihari, Z.; Lakatos, M.; Szentimrey, T.; et al. *Climate of the Greater Carpathian Region*; Final Technical Report; European Commission, JRC: Brussels, Belgium, 2013.
51. Tóth, E.G.; Bede-Fazekas, Á.; Vendramin, G.G.; Bagnoli, F.; Höhn, M. Mid-Pleistocene and Holocene demographic fluctuation of Scots pine (*Pinus sylvestris* L.) in the Carpathian Mountains and the Pannonian Basin: Signs of historical expansions and contractions. *Quat. Int.* **2017**, 1–12. [CrossRef]
52. Carrer, M.; Unterholzner, L.; Castagneri, D. Wood anatomical traits highlight complex temperature influence on *Pinus cembra* at high elevation in the Eastern Alps. *Int. J. Biometeorol.* **2018**, *62*, 1745–1753. [CrossRef]
53. Carrer, M.; von Arx, G.; Castagneri, D.; Petit, G. Distilling allometric and environmental information from time series of conduit size: The standardization issue and its relationship to tree hydraulic architecture. *Tree Physiol.* **2015**, *35*, 27–33. [CrossRef]
54. Cook, E.R.; Kairiukstis, L.A. (Eds.) *Methods of Dendrochronology—Applications in the Environmental Sciences*; Kluwer Academic Publishers: Dordrecht, The Netherlands, 1990; ISBN 0792305868.
55. Schneider, C.A.; Rasband, W.S.; Eliceiri, K.W. NIH Image to ImageJ: 25 years of image analysis. *Nat. Methods* **2012**, *9*, 671–675. [CrossRef]
56. Grissino-Mayer, H.D. Evaluating crossdating accuracy: A manual and tutorial for the computer program COFECHA. *Tree Ring Res.* **2001**, *57*, 205–211.

57. Van der Werf, G.W.; Sass-Klaassen, U.G.W.; Mohren, G.M.J. The impact of the 2003 summer drought on the intra-annual growth pattern of beech (*Fagus sylvatica* L.) and oak (*Quercus robur* L.) on a dry site in the Netherlands. *Dendrochronologia* **2007**, *25*, 103–112. [CrossRef]
58. Von Arx, G.; Carrer, M. ROXAS—A new tool to build centuries-long tracheid-lumen chronologies in conifers. *Dendrochronologia* **2014**, *32*, 290–293. [CrossRef]
59. Yang, K.C.; Benson, C.A.; Wong, J.K. Distribution of juvenile wood in two stems of *Larix laricina*. *Can. J. For. Res.* **1986**, *16*, 1041–1049. [CrossRef]
60. Vaganov, E.A. The tracheidogram method in tree-ring analysis and its application. In *Methods of Dendrochronology: Applications in the Environmental Sciences*; Cook, E.R., Kairiukstis, L.A., Eds.; Kluwer Academic Publishers: Dordrecht, The Netherlands, 1990; pp. 63–76. ISBN 0792305868.
61. Denne, M.P. Definition of latewood according to Mork (1928). *IAWA Bull. New. Ser.* **1988**, *10*, 59–62. [CrossRef]
62. Harrison, X.A.; Donaldson, L.; Correa-Cano, M.E.; Evans, J.; Fisher, D.N.; Goodwin, C.E.D.; Robinson, B.S.; Hodgson, D.J.; Inger, R. A brief introduction to mixed effects modelling and multi-model inference in ecology. *PeerJ* **2018**, *6*, e4794. [CrossRef] [PubMed]
63. De Micco, V.; Campelo, F.; De Luis, M.; Bräuning, A.; Grabner, M.; Battipaglia, G.; Cherubini, P. Intra-annual Density Fluctuations in Tree Rings: How, When, Where, and Why? *IAWA J.* **2016**, *37*, 232–259. [CrossRef]
64. Sperry, J.S. Evolution of Water Transport and Xylem Structure. *Int. J. Plant Sci.* **2003**, *164*, S115–S127. [CrossRef]
65. Sperry, J.S.; Hacke, U.G.; Pittermann, J. Size and function in conifer tracheids and angiosperm vessels. *Am. J. Bot.* **2006**, *93*, 1490–1500. [CrossRef] [PubMed]
66. Hacke, U.G.; Sperry, J.S.; Pockman, W.T.; Davis, S.D.; McCulloh, K.A. Trends in wood density and structure are linked to prevention of xylem implosion by negative pressure. *Oecologia* **2001**, *126*, 457–461. [CrossRef] [PubMed]
67. Maherali, H.; DeLucia, E.H. Xylem conductivity and vulnerability to cavitation of ponderosa pine growing in contrasting climates. *Tree Physiol.* **2000**, *20*, 859–867. [CrossRef] [PubMed]
68. Majer, A. *Fenyves a Bakonyalján*; Akadémiai Kiadó: Budapest, Hungary, 1988; ISBN 963-05-4461-x.
69. Kalliokoski, T.; Makinen, H.; Jyske, T.; Nojd, P.; Linder, S. Effects of nutrient optimization on intra-annual wood formation in Norway spruce. *Tree Physiol.* **2013**, *33*, 1145–1155. [CrossRef] [PubMed]
70. Jolly, W.M.; Dobbertin, M.; Zimmermann, N.E.; Reichstein, M. Divergent vegetation growth responses to the 2003 heat wave in the Swiss Alps. *Geophys. Res. Lett.* **2005**, *32*, L18409. [CrossRef]
71. Piermattei, A.; Campelo, F.; Büntgen, U.; Crivellaro, A.; Garbarino, M.; Urbinati, C. Intra-annual density fluctuations (IADFs) in *Pinus nigra* (J. F. Arnold) at high-elevation in the central Apennines (Italy). *Trees* **2020**, *34*, 771–781. [CrossRef]
72. Theodorou, C. Soil moisture and the mycorrhizal association of *Pinus radiata* D. don. *Soil Biol. Biochem.* **1978**, *10*, 33–37. [CrossRef]
73. Lieffers, V.J.; Rothwell, R.L. Rooting of peatland black spruce and tamarack in relation to depth of water table. *Can. J. Bot.* **1987**, *65*, 817–821. [CrossRef]
74. Rigling, A.; Waldner, P.O.; Forster, T.; Bräker, O.U.; Pouttu, A. Ecological interpretation of tree-ring width and intraannual density fluctuations in *Pinus sylvestris* on dry sites in the central Alps and Siberia. *Can. J. For. Res.* **2001**, *31*, 18–31. [CrossRef]
75. Körner, C.; Paulsen, J. A world-wide study of high altitude treeline temperatures. *J. Biogeogr.* **2004**, *31*, 713–732. [CrossRef]
76. Tyree, M.T.; Davis, S.D.; Cochard, H. Biophysical Perspectives of Xylem Evolution: Is there a Tradeoff of Hydraulic Efficiency for Vulnerability to Dysfunction? *IAWA J.* **1994**, *15*, 335–360. [CrossRef]
77. Pittermann, J.; Sperry, J. Tracheid diameter is the key trait determining the extent of freezing-induced embolism in conifers. *Tree Physiol.* **2003**, *23*, 907–914. [CrossRef] [PubMed]
78. Cavender-Bares, J. Impacts of Freezing on Long Distance Transport in Woody Plants. In *Vascular Transport in Plants*; Elsevier: Amsterdam, The Netherlands, 2005; pp. 401–424. ISBN 9780120884575.
79. Mihalca, I.A. The landscape of marshes in the Land of Dorna. *Analele Univ. Oradea Ser. Geogr.* **2012**, *22*, 66–71.
80. Aurelia, D.M.; Toader, C. Contributions to the Study of Vegetation in Tinovul Mare (Poiana Stampei—District of Suceava). *Bul. Grădinii Bot. Iaşi* **2007**, *14*, 75–80.
81. Barker, J.E. Growth and Wood Properties of *Pinus radiata* in Relation to Applied Ethylene. *N. Z. J. For. Sci.* **1979**, *9*, 15–19.
82. Yamamoto, F.; Kozlovwski, T.T. Effect of ethrel on growth and stem anatomy of *Pinus halepensis* seedlings. *IAWA Bull. New Ser.* **1987**, *8*, 11–19. [CrossRef]
83. Yamamoto, F.; Kozlowski, T.T. Effects of Flooding of Soil on Growth, Stem Anatomy, and Ethylene Production of *Thuja Orientalis* Seedlings. *IAWA J.* **1987**, *8*, 21–29. [CrossRef]
84. Kozlowski, T.T. Responses of woody plants to flooding and salinity. *Tree Physiol.* **1997**, *17*, 490. [CrossRef]
85. Gindl, W.; Grabner, M.; Wimmer, R. Effects of altitude on tracheid differentiation and lignification of Norway spruce. *Can. J. Bot.* **2001**, *79*, 815–821. [CrossRef]
86. Hevia, A.; Campelo, F.; Chambel, R.; Vieira, J.; Alía, R.; Majada, J.; Sánchez-Salguero, R. Which matters more for wood traits in *Pinus halepensis* Mill., provenance or climate? *Ann. For. Sci.* **2020**, *77*, 55. [CrossRef]
87. Borbély, E.; Indreica, A. Conservation status of the rare species *Betula nana* in the peatbog Tinovul Luci (the Harghita Mountains, Romania). *Stud. Univ. Babeş-Bolyai Biol.* **2019**, *64*, 5–12. [CrossRef]

88. Rosner, S.; Světlík, J.; Andreassen, K.; Børja, I.; Dalsgaard, L.; Evans, R.; Luss, S.; Tveito, O.E.; Solberg, S. Novel Hydraulic Vulnerability Proxies for a Boreal Conifer Species Reveal That Opportunists May Have Lower Survival Prospects under Extreme Climatic Events. *Front. Plant Sci.* **2016**, *7*, 1–14. [CrossRef]
89. Diaconu, A.-C.; Tanţău, I.; Knorr, K.-H.; Borken, W.; Feurdean, A.; Panait, A.; Gałka, M. A multi-proxy analysis of hydroclimate trends in an ombrotrophic bog over the last millennium in the Eastern Carpathians of Romania. *Palaeogeogr. Palaeoclimatol. Palaeoecol.* **2020**, *538*, 109390. [CrossRef]
90. Jelonek, T.; Tomczak, A.; Remlein, A.; Katrusiak, A.; Arasimowicz-Jelonek, M.; Gzyl, J. The novel relationship between the morphological characteristics of trees and ultrastructure of wood tissue in scots pine (*Pinus sylvestris* L.). *Wood Res.* **2015**, *60*, 519–530.
91. Tóth, E.G. Phylogeography and Adaptive Genetic Variation of Scots Pine (*Pinus sylvestris* L.) Populations from the Carpathians and the Pannonian Basin. Ph.D. Thesis, Szent István University, Budapest, Hungary, 2017.
92. Gruber, A.; Strobl, S.; Veit, B.; Oberhuber, W. Impact of drought on the temporal dynamics of wood formation in *Pinus sylvestris*. *Tree Physiol.* **2010**, *30*, 490–501. [CrossRef] [PubMed]
93. Novak, K.; de Luís, M.; Raventós, J.; Čufar, K. Climatic signals in tree-ring widths and wood structure of *Pinus halepensis* in contrasted environmental conditions. *Trees* **2013**, *27*, 927–936. [CrossRef]
94. Carvalho, A.; Nabais, C.; Vieira, J.; Rossi, S.; Campelo, F. Plastic Response of Tracheids in *Pinus pinaster* in a Water-Limited Environment: Adjusting Lumen Size instead of Wall Thickness. *PLoS ONE* **2015**, *10*, e0136305. [CrossRef] [PubMed]
95. Pacheco, A.; Camarero, J.J.; Carrer, M. Linking wood anatomy and xylogenesis allows pinpointing of climate and drought influences on growth of coexisting conifers in continental Mediterranean climate. *Tree Physiol.* **2016**, *36*, 502–512. [CrossRef] [PubMed]

forests MDPI

Article

Early Performance of Tree Species in a Mountain Reforestation Experiment

Robert Jandl *, Georg Kindermann, Cecilie Foldal, Silvio Schüler and Christina Bouissou

Austrian Forest Research Center, Seckendorff Gudent Weg 8, A-1131 Vienna, Austria;
georg.kindermann@bfw.gv.at (G.K.); cecilie.foldal@bfw.gv.at (C.F.); silvio.schueler@bfw.gv.at (S.S.);
christina.bouissou@bfw.gv.at (C.B.)
* Correspondence: robert.jandl@bfw.gv.at; Tel.: +43-664-826-9907

Abstract: Climate change requires forest managers to explore new concepts in reforestation. High-elevation sites are posing challenges because the range of tree species that can cope with present and future conditions is small and limited experience with candidate species is available. Methods: We selected a mountain site with nutrient-poor silicatic soils. The previous Norway spruce (*Picea abies*) stand performed poorly. We established a reforestation experiment with 27 tree species that were planted in different combinations in order to evaluate silvicultural options. Site preparation activities and planting techniques reflected the locally applied regular procedures. After planting, we monitored height growth and phenological characteristics of needle/leaf development in spring. The presently dominant Norway spruce was genetically characterized. Results: Tree seedlings planted at high elevation are highly vulnerable. The temporal course of needle/leaf sprouting varies widely. Early developers are vulnerable to frost, impairing tree development. Biotic stressors such as high population densities of weevils or mice can cause high mortality. Conclusion: we suggest a conservative approach to tree species selection because present site conditions in mountain areas may impair the development of many tree species that could be viable options in a considerably warmer climate.

Keywords: mountain forest; climate change; reforestation; tree species selection

Citation: Jandl, R.; Kindermann, G.; Foldal, C.; Schüler, S.; Bouissou, C. Early Performance of Tree Species in a Mountain Reforestation Experiment. *Forests* **2021**, *12*, 256. https://doi.org/10.3390/f12020256

Academic Editor: Csaba Mátyás

Received: 27 December 2020
Accepted: 15 February 2021
Published: 23 February 2021

1. Introduction

Climate change requires adaptive forest management strategies with considerable foresight. Reforestation activities after harvesting operations offer the opportunity to establish forest types that would not develop under present site conditions and that comprise tree species and tree-species mixtures that could be relevant in a future climate. Knowledge on the performance of many tree species in experiments increases the options for climate-smart forestry during stand development. Higher air temperatures in the future will allow the use of tree species in mountain forests that are currently predominantly found at warmer sites, e.g., at lower elevations. The challenge is finding tree species that can cope with currently harsh and future climatic conditions in mountain areas. As a consequence of climate change, many temperate forests are increasingly damaged by abiotic stressors such as frost and drought, and biotic stressors such as newly invading pests and pathogens [1]. The emerging problems are widely discussed, yet difficult to capture in forest growth models, because they require detailed and often unavailable information on site conditions. Moreover, planning of forest management strategies is hampered by the wide range of possible futures. In the case that worldwide climate-change mitigation strategies are successfully implemented, the future warming will be small, yet pessimistic scenarios indicate a stronger warming trend. Forests that are established now would ideally be able cope with a wide range of possible future climates. Climate scenarios are used in order to approximate the warming trend. However, the results from different climate models vary widely. Foresters are encouraged to interpret locally and regionally downscaled climate scenarios with climatology experts in order to get a good understanding of future conditions [2,3].

Change is the known unknown in decadal forest ecosystem development. The use of potential natural vegetation as guidance for the choice of tree species in forestry is partially compromised due to climate change effects [4]. Moreover, management intensity, air pollution, climate change, and increased nitrogen availability in recent decades has had some unexpected consequences on forests [5–9]. Using low-cost natural regeneration is not necessarily a promising method of sustainable forestry. Some tree species may already be underperforming as a consequence of climate change, as site conditions have already changed during the lifetime of existing forests. Some tree species shift their habitat range and change the competition compared to the present forest types [10–12]. An obvious approach to the selection of tree species in reforestation projects is the analysis of the performance of regionally encountered tree species. Yet, in some areas, the diversity of tree species is narrow, both due to natural constraints and management strategies that have favored monospecies forests. Climate conditions in mountain regions narrow the options of tree species selection. Frost episodes in late spring and early autumn render the use of tree species that develop their needles or leaves early in the growing season impossible. Climate change may lead to unprecedented water shortage and drought periods in summer. Understandably, silvicultural experiments have been focused on economically relevant and abundant tree species [13–15]. The performance of some tree species that may play an important role in mountain regions in the Alps is not yet sufficiently investigated. Textbook knowledge and the interpretation of observations from other regions are an unsatisfying basis for knowledge-based silvicultural concepts to cope with climate change. Field experiments are needed in order to analyze the productivity of different tree species and to understand their resilience to biotic and abiotic disturbances.

Forest managers have a range of adaptation options for climate change effects. A large intraspecific variation was found for Norway spruce in the Alps and its surrounding central and southeastern European range [16–18]. This high variation can support adaptation when forest practitioners select seed provenances that are better suited to future conditions [19]. Another option is often referred to as 'assisted migration', i.e., the intentional movement of tree species in response to anticipated climate change into new habitats. It is applied when tree species are feared to become maladapted upon climate change, but potentially relevant tree species are not able to disperse as quickly as the climatic conditions are changing [20–22]. Assisted migration can be applied to native tree species; unprecedented tree species combinations; or non-native, yet potentially relevant, tree species. The choice of non-native trees for assisted migration efforts is controversial. In particular, nature conservationists are discussing whether non-native plants are a benefit or threat for ecosystems [23].

Our experiment is included in the long-term experiment program of the Austrian Forest Research Center [24]. The intention is monitoring the stand development for several decades. The objectives of the experiment are as follows:

- Benchmark the performance (productivity, mortality) of a variety of tree species in comparison to the already encountered local tree population;
- Identify the candidate tree species for forests that need to cope with a changed climate in mountain regions of the Eastern Alps;
- In the long term, identify threats and challenges for tree species in a future climate;
- Characterize the performance of trees during their development;
- Identify the experimental challenges for the comparison of tree species with respect to adaptation of climate change.

In this paper, we describe the experimental setup, provide the rationale for selecting the chosen tree species, and document the planting process. Survival rate, height increase, and phenology are used as the first available indicators of the performance of different tree species. As the experiment advances we expect useful information for climate-smart reforestation projects at high-elevation sites on silicatic bedrock in the Eastern Alps.

2. Materials and Methods

2.1. Site Characteristics

The experimental site, Wechsel (47.9999° N, 15.9741° E), is located at an elevation of 1340 m a.s.l. in central Austria. The southwest-facing slope has an inclination of about 20%. The forests, according to the potential natural vegetation, are spruce-fir forests (*Luzulo nemorosae-Piceetum*) and Norway spruce is by far the dominating tree species [25]. The forests have low productivity. The yield class of the previous forest was 6, i.e., a mean annual growth of 6 m^3 stem wood during a production time of 100 years. This is far below the Austrian average, which is presently at 9 m^3 (http://waldinventur.at, accessed on 9 February 2021). Soils are derived from gneiss and schists and are sandy, rather shallow, and poor in nutrients. The water infiltration potential is high due to the abundance of weathered rocks, however, the water retention capacity in the rooted zone is low. The C:N ratio is wide and both the cation exchange capacity and the base saturation are low (Table 1).

Table 1. Chemical soil characterization. K, Ca, Mg, Al—exchangeable cations (unbuffered $BaCl_2$ extract), CEC—cation exchange capacity, BSat—base saturation.

Depth	pH	C	N	C:N	K	Ca	Mg	Al	CEC	BSat
cm	$CaCl_2$	mg/g		-			mmol$_c$/gram			%
−10–0	3.97	450	12	38						
0–10	3.83	46	1.8	26	1.4	15	2.4	53	76	24.1
10–20	4.06	19	0.9	21	0.6	2.8	1	34	40	11.0
20–30	4.18	8	0.6	13	0.5	3.6	1.3	24	31	17.3

The climatic conditions are shown in Figure 1. The Walter–Lieth diagram is based on a 60-year record of climate data collected at the closest climatological monitoring site, which is located 600-m below the experimental site (Figure 1). The diagram shows permanent snow cover from November to April and a chance of low temperatures until May. The growing season is consequently very short. Rain is quite abundant and drought is apparently a minor threat at the site. Climate change scenarios do not show a clear trend in precipitation patterns but a warming between 2 °C and 3.5 °C, depending on whether a path of RCP 4.5 or RCP 8.5 is followed [3,26].

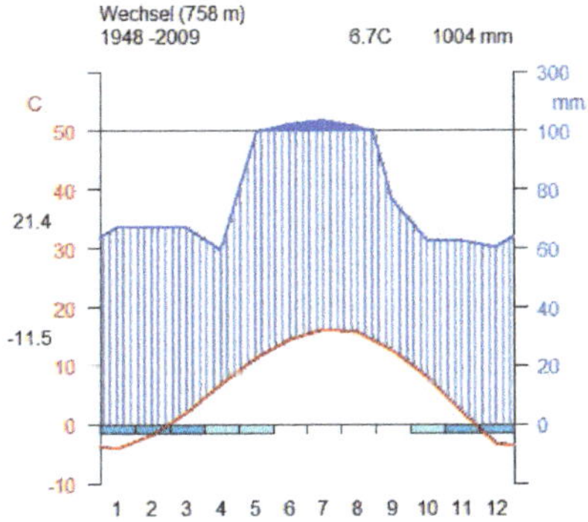

Figure 1. Walter–Lieth diagram, characterizing the climatic conditions at the experimental site, Wechsel. The red line shows the monthly average air temperatures, the blue line shows the monthly sums of precipitation. The horizontal bar shows the duration of permanent (dark blue) and intermittent (light blue) snow cover. The mean maximum daily temperature (21.4 °C) and the mean minimum daily temperature (−11.5 °C) are also shown.

The site was chosen for the experiment because the local forest owner, who had managed the forest enterprise during four decades, observed poor performance of his high-elevation spruce forests. Thinning operations that were expected to make more light, nutrients, and water available to the remaining trees had little effect on the productivity. The forest owner sought advice from scientists in order to make knowledge-based decisions when establishing the next forest generation. The previous Norway spruce-dominated forest was planted early in the 20th century. At that time, knowledge on tree genetics and provenances was in its infancy [27]. It is not documented whether site-adapted provenances have been chosen upon planting or seeding. Yet, it is well known that the forests in the region have been unsustainably used for centuries. Litter raking has deprived the soil nutrient pool. The timber and fuel-wood demand of the local population and the support of a small-scale glass industry, evidenced by the name of the adjacent village, 'Glashütten', led to exploitative forest use and degradation.

The previous Norway spruce stand was harvested in 2016. The stems were removed, roots and stumps were left in the soil, and logging residues (twigs and branches) were piled up at several stripes within the experimental site. No mulching or other soil treatment was performed. Both the harvesting technique and the dealings with branch and needle biomass reflect the normal modus operandi of the forest enterprise and its partners. The experimental site was fenced in order to keep out ungulates because the population density of ungulates is high and browsing would destroy the experiment. Just in the first year, a dense grass cover (mostly *Calamagrostis arundinacea*) developed and competed with tree seedlings for light and other resources. We did not take measures to control grasses, although they compete for light and nutrients in the early phase of stand development. The decision was taken because the managers of the forest enterprises in the region do without grass control for economic reasons and accept potential growth reductions of trees in the first couple of years. In order to ensure that our experimental site is not treated differently from other reforestations in the region, we adopted the same strategy.

2.2. Choice of Tree Species

The present dominance of Norway spruce is driven both by the biogeographical conditions and by forest management decisions of the past. Admixed species such as Silver fir (*Abies alba*), European larch (*Larix decidua*), and deciduous trees were often eliminated to follow forest concepts that were favoring Norway spruce. A further reduction of tree-species diversity was caused by selective browsing by ungulates [28]. Climate change will reduce the share of Norway spruce [10]. Yet, even in a warmer world, Norway spruce will be an important tree species in Austrian mountain forests. In high-elevation regions, even warming transiently increases the productivity of forests due to longer growing seasons. This is particularly the case in montane and subalpine, inner-alpine forests, where forest site conditions favor Norway spruce and exclude most competing tree species [25,29,30].

For our experiment, we used 27 tree species that are partially characterized in the European tree atlas [31]. Their expected growth performance and their tolerance towards frost and drought are shown in Table 2. For Norway spruce, we used different provenances: the presently recommended provenance for the forest district (*Picea abies* (high elevation)) and a provenance that is recommended for lower elevation (*Picea abies* (low elevation)) [32]. In addition, we collected local seedlings and left space for natural regeneration (*Picea abies* (natural regeneration)). Thereby, we can compare the performance of the planted Norway spruces with the locally occurring trees. The choice of tree species for our experiment was based on a discussion between regional forest managers, the regional forest authorities, and scientists. The size restriction at the experimental site limited the choice of tree species. We chose species that are expected to have relevance for timber production in the future and, in addition, some species that may have little relevance for forest enterprise and that are under-researched (Table 2).

Table 2. The used tree species in the reforestation experiment together with a brief characterization of growth (3—high to 1—low) and the a priori knowledge on their tolerance towards frost and drought (3—high to 1—low). Moreover, the table shows whether the tree species are native or non-native. 'N.I' stands for 'no reliable information available'.

Scientific Name	Common Name	Growth	Tolerance		Native/Non-Native
			Frost	Drought	
Picea abies	Norway spruce	3	3	1	native
Larix decidua	European larch	3	1	1	native
Abies alba	Silver fir	3	3	1	native
Pinus sylvestris	Scots pine	3	NI	3	native
Pinus cembra	Cembran pine	1	3	3	native
Pinus nigra	Black pine	3	NI	3	native
Pseudotsuga menziesii	Douglas fir	3	1	1	non-native
Pinus contorta	Lodgepole pine	3	NI	3	non-native
Picea engelmannii	Engelmann spruce	2	3	1	non-native
Abies grandis	Grand fir	3	3	1	non-native
Abies nordmanniana	Nordmann fir	2	NI	3	non-native
Larix x eurolepis	Hybrid larch	NI	NI	NI	non-native
Fagus sylvatica	European beech	1	1	3	native
Acer pseudoplatanus	Mountain maple	3	NI	1	native
Sorbus aucuparia	Rowan	NI	3	1	native
Sorbus aria	Common whitebeam	1	NI	3	native
Ulmus glabra	Scots elm	2	NI	3	native
Populus tremula	Eurasian aspen	3	3	2	native
Betula pendula	Silver birch	3	3	1	native
Pyrus austriaca	Pear	NI	NI	3	native
Prunus avium	Wild cherry	3	NI	1	native
Salix viminalis	Willow	NI	1	1	native
Tilia cordata	Small-leaved lime	NI	1	3	native
Quercus robur	Common oak	1	1	2	native
Alnus incana	Grey alder	3	3	1	native
Populus tremuloides X	Hybrid aspen	NI	NI	NI	non-native
Betula maximowicziana	Monarch birch	3	3	1	non-native
Sorbus intermedia	Swedish whitebeam	NI	NI	1	non-native

All trees were produced as containerized seedlings from two regional providers who regularly do business with the forest enterprise. The planting material was delivered in September 2017 and was immediately planted. Holes were made by a custom-made corer with an inner diameter exactly fitting the size of the root balls of the seedlings. The corer also had marks indicating the planting depth, as prescribed by the producer. After inserting the seedlings in the holes, the soil was cautiously tightened and covered with the previously removed soil. Locally collected spruce seedlings were treated differently. We selected naturally regenerating spruce specimens of similar height as the containerized seedlings, carefully excavated them with a spade in order to prevent root damage, inserted them in small open pits, and refilled the holes with the previously excavated soil material. The transfer of the naturally regenerated trees to their new growing spot was finished in approximately 30 minutes. In the experiment, the plots with transplanted seedlings from natural regeneration serve as a benchmark for the performance of spruce without forest management interference.

The number of used tree seedlings reflects the trade-off between the anticipated commercial relevance of trees in forests that are expected in the future, and scientific curiosity on the performance of tree species. The most abundant species are the local provenances of Norway spruce ('high elevation', 'low elevation'), followed by European larch, Silver fir, aspen (*Populus tremula*), Douglas fir (*Pseudotsuga menziesii*), pines (*Pinus sp.*), beech, and maple (*Acer pseudoplatanus*), shown in Table 3.

Table 3. Total number of tree seedlings that are used in the reforestation experiment at Wechsel.

Scientific Name	Number	Scientific Name	Number	Scientific Name	Number
Abies alba	216	Picea abies (high elevation)	432	Prunus avium	48
Abies grandis	48	Picea abies (local regeneration)	96	Pseudotsuga menziesii	216
Abies nordmanniana	48	Picea abies (low elevation)	216	Pyrus austriaca	48
Acer pseudoplatanus	144	Picea engelmannii	48	Quercus robur	48
Alnus incana	216	Pinus cembra	48	Salix viminalis	48
Betula maximowicziana	50	Pinus contorta	144	Sorbus aria	48
Betula pendula	48	Pinus nigra	48	Sorbus aucuparia	48
Fagus sylvatica	144	Pinus silvatica	144	Sorbus intermedia	48
Larix decidua	216	Populus tremula	216	Tilia cordata	47
Larix x eurolepis	144	Populus tremuloides	192	Ulmus glabra	47

The Austrian Forest Act limits the size of clear-cut areas and extra permission is required to justify a larger open area. We aimed at accommodating 31 treatments (single-species plots and plots with tree species combinations, with up to 3 replicates for each treatment; Table 4).

Table 4. Treatments in the reforestation experiment, Wechsel. 'n' denotes the number of replications.

Code	n	Description
	monospecies stands of coniferous trees	
Fi	3	pure Norway spruce; provenance for high elevation
FiT	3	pure Norway spruce; provenance for low elevation
Fi+FiT	3	mixture of high and low elevation provenances of Norway spruce
FiW	2	pure Norway spruce; locally collected seedlings
Ta	3	pure Silver fir
Lä	3	pure European larch
HLä	3	pure hybrid larch
Do	3	pure Douglas fir
DKi	3	pure Logepole pine
WKi	3	pure Scots pine
	monospecies stands of deciduous trees	
Bu	3	pure European beech
BAh	3	pure Mountain maple
Er	3	pure grey alder
As	3	pure Eurasian aspen
HAs1	1	pure aspen clone number 1
HAs2	1	pure aspen clone number 2
HAs3	1	pure aspen clone number 3
HAs123	1	mixture of aspen clones number 1, 2, 3
	mixed-species stands	
Fi+Lä	3	mixture of Norway spruce (provenance high elevation) and European larch
Do+Fi	3	mixture of Douglas fir and Norway spruce (provenance high elevation)
Er+Fi	3	mixture of grey alder and Norway spruce (provenance high elevation)
As+Ta	3	mixture of Eurasian aspen and Silver fir
	other characterization	
other species	variable	several species of interest that were planted on otherwise unused space
0	1	natural regeneration of any tree species

Size restrictions called for a space-efficient experimental setup. We applied a beehive-shapeddesign that minimizes the edge effect and densely packs a maximum number of plots in the available area. A grid of 5.4 m was set up across the entire site. Each grid point was marked with a wooden pole. Around each pole, a group of 4 seedlings was planted. In the upper-left corner of Figure 2, the concept is shown: one tree is planted in the center (number 2 in Figure 2), the next one is 1.5 m to the left along the contour line (number 1), and two more trees are 1.5 m to the right at an angle of +60° and −60° from the contour line (numbers 3 and 4). In the treatments involving two tree species, we alternated the planting pattern. In one 4-specimen group, the first tree species was planted in the center and the three surrounding positions were populated with the second tree species. In the next group, the pattern was reversed, so that we obtained a mixture ration of 50%. In the treatment that combines 3 clones of hybrid aspen (*Populus tremuloides X*), we alternated the pattern within each group in order to use the same number of seedlings from each clone.

Twelve cells are clustered to form a single plot containing 48 tree seedlings on 303 m^2. This represents a planting density of 1584 plants per hectare.

At the edges of the almost rectangular experimental site, no full plots could be fitted. At these spots, more tree species were planted in irregularly shaped plots. We used that space for planting several underinvestigated tree species in order to satisfy our scientific curiosity, but not under the assumption that these tree species will be commercially relevant in the future. The gray cell in the upper center ('0') was left open and allows the monitoring of natural regeneration of any tree species.

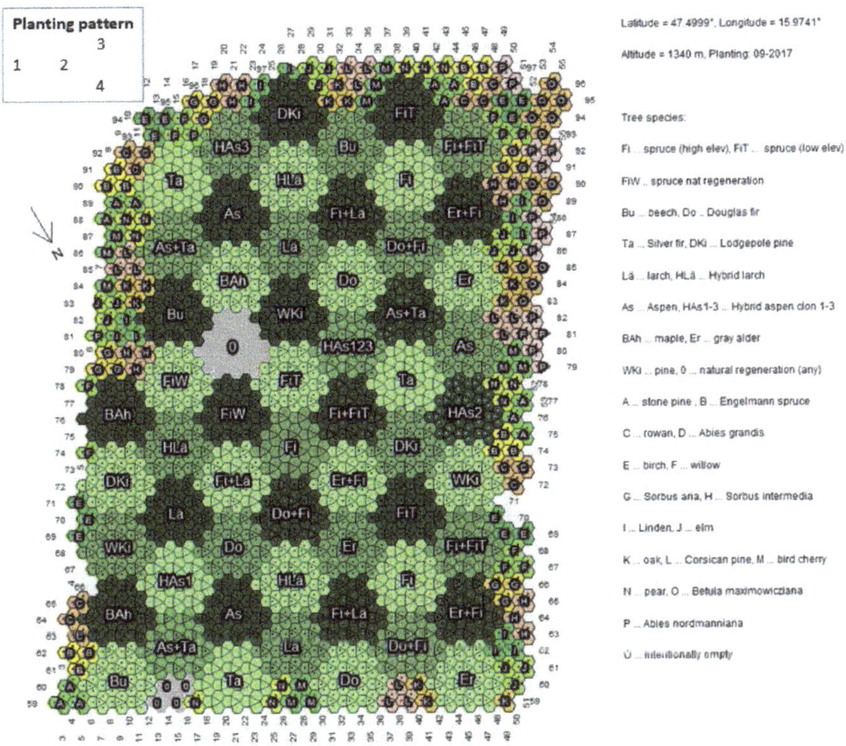

Figure 2. The space-saving experimental design minimizes the edge effect and allows for the maximum number of plots in a given area. The letters represent the tree species combinations on each plot. The arrangement of tree seedlings in one individual cell is shown in the upper-left corner.

The individual plots were spatially distributed according to a randomized process. The choice of tree species combinations was made in a discussion with the regional forest managers.

2.3. Phenology, Growth, Mortality

Beginning in spring 2017, we monitored the phenology of the tree seedlings. We distinguished between needles/leaves not-yet developed, partially developed, and fully developed. In addition, we assessed whether tree seedlings were damaged and identified the reasons for the damage by expert judgment. Biotic damages were inflicted by weevils (*Hylobius abietis*) that were unfortunately quite active in 2017 and 2018, and in 2020 by mice. Measures to control weevils by chemical measures were only partially successful. Abiotic damages were caused by frost. Dead seedlings were annually replaced, and the replacement was documented. The growth performance was assessed by measuring tree height in the autumns of 2017 to 2020. Here, we report the results of the last two years. The relative height increment was calculated as the ratio of the height difference between the years 2020 and 2017 and the initial height (2017). An ANOVA was used to test the significance of differences between species and the means were compared with a Tukey test in R version 4.0.3.

2.4. Genetics

A genetic analysis of Norway spruce was instigated in order to characterize the genetic variability of tree seedlings from a purportedly homogenous batch, because all seeds were collected from the provenance of Norway spruce in high elevation. Seeds were collected from different forest stands in the same growth region [25,32]. Needles of 65 Norway spruce specimen were picked from seedlings in summer 2019. A well-established Single Nucleotide Polymorphism Chip for Norway spruce was used with loci characterizing the climate, phenology, and drought stress [33]. DNA was extracted and a quality and quantity check was done by gel electrophoresis. Data were classified with respect to drought tolerance using machine learning [34]. A prediction accuracy of approximately 40% was achieved.

3. Results

The phenology in spring is shown in Figure 3. The early development varies widely. The three aspen clones, alder (*Alnus incana*), silver birch (*Betula pendula*), cherries (*Prunus sp.*), and rowan (*Sorbus aucuparia*) developed quickly, obviously rapidly responding to increasing temperatures. The conifers (pines, firs, larches, and spruces) are showing a slower start in spring. The differences within the Norway spruce provenances are minor. The gradual development in the two compared years is similar. In 2019, the phenology was monitored until the end of June. In 2020, the monitoring period was greatly extended in order to fully capture the needle-unfolding of the particularly slowly developing Corsican pine (*Pinus nigra*) and Scots pine (*Pinus sylvestris*).

The height of the trees in the second and third year of the experiment is shown in Figure 4. Considering that the grasses were more than half a meter high, many tree species were overgrown by grasses and primarily deciduous trees, which were delivered as taller seedlings reached over the canopy of the grasses. The measured differences in heights of different tree species were consistent between years. The 'number of specimen' at the top of the graph indicates the number of measured tree seedlings. It reflects and takes into account that many tree seedlings have died off for a number of reasons (compare Table 3).

The annual increase of height of different tree species is shown in Figure 5. In both years, the average height growth was modest and, in most cases, below 30 cm. Negative height increments indicate the effect of damages, when leading shoots were either damaged by frost or high individuals were damaged due to biotic factors such as weevils and mice. An ANOVA showed that the differences of the relative height increment were highly significant (Table 5). A Tukey test identified 6 groups with a considerable overlap between the different tree species.

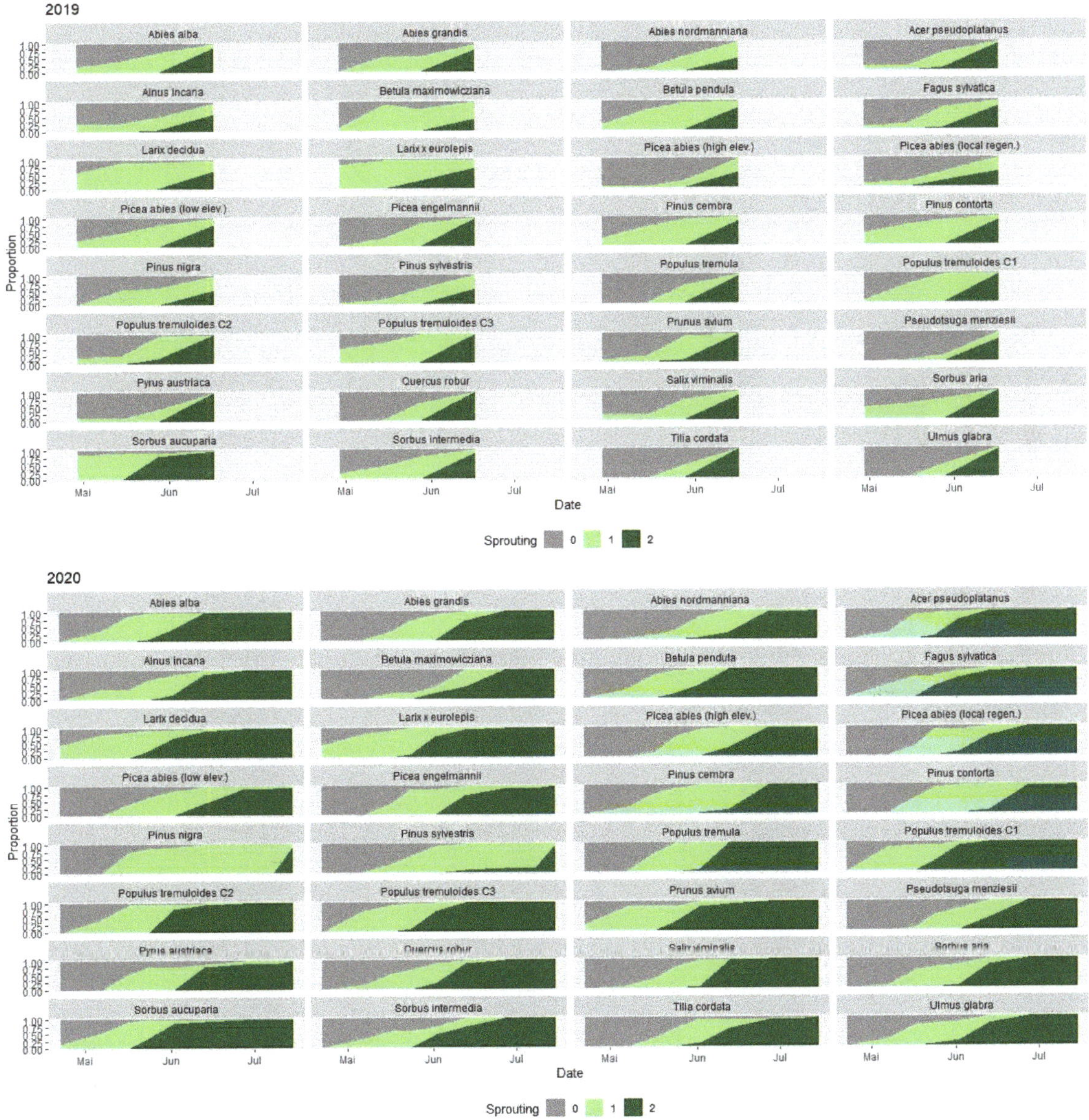

Figure 3. Temporal course of needle/leaf development at the experimental site of Wechsel in the springs of 2019 and 2020. Sprouting is classified as 0—no needles/leaves emerged, 1—needles/leaves partially developed, 2—needles/leaves fully developed.

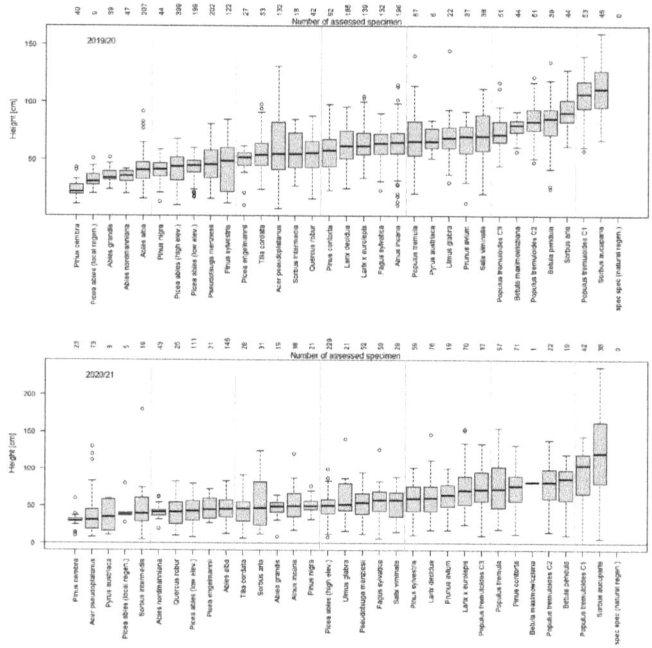

Figure 4. Height of trees in the reforestation experiment site, Wechsel, in 2019 and 2020. The number of assessed seedlings is shown at the top.

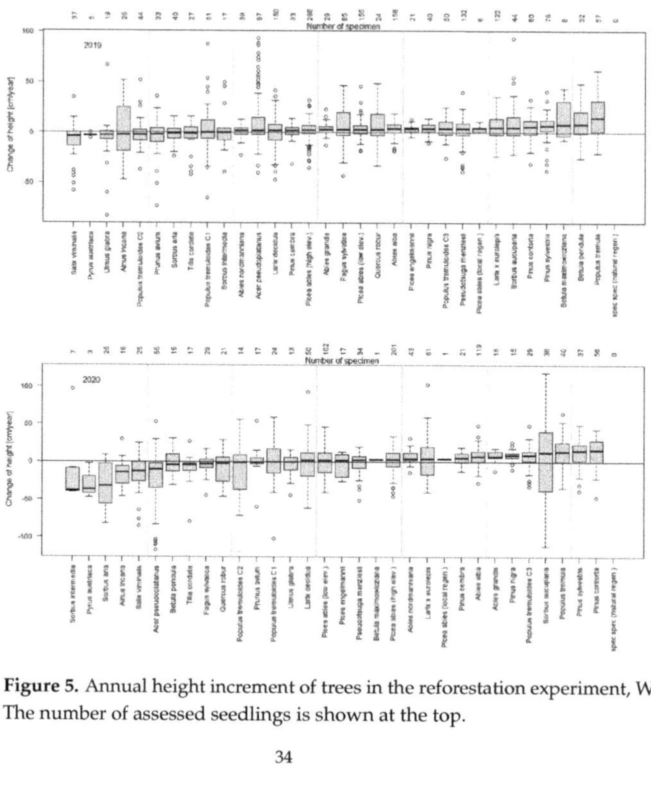

Figure 5. Annual height increment of trees in the reforestation experiment, Wechsel, in 2019 and 2020. The number of assessed seedlings is shown at the top.

Table 5. Analysis of variance for the height increment of different tree species in the reforestation experiment, Wechsel, between 2017 and 2020.

	Effect of Tree Species	Residuals
Sum of squares	221.2	617.2
Degrees of freedom	27	1279
F-value	16.9	

The mortality rate of tree seedlings was very high in both seasons. Monarch peak and alder was almost completely lost in both years and Douglas fir also did not develop well. We observed substantial differences between the years. Beech was mostly lost in the second year. Other tree species such as fir, pine, and rowan mostly survived (Figure 6).

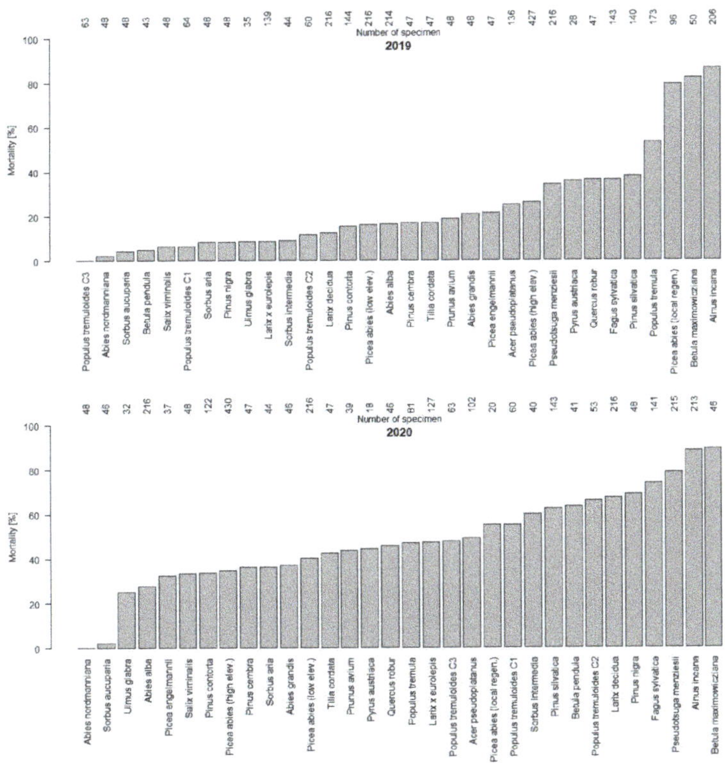

Figure 6. Mortality of tree seedlings in the reforestation experiment, Wechsel, in 2019 and 2020.

The genetic analysis showed the wide variation of drought tolerance within Norway spruce. Among the 65 investigated specimen, 30 were intolerant to drought, 22 were tolerant, and the remaining 13 assumed an intermediate position.

4. Discussion

The presently dominant tree species in the mountain forests of Austria is Norway spruce. The relevance of different spruce provenances and the options for forest managers to successfully deal with climate change have been demonstrated in several investigations [16,17]. Comparing the spruce provenances for low and high elevation, and the growth performance of locally collected seedlings, we observed only minor differences. The timing of sprouting and the mortality rate are similar and the annual height growth is around the average of the compared tree species (Figures 4 and 5). The high genetic variability of spruce within the same provenance indicates that large flexibility with respect to an eventually changing water supply. We preliminary conclude that low-elevation provenances of Norway spruce are suitable options for reforestation projects in high altitudes.

The high mortality of transplanted spruce seedlings that have been collected from adjacent stands is in agreement with observations from the local forest manager and was therefore expected. Transplanted tree seedlings surviving the transfer to a new location are potentially successful. However, we do not expect that the transplanted seedlings are a viable future option for forest managers, because the recently harvested forest was not growing well. The transplant seedlings rather complement the experiment when we benchmark their growth with the development of the purchased seedlings after the experiment. It allows us to establish whether the tree species used in the experiment indeed increase the options of forest managers in comparison to the expected natural regeneration (Figures 3–5).

A lot of research in the context of climate-smart forestry is done on Silver fir. It is a common native member of many forest types in the Alps and highly appreciated by nature conservationists. Its slow, juvenile development; slightly poorer technological quality compared to Norway spruce; susceptibility to damages caused by air pollution; and low tolerance to deer browsing have contributed to the decline of the population in productive forests in Austria. These adverse factors are now compensated by a higher tolerance to drought and a higher optimum temperature compared to Norway spruce [30,35]. In our experiment, the mortality of Silver fir was low. The performance with respect to height growth was average, but not particularly weak (Figures 3–5). The non-native *Abies grandis* was showing similar characteristics to Silver fir. Although the elevation range of the experimental site is well within the range of the natural occurrence of *Abies grandis*, the low soil fertility of the site may be an obstacle [36] (Table 1). Similar considerations apply for *Abies nordmanniana*, which performed very well in the first years of the experiment.

Many pines are tolerant to a wide range of site conditions. In our experiment, we included Scots pine, Stone pine, and Black pine as native, and Lodgepole pine as non-native species (Figure 2 and Table 3). Scots pine is a typical late developer in spring and is therefore believed to be especially tolerant to late frost events in spring (Figure 2). In our experiment, the mortality rate was low and the height growth was above average. Black pine, which grows naturally at low elevations, performed poorly. A relevant factor may be that the delivered first batch of seedlings was tiny and had little chance to cope with the transfer from the production site to a harsh environment. The compared Stone pine is typically encountered in high-elevation ecosystems [30]. It is usually not seen as an option for mountain forests in a warmer world [37,38]. We introduced Stone pine in order to properly reflect both present and future site conditions. Stone pine is presently developing slowly, undeterred by mortality for biotic or abiotic factors. Under future climatic conditions the experimental site may well be far out off the range of site conditions where Stone pine belongs. Eventually, Stone pine may be pushed out of its natural habitat, as it has been demonstrated for herbaceous plants [39,40]. The non-native lodgepole pine was developing in a similar fashion to Scots pine (Figures 3–5).

The non-native Douglas fir is considered as an economically viable alternative for Norway spruce. It is particularly interesting because it is a native species sensu lato that was 'lost' during the Pleistocene and could be reintegrated in the flora of Central Europe [41]. Among the many provenances of Douglas fir, only some may be relevant for mountain forests in Austria in a future climate. However, in our experiment Douglas fir was performing poorly so far. The mortality was exceptionally high (>60%). Frost damaged the tree species, although did not distinctly belong to early developers. The height growth was above the average of the compared species (Figures 3–5). Due to its high mortality, Douglas fir is not considered as additional 'treatment' in the future of the experimental site.

Deciduous trees were quite different to the conifers. When delivered and planted, the specimens were taller than coniferous trees. An obvious choice for our experiment was birch and we included two species (compare Table 2), i.e., Silver birch and Monarch birch. Birch is a common tree species in Nordic countries and was expected to perform well at the experimental site. However, both species showed a high mortality rate. The harsh climate conditions may be an important factor, but we mostly consider damages inflicted by weevils and mice as the main reasons. Birch is definitely a late-developing species (Figure 3), thereby avoiding the deleterious effect of late frost. Particularly successful in our experiment were, until now, the members of the *Sorbus* genus. We included Rowan and Common whitebeam as native trees in mountain forests and Swedish whitebeam as the non-native species. *Sorbus* had similar sprouting characteristics (Figure 3) and a low mortality rate. Rowan withstood biotic stress better than other species of the genus.

Maple was expected to be a sure high-performer. In the vicinity of the experimental site, we encountered patches of maple in the spruce-dominated forests. However, maple suffered from damages inflicted by weevils and mice. The absence of maples in the presently encountered forests gives reason to believe that under current management strategies and site constraints, the species is not competitive.

The poor performance of typical low-elevation species such as oak, cherry, and pear indicates a major challenge of 'assisted tree migration'. These species may well thrive under future climatic conditions, but it is risky to include them in reforestation experiments. Overall, they are suffering from frost in high elevation, and a small proportion of the planted specimen has a chance to survive the juvenile phase.

We included additional species in our experiment, but we are reluctant to express a final verdict on their future performance. Although the phenology in spring shows clear patterns in both years of observations, the height growth has been less consistent. Even when tree species are quite successful during several years, they may be vulnerable to extreme climatic episodes that are not reflected in long-term climatic averages. Frost events in late spring and early summer can confine the range of options when choosing tree species for the next forest generation. Species that are developing their needles/leaves early in spring require a high frost tolerance, because late frost will inevitably happen eventually in mountain forests of the Eastern Alps.

The mortality rate of trees was frustratingly high. Yet, we learned from experience that experiments in high-elevation forests need to be tightly controlled in order to reduce the potential stress factors. Biotic damages due to weevil attacks or mice are totally unrelated to site conditions and climate change, and the performance of trees affected by them does not contribute greatly to the knowledge on future forest management options. However, every effort is made to minimize the effect of undesired and uncontrolled effects on experimental outcomes. The experimental team is required to minimize external and random effects such as damages due to weevil or mice populations in order to isolate the response factor, i.e., the response of tree species to future site conditions. In order to strengthen the power of the experiment, additional activities are recommended. Fencing the experimental plot was already justified in order to minimize the damages inflicted by deer. Defense measures against pests (e.g., weevils and mice) must be considered in order to ensure maximum effectiveness of the conducted experiment.

The future of forests in mountain environments is at stake, and forestry relies on many controlled experiments in order to draw reproducible conclusions for climate-smart forestry. The described experiment here is one of many required building blocks for establishing knowledge for science-based solutions. The design of the experiment is capable of maintaining a high tree species diversity, even after thinning operations. The alternation of tree species in each 4-tree group, as described in the experimental setup, ensures that the main tree species are continuously represented. With one tree remaining in each 4-tree group after thinning, the forest will maintain a high tree-species diversity (Figure 2). An element that we could not yet exploit in our analysis is the natural regeneration of different tree species at the experimental site. Monitoring in the future will show how transplanted spruce seedlings and natural regeneration of any tree species will perform in comparison to the planted trees. Thereby, we will be able to determine effects of the natural process compared to active reforestation.

Author Contributions: Conceptualization, R.J., S.S., G.K., C.F.; methodology, R.J., S.S., G.K., C.F.; software, G.K., C.F.; validation, R.J., S.S., G.K., C.F.; formal analysis, R.J., S.S., G.K., C.F.; investigation, R.J., S.S., G.K., C.F., C.B.; data curation, C.F., G.K.; writing—original draft preparation, R.J.; writing—review and editing, R.J., S.S., G.K., C.F.; visualization, G.K., C.F.; supervision, R.J.; project administration, C.F.; funding acquisition, R.J. All authors have read and agreed to the published version of the manuscript.

Funding: This research was funded by Austrian Research Promotion Agency (FFG) under grant number 858539.

Institutional Review Board Statement: Not applicable.

Informed Consent Statement: Not applicable.

Data Availability Statement: The data presented in this study are available on request from the corresponding author. The data are not publicly available due to their complexity.

Acknowledgments: The forest enterprise Schenker Forest provided technical support and in kind contributions to field work (e.g., materials used for experiments). Eva-Maria Sehr from the Austrian Institute of Technology (AIT) provided the genetic analysis. James Connell investigated plant damages and interpreted the impact of biotic and abiotic stressors. The research was financially supported by the Austrian Research Promotion Agency (FFG).

Conflicts of Interest: The authors declare no conflict of interest.

References

1. Seidl, R.; Thom, D.; Kautz, M.; Martin-Benito, D.; Peltoniemi, M.; Vacchiano, G.; Wild, J.; Ascoli, D.; Petr, M.; Honkaniemi, J.; et al. Forest disturbances under climate change. *Nat. Clim. Chang.* **2017**, *7*, 395–402. [CrossRef] [PubMed]
2. Collins, M.; Knutti, R.; Arblaster, J.; Dufresne, J.L.; Fichefet, T.; Friedlingstein, P.; Gao, X.; Gutowski, W.; Johns, T.; Krinner, G.; et al. Long-term Climate Change: Projections, Commitments and Irreversibility. In *Climate Change 2013: The Physical Science Basis. Contribution of Working Group I to the Fifth Assessment Report of the Intergovernmental Panel on Climate Change*; Cambridge University Press: Cambridge, UK; New York, NY, USA, 2013; Chapter 12; pp. 1029–1136.
3. Chimani, B.; Heinrich, G.; Hofstätter, M.; Kerschbaumer, M.; Kienberger, S.; Leuprecht, A.; Lexer, A.; Pessenteiner, S.; Poetsch, M.; Salzmann, M.; et al. Endbericht ÖKS15–Klimaszenarien für Österreich. Available online: https://hdl.handle.net/20.500.11756/06edd0c9 (accessed on 5 February 2018).
4. Hickler, T.; Vohland, K.; Feehan, J.; Miller, P.A.; Benjamin, S.; Costa, L.; Giesecke, T.; Fronzek, S.; Carter, T.R.; Cramer, W.; et al. Projecting the future distribution of European potential natural vegetation zones with a generalized, tree species-based dynamic vegetation model. *Glob. Ecol. Biogeogr.* **2012**, *21*, 50–63. [CrossRef]
5. Schleppi, P.; Körner, C.; Klein, T. Increased Nitrogen Availability in the Soil Under Mature Picea abies Trees Exposed to Elevated CO_2 Concentrations. *Front. For. Glob. Chang.* **2019**, *2*, 59. [CrossRef]
6. Spiecker, H.; Mielikäinen, K.; Köhl, M.; Skovsgaard, J. *Growth Trends in Europe-Studies from 12 countries*; EFI Research Reports; Springer: Heidelberg, Germany, 1996; Volume 5, pp. 275–289.
7. Seidl, R.; Schelhaas, M.J.; Lexer, M.J. Unraveling the drivers of intensifying forest disturbance regimes in Europe. *Glob. Chang. Biol.* **2011**, *17*, 2842–2852. [CrossRef]
8. Seidl, R.; Rammer, W.; Lexer, M.J. Adaptation options to reduce climate change vulnerability of sustainable forest management in the Austrian Alps. *Can. J. For. Res.* **2011**, *41*, 694–706. [CrossRef]
9. Thom, D.; Seidl, R.; Steyrer, G.; Krehan, H.; Formayer, H. Slow and fast drivers of the natural disturbance regime in Central European forest ecosystems. *For. Ecol. Manag.* **2013**, *307*, 293–302. [CrossRef]
10. Hanewinkel, M.; Cullmann, D.A.; Schelhaas, M.J.; Nabuurs, G.J.; Zimmermann, N.E. Climate change may cause severe loss in the economic value of European forest land. *Nat. Clim. Chang.* **2012**, *3*, 203–207. [CrossRef]
11. Millar, C.I.; Stephenson, N.L. Temperate forest health in an era of emerging megadisturbance. *Science* **2015**, *349*, 823–826. [CrossRef] [PubMed]
12. Casalegno, S.; Amatulli, G.; Bastrup-Birk, A.; Durrant, T.H.; Pekkarinen, A. Modelling and mapping the suitability of European forest formations at 1-km resolution. *Eur. J. For. Res.* **2011**, *130*, 971–981. [CrossRef]
13. Johann, K. Ergebnisse von Düngungsversuchen nach 30 Jahren ertragskundlicher Beobachtung. *Berichte FBVA* **2000**, *114*, 1–93.
14. Pretzsch, H. *Forest Dynamics, Growth and Yield: From Measurement to Model*; Springer: Berlin/Heidelberg, Germany, 2010.

15. Pretzsch, H.; Forrester, D.I.; Bauhus, J. (Eds.) *Mixed-Species Forests: Ecology and Management*; Springer: Berlin/Heidelberg, Germany, 2017.
16. Kapeller, S.; Lexer, M.J.; Geburek, T.; Hiebl, J.; Schueler, S. Intraspecific variation in climate response of Norway spruce in the eastern Alpine range: Selecting appropriate provenances for future climate. *For. Ecol. Manag.* **2012**, *271*, 46–57. [CrossRef]
17. Chakraborty, D.; Jandl, R.; Kapeller, S.; Schueler, S. Disentangling the role of climate and soil on tree growth and its interaction with seed origin. *Sci. Total. Environ.* **2019**, *654*, 393–401. [CrossRef] [PubMed]
18. Trujillo-Moya, C.; George, J.P.; Fluch, S.; Geburek, T.; Grabner, M.; Karanitsch-Ackerl, S.; Konrad, H.; Mayer, K.; Sehr, E.M.; Wischnitzki, E.; et al. Drought Sensitivity of Norway Spruce at the Species' Warmest Fringe: Quantitative and Molecular Analysis Reveals High Genetic Variation Among and Within Provenances. *G3 Genes Genomes Genet.* **2018**, *8*, 1225–1245. [CrossRef]
19. Frank, A.; Howe, G.T.; Sperisen, C.; Brang, P.; Clair, J.B.S.; Schmatz, D.R.; Heiri, C. Risk of genetic maladaptation due to climate change in three major European tree species. *Glob. Chang. Biol.* **2017**, *23*, 5358–5371. [CrossRef]
20. Dumroese, R.K.; Williams, M.I.; Stanturf, J.A.; Clair, J.B.S. Considerations for restoring temperate forests of tomorrow: Forest restoration, assisted migration, and bioengineering. *New For.* **2015**, *46*, 947–964. [CrossRef]
21. Gömöry, D.; Krajmerová, D.; Hrivnák, M.; Longauer, R. Assisted migration vs. close-to-nature forestry: What are the prospects for tree populations under climate change? *Cent. Eur. For. J.* **2020**, *66*, 63–70. [CrossRef]
22. Chakraborty, D.; Gaviria, J.; Bednárová, D.; Bolte, A.; Bouissou, C.; Buchacher, R.; Hazarika, R.; Henning, L.; Kowalczyk, J.; Longauer, R.; et al. Implementing assisted migration. Output of the Interreg Central Europe Programme 2014–2020. *SUSTREE Policy Brief* **2019**, *2*. [CrossRef]
23. Schlaepfer, M.A.; Sax, D.F.; Olden, J.D. The Potential Conservation Value of Non-Native Species. *Conserv. Biol.* **2011**, *25*, 428–437. [CrossRef]
24. Ledermann, T. (Ed.) *Long-Term Experimental Plots–Valuable Heritage or Heavy Burden?* BFW-Berichte; Austrian Forest Research Center: Vienna, Austria, 2017; Volume 153.
25. Kilian, W.; Müller, F.; Starlinger, F. Die Forstlichen Wuchsgebiete Österreichs. *FBVA Berichte* **1994**, *82*, 1–58.
26. Jacob, D.; Petersen, J.; Eggert, B.; Alias, A.; Christensen, O.B.; Bouwer, L.M.; Braun, A.; Colette, A.; Déqué, M.; Georgievski, G.; et al. EURO-CORDEX: New high-resolution climate change projections for European impact research. *Reg. Environ. Chang.* **2014**, *14*, 563–578. [CrossRef]
27. Finkeldey, R. Forschung zur Vielfalt, vielfältige Forschung: Ziele und Wege der Forstgenetik. *Schweiz. Z. FüR Forstwes.* **2001**, *152*, 162–168. [CrossRef]
28. Schodterer, H. (Ed.) *Bundesweites Wildeinflussmonitoring 2016–2018*; BFW Praxisinformation; Austrian Forest Research Center: Vienna, Austria, 2019; Volume 48.
29. Lexer, M.J.; Jandl, R.; Nabernegg, S.; Bednar-Friedl, B. Forestry. In *Economic Evaluation of Climate Change Impacts-Development of a Cross-Sectoral Framework and Results for Austria*; Steininger, K.W., König, M., Bednar-Friedl, B., Kranzl, L., Loibl, W., Prettenthaler, F., Eds.; Springer: Berlin/Heidelberg, Germany, 2015; Chapter 9; pp. 145–165.
30. Mayer, H. *Wälder des Ostalpenraumes–Standort, Aufbau und Waldbauliche Bedeutung der Wichtigsten Waldgesellschaften in den Ostalpen Samt Vorland*; G. Fischer: Stuttgart, Germany, 1974.
31. San-Miguel-Ayanz, J.; de Rigo, D.; Caudullo, G.; Houston Durrant, T.; Mauri, A. (Eds.) *European Atlas of Forest Tree Species*; Publication Office of the EU: Luxembourg, 2016.
32. BFW. Herkunftsberatung. 2020. Available online: http://bfw.ac.at/hkd/herkauswahl.einstieg (accessed on 26 January 2021).
33. Heer, K.; Ullrich, K.; Liepelt, S.; Rensing, S.; Zhou, J.; Ziegenhagen, B.; Opgenoorth, L. Detection of SNPs based on transcriptome sequencing in Norway spruce (*Picea abies* (L.) Karst.). *Conserv. Genet. Resour.* **2016**, *8*, 105–107. [CrossRef]
34. Liaw, A.; Wiener, M. Classification and Regression by randomForest. *R News* **2002**, *2*, 18–22.
35. George, J.P.; Schueler, S.; Karanitsch-Ackerl, S.; Mayer, K.; Klumpp, R.T.; Grabner, M. Inter- and intra-specific variation in drought sensitivity in Abies spec. and its relation to wood density and growth trait. *Agric. For. Meteorol.* **2015**, *214–215*, 430–443. [CrossRef]
36. Ruhm, W.; Schuster, K.; Schönauer, J. Mögliche Gastbaumarten für die österreichische Forstwirtschaft. *Die Landwirtsch.* **2015**, *20*, 1–20.
37. Jandl, N.; Jandl, R.; Schindlbacher, A. Future management options for cembran pine forests close to the alpine timberline. *Ann. For. Sci.* **2018**, *75*, 81. [CrossRef]
38. Bircher, N. To Die or Not to Die: Forest Dynamics in Switzerland under Climate Change. Ph.D. Thesis, ETH Zurich, Zurich, Switzerland, 2015. [CrossRef]
39. Grabherr, G.; Gottfried, M.; Pauli, H. Climate effects on mountain plants. *Nature* **1994**, *369*, 448. [CrossRef]
40. Pauli, H.; Gottfried, M.; Dullinger, S.; Abdaladze, O.; Akhalkatsi, M.; Alonso, J.L.B.; Coldea, G.; Dick, J.; Erschbamer, B.; Calzado, R.F.; et al. Recent Plant Diversity Changes on Europe's Mountain Summits. *Science* **2012**, *336*, 353–355. [CrossRef] [PubMed]
41. Schulze, E. Effects of forest management on biodiversity in temperate deciduous forests: An overview based on Central European beech forests. *J. Nat. Conserv.* **2018**, *43*, 213–226. [CrossRef]

forests

MDPI

Article

Performance and Phenotypic Stability of Norway Spruce Provenances, Families, and Clones Growing under Diverse Climatic Conditions in Four Nordic Countries

Tore Skrøppa * and Arne Steffenrem

Norwegian Institute of Bioeconomy Research (NIBIO), Division of Forestry and Forest Resources, P.O. Box 115, NO-1431 Ås, Norway; arne.steffenrem@nibio.no
* Correspondence: tore.skroppa@nibio.no; Tel.: +47-930-07-710

Abstract: Genetic variation and phenotypic stability in Norway spruce were studied based on provenances, families, and clones planted in trials at 12 sites in four Nordic countries. The families were generated in a factorial cross between 10 parents of Norwegian origin and 10 parents of Eastern European origin, and the clones were propagated from seedlings within 20 of the same families. Traits analyzed were survival, proportion of trees with stem defects, and tree heights. Stability was analyzed by regression analyses with the genetic entries' annual shoot increment as the dependent variable and the total site mean as an environmental index. Information about growth and phenology traits were available from short-term tests. For tree heights, significant variance components were present both among female and male parents, but not for their interactions, indicating that non-additive genetic effects are small. Genotype × environment interactions were significant at all three genetic levels, but their variance components had considerably lower values than the variance components estimated for the effects of families and clones. For the set of families of Norwegian origin, strong relationships were observed between the timing of annual shot elongation, mortality, and height growth. Large variation was found at all three genetic levels for phenotypic stability measured by regression coefficients. A positive relationship was present between the regression coefficient and the timing of annual shoot growth for families, indicating that later flushing families responded more to a high site index. The regression coefficient can be a useful supplement to the breeding value when selecting for superior and stable genotypes.

Keywords: *Picea abies*; field trials; provenances; families; clones; height growth; phenotypic stability; phenotypic plasticity

Citation: Skrøppa, T.; Steffenrem, A. Performance and Phenotypic Stability of Norway Spruce Provenances, Families, and Clones Growing under Diverse Climatic Conditions in Four Nordic Countries. *Forests* **2021**, *12*, 230. https://doi.org/10.3390/f12020230

Academic Editor: Csaba Mátyás

Received: 13 January 2021
Accepted: 13 February 2021
Published: 17 February 2021

1. Introduction

The norm of reaction of a genotype describes its pattern of phenotypic response to changing environmental growth conditions. It characterizes its phenotypic plasticity, which is the ability of the genotype to form varying phenotypes. A related concept, more appropriate when considering the growth performance of forest trees in a series of trials, is phenotypic stability, which is here defined in the dynamic sense [1]: a genetic unit is said to be stable if its growth performance is similarly related to the mean production level at sites with different environmental conditions for growth. Deviations from the general response function of all genetic units tested are considered as contributions to instability. If norms of reaction of different genotypes are not parallel, then genotype × environment (G × E) interactions are present. Such instability causes changes in the ranking of the genotypes growing at different environments. This has implications in tree breeding, where one objective is to produce reproductive materials that have a stable and high growth performance across a wide range of environmental conditions. The materials should also be stable with temporal environmental variation at the same site (e.g., climate change). If the interactions are related to predictable factors such as site and climate, they can to some extent be adjusted for by subdivision of breeding and deployment zones [2]. However, the breeders of Norway spruce in the Nordic countries must consider that environmental heterogeneity is substantial within actual zones for factors such as soil productivity and local climatic conditions. More knowledge of the genetic components of phenotypic stability in Norway spruce and its implications in selection should therefore be provided.

In Norway spruce (*Picea abies (L.) Karst.*) trials in the Nordic countries some G × E interactions for height growth have been demonstrated both for provenances [3], families [4,5], and sets of clones [6,7]. In the lowland of southern Scandinavia, the main cause of the interactions in this species seems to be damage caused by late spring frosts [7]; clones with an early bud flush contributed most to the G × E interactions. In an analysis of G × E interactions for height growth in breeding populations in southern Sweden, the interactions were low to moderate and largely unpredictable [6], and with spring frost damage as an important factor. It has been suggested that the same genetic material could be used over large areas in southern Sweden [2,5,7]. However, in a recent study of Norway spruce progeny trials in southern and central Sweden [8], quite strong G × E interactions were found for tree height, and climate indices of spring and autumn temperature could account for a large amount of these interactions, with frost damage as the important cause. Substantial non-additive genetic effects for height and diameter in Norway spruce were found for clones from full-sib and half-sib families [9], and G × E interactions were strong for non-additive genetic effects.

The present study is based on assessments and measurements made in 15 field trials in four Nordic countries comprising provenances, full-sib families, and clones of Norway spruce. The origin and selection history varied between the parents of the full-sib families. Our objective is to present the variability and inheritance patterns at three genetic levels and the implications of different genetic backgrounds of the full-sib families and clones. Furthermore, as significant G × E interactions were found for tree height, we wanted to characterize the genetic variation in phenotypic stability of performance across multiple sites and show how this response parameter is related to phenology traits. Selection for stability in breeding of Norway spruce will be discussed.

2. Materials and Methods

2.1. Materials, Trials, and Measurements

Seedlings from 20 Norway spruce provenances, 100 full-sib families from a 10 × 10 factorial cross and 240 clones from a subset of 20 of the families, were planted in 17 field trials at 12 sites in Norway, Sweden, Finland, and Denmark in 1988 and 1989 (Table 1, Figure 1). Seedlings from the 100 full-sib families had earlier been planted in a short-term trial in Norway, and data from early height growth and phenology traits are available from this trial [10].

The materials planted in 1988 at seven sites comprise seedlings from 100 full-sib families after controlled crosses made in 1983 at Stange Seed Orchard in Norway (lat. 60°43′ N, long. 11°11′ E, alt. 180 m) and from 20 provenances. Eight of the provenances were from Sweden and Finland, six were from Eastern Europe, four were from the Carpathian Mountains, and two were from Harz, Germany. At the site Rostorp (trial 3), only six provenances were planted. The full-sib families had been generated in a complete 10 × 10 factorial cross. One group of ten parents were plus-trees originating from natural stands in a rather limited geographical area between altitudes 40 and 320 m in southern Norway (N). The other group of ten parents were young plus-trees selected from five outstanding Eastern European (E) provenances from latitudes between 47° and 57° and altitudes between 80 and 720 m in Romania, Belarus, North-East Poland, and Latvia, in the international Norway spruce provenance trial of 1938 planted at Södra Bäcksjö, Sweden, latitude 63°56′ [11]. The selections were done in 1967, at age 25 years from planting. These trees were phenotypically selected for superior height growth and absence of stem defects such as double stems, forking, and ramicorn branches [10]. In the factorial cross, five parents from each of the two groups of origin were used as maternal and five as paternal parents (Table 2). Thus, the 100 full-sib families can logically be sub-divided into four separate groups of crosses based on the origin of the parents: N × N, N × E, E × N, and E × E. Eight of the field trials were in 1989, planted with rooted cuttings from a selected set of 20 of the full-sib families. These 20 families were pairwise related with one parent in common, as shown in Table 2.

Table 1. Locations of field trials with average survival (%), tree height, and proportion of trees with stem defects. Trials 1, 2, 11, and 12 are in Norway; trials 3, 4, 13, and 14 in Sweden; trials 5, 6, 7, 16, 17, and 18 in Finland; and trial 15 is in Denmark. Trials 1 to 7 comprise provenances and families, while trials 11 to 18 comprise clones propagated from 20 of the families. The average values are obtained from measurements nine growing seasons after planting in trials 17 and 18 (*), eleven in trials 1 and 4 (**), and ten in the remaining trials.

Trial and Location	Latitude °N	Longitude °E	Altitude M	Mortality %	Height cm	Stem Defects %
1 Ølve	60°00′	5°50′	50	10.2	260 **	8.9
2 Skiptvedt	59°28′	11°11′	100	24.8	224	15.5
3 Rostorp	57°47′	15°47′	165	66.0	235	18.1
4 Lappkuliden	64°16′	19°37′	190	52.8	174 **	19.3
5 Paimio	60°27′	22°44′	40	4.1	279	21.7
6 Janakkala	61°00′	24°45′	130	27.6	178	16.0
7 Imatra	61°10′	28°53′	80	42.9	152	26.9
11 Kaupanger	61°11′	7°05′	80	8.0	183	10.0
12 Nannestad	60°19′	11°05′	200	22.1	256	14.5
13 Ribbingelund	59°19′	16°42′	50	31.9	206	11.5
14 Lappkuliden	64°1′	19°37′	190	33.2	147	37.0
15 Hørsholm	55°52′	12°04′	50	14.7	377	31.9
16 Finström	60°15′	19°57′	5	24.3	227	27.4
17 Paimio	60°27′	22°44′	40	4.5	257 *	36.1
18 Imatra	61°10′	28°53′	80	28.9	126 *	39.0

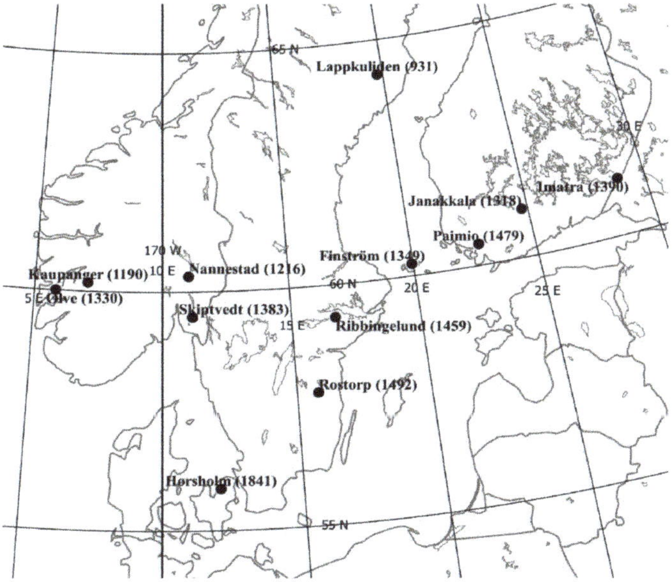

Figure 1. The field test sites in four Nordic countries with the annual temperature sum for mean daily temperature above 5 °C in brackets.

Table 2. The 100 factorial crosses made between 10 Norwegian (N) and 10 Eastern European (E) parents. The N × N and E × E families are indicted as shaded squares. The 20 families that were propagated by rooted cuttings are denoted by <u>X</u>.

Famale Parents		**Male Parents**									
		Norwegian (N)					**Eastern European (E)**				
		11	**12**	**13**	**14**	**15**	**16**	**17**	**18**	**19**	**20**
Norwegian (N)	1	X	X	X	<u>X</u>	X	X	X	X	<u>X</u>	X
	2	X	X	X	X	<u>X</u>	X	X	X	X	<u>X</u>
	3	<u>X</u>	X	X	X	X	<u>X</u>	X	X	X	X
	4	X	<u>X</u>	X	X	X	X	<u>X</u>	X	X	X
	5	X	X	<u>X</u>	X	X	X	X	<u>X</u>	X	X
Eastern European (E)	6	<u>X</u>	X	X	X	X	<u>X</u>	X	X	X	X
	7	X	<u>X</u>	X	X	X	X	<u>X</u>	X	X	X
	8	X	X	X	<u>X</u>	X	X	X	X	<u>X</u>	X
	9	X	X	<u>X</u>	X	X	X	X	<u>X</u>	X	X
	10	X	X	X	X	<u>X</u>	X	X	X	X	<u>X</u>

The field test sites cover a wide range of environmental conditions in the Nordic countries (Table 1, Figure 1) regarding temperature climate, rainfall, and distance from the coast. Several sites are located at approximately the same latitude (60° N), being situated from the Atlantic oceanic coastal climate of West Norway, in East Norway, in Central Sweden, on the Åland Islands, and in the coast and interior of Finland. In addition, one southern site, Hørsholm, Denmark, and one northern, Lappkuliden, Sweden, are included. Both types of trials were planted at the same location at Lappkuliden, Paimio, and Imatra. The sites Lappkuliden, Imatra, and Hørsholm are outside the recommended deployment area for seedlings from seed produced in Stange Seed Orchard.

Climate reference data for the sites were obtained for the sites in Norway based on weighted mean values of the 1961–1990 period from the closest meteorological stations [12]. For the sites in Sweden, Finland, and Denmark, the data were derived from meteorological observations interpolated into high resolution gridded data in climate models [13]. The range of mean annual temperature is from 2.0 to 8.8 °C for the 12 sites, and the temperature sum, calculated as the accumulated daily mean temperatures above 5 °C, varies from 931 to 1841 degrees (Figure 1).

The seedlings for the family and provenance trials were produced at Biri Nursery (60.95° N, 10.60° E, 160 m elevation), Norway, and were one year old when they were planted in the spring of 1988. Six of the seven trials were planted at forest sites at a spacing of either 1.8 or 2 m. One of the trials (Paimio) was planted at agricultural soils at 1 m spacing. The experimental design was, in all trials, single tree plots with families and provenances mixed and randomized in 40 replicates.

The rooted cuttings for clonal trials were also produced at Biri Nursery and were planted in 1989 two growth seasons after propagation. From each of the 20 full-sib families, 18 clones were originally propagated, and 12 of these were selected to be included in the trials. One ramet of each clone was planted in single tree plots in a randomized block design with six replicates. Six of the eight trials were planted at forest sites and two on agricultural soils (Paimio and Hørsholm).

Tree heights were measured in the field trials nine (two trials), ten (eleven trials), or eleven (two trials) growing seasons after planting, and assessments were made of each tree whether it had double stems, forks, or ramicorn branches, hereafter called stem defects. Assessments of bud burst in scoring classes [14] were made in 1999 in the clone trial at Kaupanger. In one of the two the short-term trials established earlier at Ås, Norway, with the 100 families, the elongation of terminal shoots was measured weekly during the growing season in 1989 and 1990, and Weibull shoot elongation curves were estimated [10]. From these curves, the days of initiation and cessation of shoot elongation, counted in days from April 1, were estimated separately for each tree. Least square family means (LS) of these traits will be related to measurements made in the field trials. Similar measurements were made in the trial at Skiptvedt during the growing season in 1992 [10], and means of provenances were calculated for the same traits.

2.2. Measures of Phenotypic Stability

Several criteria are available for characterizing phenotypic stability [1,4,15]. Here, regression coefficients were calculated for each genetic unit using its mean annual shoot growth at each site as the dependent variable and the site mean as an environmental index and explanatory variable in linear regression analyses. Due to unequal ages, when measurements were made, regression analyses were done on tree heights divided by the age when measured, here denoted by height increment. The regression coefficient b of each genetic unit characterizes its specific response in height growth to the changing environmental conditions defined by the mean performance of all genotypes and is an indirect measure of its phenotypic stability. It is a measure of the relationship between the interaction effects and the environmental values [15]. If the estimated regression coefficient b is close to 1, then the genetic unit has an average response to the site conditions and its growth performance parallels that of the site means. A $b > 1.0$ indicates relative high performance at the most productive sites, while a $b < 1.0$ indicates relative better growth on low yielding sites. Large deviations from the regression line characterize a low fit to the estimated response and instability in height growth relative to the mean performance, caused by sampling errors or non-linearity, and can be measured by the deviation mean square in the regression or the coefficient of determination (R^2) [1]. These two measures are strongly related, and the latter will be used here. Both b and R^2 are independent of the units of measurements.

An alternative index of the environmental conditions in each trial could be the site mean height of the 10% tallest trees in each replicate, regardless of their provenance or family relationships. This parameter should estimate the production potential of a forest site and is like the "dominating tree height" [16], which often is used as a site index. Regressions analyses were here also performed with this index and with annual mean temperature and the degree days at each site as explanatory variables.

2.3. Calculations and Statistical Analyses

The percentage of dead trees (mortality) and the percentage of trees with at least one type of stem defects were calculated for each genetic entry in each trial, and the percentages were transformed by the log square-root transformation before statistical analyses were made.

Mean provenance and family values were available from earlier analyses [10] for the days of shoot growth initiation (Day1) and cessation (Day2). Mean values for flushing scores of clones were available from the clone trial at Kaupanger.

Separate analyses of variance of heights were made within each site both in the provenance and family trials and in the clonal trials using SAS Proc Mixed and Proc GLM [17]. The significance of variance components was tested by *F*-tests, specifying Method = Type 3. As there were some differences in the root mean square errors (RMSE) among sites from these analyses, transformations were made by dividing the individual tree heights by the RMSE in each trial. Statistical analyses across sites were then performed based both on the original heights and on their transformed values. Least square (LS) height means of all genetic entries at each site were estimated by Proc GLM.

2.3.1. Provenance Data

Analyses of variance of height growth across site for the provenances were based on the model

$$Y_{ijk} = \mu + P_i + S_j + PS_{ij} + B_{jk} + E_{ijk} \quad (1)$$

where μ is the general mean, P_i is the fixed effect of provenance i, S_j is the fixed site effect, PS_{ik} is the interaction between provenance and site, B_{kl} is the random block effect within site, and E_{ijk} is the random error term.

2.3.2. Family Data

The two groups of parents, Norwegian (N) and Eastern European (E), belong to populations with different origins and different selection histories and show different patterns of variability [10]. It was therefore found necessary to make quantitative genetic analyses in each of the small 5 × 5 factorials separately, a fact that limits the generalizations that are drawn from estimated genetic parameters [10]. In the analyses and presentations focus has been on the N × N and E × E family groups. The low number of parents and families is also the reason why no attempts were made to separate the total genetic variance into additive, dominance, and epistatic components, which can be done when measurements are available from trials with families and clones within families [9,18,19].

The analyses of variance of height across sites were therefore done separately within each of the four groups of crosses, according to the model

$$Y_{ijkl} = \mu + F_i + M_j + FM_{ij} + S_k + FS_{ik} + MS_{jk} + FMS_{ijk} + B_{kl} + E_{ijkl} \quad (2)$$

where μ is the general mean, F_i is the effect of female parent i, M_j that of male parent j, FM_{ij} is the interaction between female parent i and male parent j, S_k is the site effect, FS_{ik}, MS_{jk}, and FMS_{ijk} are the three interactions with parents and site, B_{kl} is the block effect within site, and E_{ijkl} is the random within plot error term. All effects, except S_k, are assumed to be random with expectations 0 and respective variance components. The analyses were made across all sites and with the two most extreme sites Lappkuliden and Imatra excluded.

2.3.3. Clonal Data

The five full-sib families that are represented by clones within each of the four crossing groups were analysed across sites according to the following model

$$Y_{ijkl} = \mu + FA_i + C_{ij} + S_k + FAS_{ik} + CS_{ijk} + B_{kl} + E_{ijkl} \quad (3)$$

where μ is the general mean, FA_i is the effect of full-sib family i, C_{ij} that of clone within family, S_k is the site effect, FAS_{ik} is the interaction between family and site, CS_{ijk} is the interaction between clones (within family) and site, B_{kl} is the block effect within site, and E_{ijkl} is the random error term. All effects, except S_k, are assumed to be random with expectations 0 and respective variance components. Analyses of variance within each site were based on similar models, but with the site effect and genetic entry by site interactions excluded. The narrow sense heritability in the family trials was calculated for the N × N and E × E families as twice the sum of the female and male variance components divided by the total phenotypic variance that was the sum of all the other variance components, except the block components. The broad sense heritability in the clonal trial was similarly calculated as the sum of the family and clone components divided by the total phenotypic variance. The standard errors of the heritabilities were calculated by the Taylor expansion for variances of ratios [15].

Pearson correlation coefficients for height between pairs of sites for the three types of material were calculated based on LS means at each site. Correlation coefficients were calculated between the family means for the day of growth initiation and growth cessation in the short-term test and mean tree height, percent mortality, and percentage of trees with stem defects in the field trials.

Height measurements were made in the trials at different ages, after nine, ten, or eleven growth seasons. Therefore, the mean annual height increment (LS) was calculated for all genetic entries to make in order to compare the phenotypic response of provenances, families, and clones to the varying site conditions. Linear regression analyses by SAS Proc Reg were performed using this mean increment at each site as dependent variable and the total site mean height increment as independent variable. For the provenances, families, and clones, a total of 20, 100, and 240 regressions coefficients, respectively, were estimated. The regression analyses were made both for the original increments and for their transformed values. As there were high correlations between the regression coefficients from the two analyses (r = 0.80–0.95), only the results from the first analyses will be presented. Regression analyses were also made for the family trials using the mean annual height increment of the 10% tallest trees per replicates as site index and explanatory variable and also based on the annual mean temperature and degree days for the trial sites.

For the family trials, a factorial analysis of variance of the regression coefficients for all 100 families was made with the female and male parents as the main effects and their interaction as the error term. A similar analysis was made for the clones, with the terms families and clones within families in the model. These analyses were based on models (4) and (5) for families and clones, respectively,

$$Y_{ij} = \mu + F_i + M_j + E_{ij} \quad (4)$$

$$Z_{ik} = \mu + FA_i + C_{ikk} \quad (5)$$

where Y_{ij} and Z_{ik} are the regression coefficients for family $_{ij}$ and clone $_{ik}$, respectively, and F and Mj are the parental effects in the factorial cross, E_{ij} is the interaction between the two parents, and FA_i and C_{ik} are the family and clone within family effects.

3. Results

3.1. Mortality and Stem Defects

The mean mortality and the percentage of trees with stem defects varied considerably among the sites (Table 1). However, there were no relationships present between these trial means and the site mean annual temperature or temperature sum. For provenances, the variation in mean mortality across sites was significant ($p < 0.01$) and varied from 23% to 43% with lowest values for the Nordic and Eastern European provenances. The same provenances also had the lowest percentage of trees with stem defects, varying from 16% to 35% among provenances, but these differences were not significant. Among the 100 full-sib families, the mean mortality ranged from 24% to 40% and for stem defects from 8% to 37%. Only minor differences in mortality were present among the four groups of crosses, whilst the mean percentage of defects varied from 16% to 23% with the lowest mean for the E × E and highest mean for the N × N families. In the clonal trials, small differences were present for mortality both for crossing types and for families. The percentages of stem defects showed larger variation, with a mean of 32% for clones from the N × N and 19% for those from the E × E crosses. The provenance hybrid families were intermediate. The mean frequencies of defects of the 20 families that were represented both with clones and seedlings were strongly related (r = 0.86).

3.2. Height

At all sites, significant differences ($p < 0.001$) were present for height among both the female and male parents in all four groups of crosses (analyses not shown). In the clonal trials, there were significant differences ($p < 0.001$) for height among families and clones within families, with a range among clones from 49% to 180% of the site mean (analyses not shown).

Analyses of variance across sites were made both of tree heights and their transformed values and showed significant variation both for provenances, families, clones within families, and their interactions with sites. As minor differences in results were present between the two types of analyses, only those made on the original tree heights are presented.

The mean height across sites for the provenances was 202 cm, with a range of variation from 158 to 230 cm. The eight Nordic provenances had a mean height of 190 cm, while the six provenances from Eastern Europe had a mean of 215 cm. The variation among provenances was significant ($p < 0.0001$), and there was a significant interaction among provenances and sites ($p < 0.0001$). In a separate analysis of the eight Nordic provenances, they differed significantly ($p < 0.0001$), but the interaction with sites was now not significant ($p = 0.09$).

The mean height for the families was 214 cm, with a range of variation from 171 to 249 cm among the 100 full-sib families. In the analyses of variance, considerably larger variance components were found for the female and male parents in the N × N than in the E × E crosses, and the estimates of heritability were 0.20 and 0.03, respectively (Table 3). The female x male variance components showed low values and were not significant in three of the four crossing groups. Significant interactions between parents and sites were present in all four groups. In additional analyses made, excluding the two most extreme sites Lappkuliden (northern) and Imatra (eastern), the interactions were still significant and with quite similar values of the variance component estimates.

Table 3. Estimates of variance components and narrow sense heritabilities (h^2) for height of 100 families after ten or eleven growing seasons at seven sites (sites 1–7) for each crossing type separately. In parentheses, *p*-values of variance components and standard error of the heritability.

Source/Cross	N × N	N × E	E × N	E × E
Mean Height (cm)	205	221	209	219
F	215.5 (0.0002)	81.3 (0.0002)	0.0	0.0
M	476.8 (<0.0001)	59.5 (0.007)	423.3 (<0.0001)	105.2 (0.001)
F × M	38.5 (0.07)	73.5 (0.0012)	22.3 (0.20)	26.0 (0.05)
F × S	0.0	69.5 (0.0009)	88.8 (<0.0001)	66.3 (<0.0001)
M × S	71.9 (0.01)	114.7 (<0.0001)	117.5 (<0.0001)	47.7 (0.004)
F × M × S	101.2 (0.002)	2.7 (0.54)	0.0	0.0
Error	5926.4	6559.1	6066.1	5861.8
h^2	0.20 (0.15)			0.03 (0.04)

F: female parent; M: male parent; S: site; N: Norwegian parent; E: Eastern European parent.

In the trials with clones, similar differences in patterns of variation were present between the N × N and E × E family groups (Table 4). The clone within family component was slightly lower than the family component in the N × N families, and slightly higher in the E × E families. Both interaction variance components (family × site and clone (family) × site) were significant in all four family groups. These were, however, in most cases considerably smaller than the family and clone components. The proportional variance of the family and clone within family components, interaction components, and the error component were 26%, 6%, and 68%, respectively, in the N × N family group, and 15%, 8%, and 77% in the E × E families. The estimates of the broad sense heritability were, as expected, higher than the narrow sense estimates from the family trials. In separate analyses, excluding the Lappkuliden and Imatra sites, or the Hørsholm site, the relative values of the interaction components were slightly reduced and those for components for families and clones within families increased, but most components were still significant at the 5% level.

Table 4. Means and estimates of variance components and the broad sense heritabilities (H^2) for height of clones within families after ten growing seasons at eight sites (sites 11–18). In parentheses, *p*-values of variance components and standard error of the heritability.

Source	N × N	N × E	E × N	E × E
Mean height (cm)	169	178	194	187
Family	768.3 (<0.0001)	651.2 (<0.0001)	191.8 (0.05)	269.6 (0.003)
Family × site	164.3 (<0.0001)	84.5 (0.01)	279.0 (<0.0001)	200.5 (<0.0001)
Clone (family)	503.5 (<0.0001)	462.3 (<0.0001)	755.1 (<0.0001)	373.1 (<0.0001)
Clone (family) × site	141.7 (0.01)	292.9 (<0.0001)	174.1 (0.04)	131.6 (0.04)
Error	3291.8	3451.2	4221.4	3333.1
H^2	0.26 (0.20)			0.15 (0.11)

The interaction between provenances and trials for height is reflected in the phenotypic correlations between pairs of trials (not shown). Only low, or even negative, values were estimated for the correlation coefficients between the mean heights of the provenances at the site on the west coast of Norway and in the other trials. The three sites in Finland have the closest relationship with pairwise correlation coefficients in the range 0.55–0.66.

For the 25 N × N families, there were in general medium to high values of the correlation coefficients between pairs of trials in the range 0.51–0.80. Rather low relationships were present among sites for the 25 families in the E × E crosses, most likely due to the lesser variation among families.

The correlation coefficients between pairs of trials for clones had generally low values in the range 0.20–0.59. They were considerably higher when they were calculated for family means, based on the 12 clones, with values between 0.46 and 0.83.

3.3. Relationships to Phenology Traits

Earlier analyses of the phenology assessments of families in the short-term trials showed heritability estimates of the days for shoot growth initiation (Day1) and cessation (Day2) equal to 0.89 and 0.88, respectively, in the N × N family group, and 0.61 and 0.47 in the E × E families [10]. The estimates of the genetic correlation between these two traits were as high as 0.92 and 0.67 for the two groups. For clones in the Kaupanger trial, the broad sense heritability estimates for the timing of bud burst were 0.84 and 0.79 in the two family groups. There were no relationships between mortality in the field tests and the phenology traits across the seven family and provenance trials, nor in the clonal trials. For provenances, there were only weak relationships between stem defects and the phenology traits Day1 and Day2. However, this relationship was stronger between height and Day1 and Day2 ($r = 0.69$ and $r = 0.77$). For the N × N families, strong associations were present between these phenology traits and stem defects and height (Figure 2), with phenotypic correlations between Day1 and stem defects as high as $r = -0.80$, and Day1 and heights of 0.76. These associations could not be found for the E × E families (Figure 2). N × N families with an early growth start and early growth cessation showed the highest frequency of stem defects and slower height growth. Similar relationships were present for the 20 families in the eight trials with clones, with correlation coefficients between Day1 and stem defects of 0.80 and Day1 and height growth of 0.60 (not shown in figure). Correlations between these traits at the clonal level were present only in the N × N family group, where the coefficients between the same pair of traits were 0.58 (Day1, stem defects) and 0.53 (Day1, height).

3.4. Phenotypic Stability Parameters

The estimates of the regression coefficients for the mean height increment against the site mean for the 20 provenances varied from 0.69 to 1.31, with quite good fit to the regression lines with R^2 values from 0.81 to 0.99. Figure 3a shows the plots of the regression lines and the corresponding values for three of the provenances studied: Åland, Finland (b = 0.69, R^2 = 0.86), Harz, Germany (b = 1.27, R^2 = 0.87), and Emmaboda, Sweden (b = 0.85, R^2 = 0.81). The Åland provenance showed inferior growth at all sites and the lowest values of the regression coefficient. The Harz provenance has a steep regression line with poor growth at the climatic most severe sites and superior performance at the sites with the best growth conditions, particularly at Ølve on the west coast of Norway. When the provenances were grouped in four regions, the Nordic region had the lowest mean regression coefficient of 0.87, the Eastern European had 1.01, the Carpathian Mountains 1.22, and Harz 1.15.

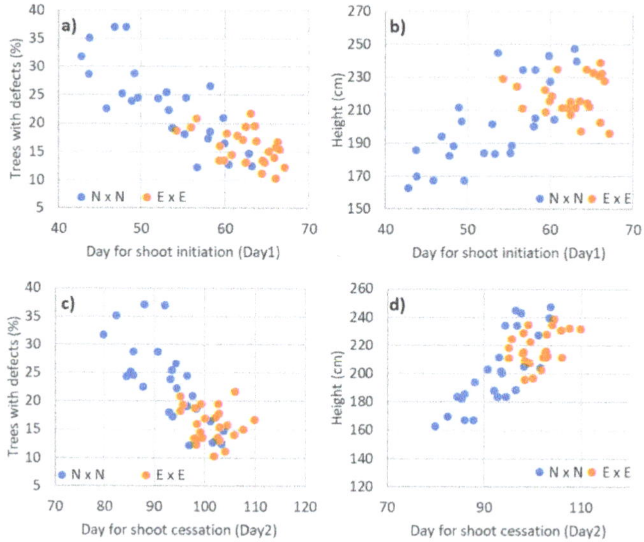

Figure 2. Plot of the relationships between the percentage of trees with stem defects (**a**,**c**) and tree height (**b**,**d**) with the day for shoot growth initiation (Day1) (**a**,**b**) and shoot growth cessation (Day2) (**c**,**d**) at the family-mean level (site 1–7). The material is grouped in N × N (blue) and E × E families (orange).

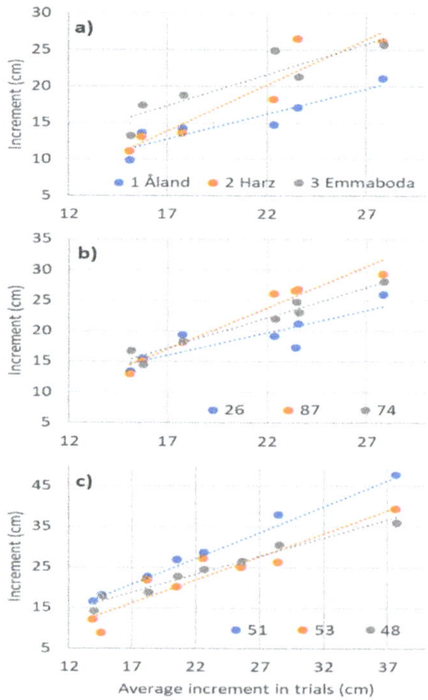

Figure 3. Estimated regressions lines for the genetic entries' height increment as a function of the average increment of the trials shown for (**a**) three selected provenances (sites 1–7), (**b**) families (sites 1–7), and (**c**) clones within family 83 (sites 11–18).

The estimated regression coefficients for the 100 full-sib families also varied considerably, with a range between 0.64 and 1.44, among families (Table 5). The fit to the regression line, as expressed by the coefficient of determination had a mean value of R^2 = 0.72 and ranged between 0.49 and 0.99. The mean regression coefficients were 0.97 and 1.02 in the N × N and E × E family groups, respectively, and with mean R^2 values of 0.81 and 0.90.

Table 5. Estimates of regression coefficients for annual height increment of 100 full-sib families planted at seven sites (sites 1–7). The regression coefficients of the N × N and E × E families are shaded.

Male / Female	11	12	13	14	15	16	17	18	19	20	Mean
1	1.05	0.94	1.03	0.88	0.98	1.00	0.93	0.86	1.14	0.75	0.96
2	1.09	1.19	0.94	0.67	1.03	1.00	1.15	1.22	0.89	0.92	1.01
3	0.75	0.82	0.91	0.85	0.82	0.74	0.86	0.98	1.03	0.79	0.86
4	1.17	1.11	1.06	0.83	1.01	0.94	1.02	1.20	1.19	1.04	1.06
5	0.98	1.12	0.97	1.05	0.90	0.69	1.05	1.15	0.87	1.04	0.98
6	0.93	0.77	0.99	1.01	0.79	0.81	0.64	1.19	0.97	0.96	0.91
7	1.07	1.05	0.85	0.97	0.99	0.98	1.14	1.25	1.04	1.05	1.04
8	1.20	1.06	0.88	1.07	0.98	0.98	1.02	1.23	1.00	1.06	1.05
9	1.44	1.36	1.01	1.20	1.01	1.08	0.99	1.19	1.11	1.03	1.14
10	1.23	1.07	0.96	0.89	0.78	0.83	0.99	1.04	1.02	0.91	0.97
Mean	1.09	1.05	0.96	0.94	0.93	0.90	0.98	1.13	1.02	0.96	1.00

The plots shown in Figure 3b present examples of the estimated regression lines and deviations for three unrelated families. Family 26 (b = 0.75), which has the largest deviations from the regression line, has inferior growth on the most productive sites compared to other two families, while family 87 (b = 1.36) has a very good growth at these sites and more average at the less productive sites. It also had a good fit to the regression line (R^2 = 0.95).

In the overall analysis of variance of regression coefficients (all families in the factorial mating), significant variation was present both among the female and male parents ($p < 0.001$), with 25% of the its variation among the female and 20% among the male parents. This clearly demonstrates the genetic component of the phenotypic response to the variation in environmental conditions at the trial sites. A similar analysis of variance of R^2 showed significant variation both among the female ($p = 0.02$) and male parents ($r = 0.002$), and with 9% and 21%, respectively, of its variation due to the two types of parents.

When the mean of the 10% tallest trees at each site was used as explanatory variable in the regression analysis instead of the mean of all trees, estimated coefficients were lower and in the range from 0.43 to 1.13. The regression coefficients from the two analyses were strongly correlated ($r = 0.91$), but a lower fit to the regression lines was obtained using this latter index in the analyses (R^2 of 0.72 compared to 0.86 in the regressions based on the total site mean). The families contributed very differently to the top 10% in the index, varying from 0 to 15 trees in each trial. One family was not represented at all in any of the indices across the seven trials, and one family was represented by 48 trees. No further analyses were done using this index.

In the regression analyses of the increments made with either the annual mean temperature or degree days as explanatory variables, a low fit was obtained in both types of analyses, with mean R^2 values equal to 0.47 and 0.32, respectively. This shows that temperature is not the only environmental factor causing differences in performance of the genetic materials at these sites.

In the clonal trials, the estimated regression coefficients for individual clones varied in the range from 0.40 to 1.64 with R^2 values from 0.46 to 0.99. For the 20 families, mean regression coefficients varied from 0.78 to 1.20. An example of the variation among the clones in one family is presented in Figure 3c. Clone 51 (b = 1.29) had the best height growth at seven sites and was superior at the two most productive sites. When the total variation of the regression coefficients was partitioned into two components, assuming unrelated families, 25% was among families ($p = 0.04$) and 75% among clones within families. Most likely a substantial part of the large variation among clones is caused by a high experimental error of the clonal means at each site.

Twenty families were represented both in the family and clonal trials (Table 2). The correlation between the mean family mean regression coefficients estimated from the two types of trials was as low as $r = 0.46$ ($p = 0.04$).

Figure 4a,b shows the plots of the regression coefficients for families in the N × N and E × E groups against Day1 and Day2, respectively. For the N × N families, there is a clear positive relationship for both phenology traits ($r = 0.61$ and $r = 0.60$); families with an early growth start have a value of the regression coefficient below 1.0 and so have families with an early growth cessation. For the families in the E × E group, there is a positive relationship between the regression coefficient and Day2 ($r = 0.60$).

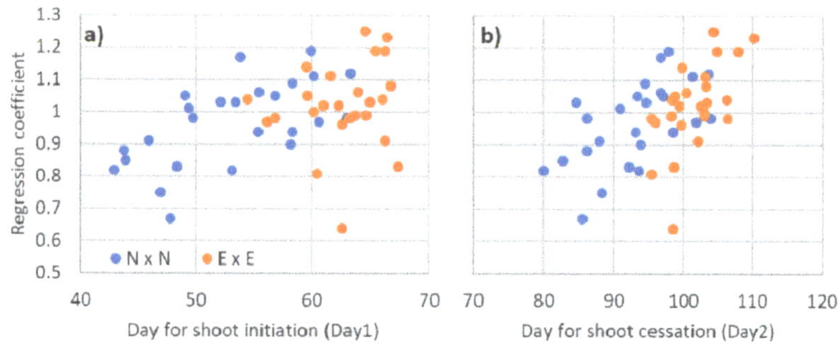

Figure 4. Plot of regression coefficients estimated for families against the days of shoot growth initiation (Day1) (**a**) and cessation (Day2) (**b**), day number after May 1 (sites 1–7). N × N families are blue and the E × E families are orange.

In the N × N family group, there was a positive relationship ($r = 0.66$) between the regression coefficient and R^2; families with the lowest value for the regression coefficient showed the largest deviation from the regression line.

There were considerable differences between the family and clonal trials in the variance of estimates of the stability parameter reflected in a mean root square error of 3.01 for the clones compared to the value of 1.94 for the regression analyses for the 100 families. In the clonal trials, six ramets of each clone were planted at each trial site. The mean mortality among sites varied from 8% to 33% and with variation among clones. Their LS means are therefore estimated with quite large and variable sampling errors, which influences the precision of the estimates of regression coefficients for the clones. Another factor contributing to differences between the regression estimates from the family and clone trials is that the site index at Hørsholm clonal trial site was high, and trees were taller than at the other sites. This certainly influences the regression coefficient estimates; clones and families that are responsive to the favorable conditions for growth will have a steep regression line. This can also be a reason for the unexpected low relationship between the mean family regression coefficients estimated for clones and seedlings (20 families that were common).

4. Discussion

There was a large variation among the 20 provenances for survival and height growth after ten years, and those from Eastern Europe and the Nordic countries generally showed the best performance. The Central European provenances performed on average poorly but were among the best ranked in the mild climate in West Norway. Evidently, provenances that are not well adapted to the harsh conditions at the sites with the most severe climatic conditions made the largest contribution to the interactions. These results confirm conclusions from reports from large provenance trials in the Nordic countries [2,20]. The different patterns of provenance performance across sites are clearly expressed by the variation of the regression coefficients. They showed that the provenances from the Carpathian Mountains and Harz had superior height growth at the most productive sites but grew poorly at the sites with the most severe climatic conditions. The Eastern European provenances showed on average good growth on all sites, while the provenances from Sweden and Finland, except that from Åland, showed relatively good growth at the less productive sites, but relatively poorer growth at the most productive sites. The variable performance of the Central European provenances across Nordic sites confirms results from earlier provenance trials [3,20].

The differences between the N × N and E × E family groups in mortality and percentage of trees with stem defects confirm the results from the short-term tests previously reported [10], as do the results from the analyses of heights showing large differences in genetic variability within the two groups families. Thus, the lower level of genetic variation among the E × E families, most likely caused by both natural selection and the artificial selection (made by the tree breeders) of the parents after growing in the climatically harsh environment for 25 years in northern Sweden, is still present after ten years in fifteen field trials.

In the analyses of the family trials (trials 1–7), the female and male variance components for height showed considerably higher values than their interaction component (F × M), which was not significant. This indicates that the additive genetic variation is the most important and that non-additive genetic effects are small for the families in the two 5 × 5 factorials. A different result, with significant F × M components, was obtained from the analyses of tree heights after five growing seasons in two short-term trials on agricultural soils with the same families [10]. Non-additive genetic effects have also been demonstrated for height and diameter growth in progeny trials of Norway spruce, comprising clones and families from the Swedish breeding program [9]. The deviating results from the present analyses with 9–11 year- old trees were confirmed in the analysis of height at age 17 years in the field trial at Skiptvedt with the same genetic material [10]. The deviations from reports of other genetic materials could be due to the small number of parents, trial site conditions, and varying G × E interactions.

The variance components for families and clones within families estimated in the clonal trials were strongly significant and showed approximately similar values. This is according to the expectation that the genetic variances among and within full-sib families are of similar size when small non-additive genetic effects are present [21].

In the short-term trials with the same 100 full-sib families planted at two sites, one with a quite mild climate and the other one with frequent frost occurrences both in the spring and in the autumn, quite large G × E interactions were found for height growth after five growing seasons, in particular for the families in the N × N group [10]. The relative magnitude of the interaction variance components was smaller in the field trials we report now, even if they were significant in the analyses of variance. In the N × N family group, they were considerably less than 50% of the family variance component, which has been considered to be a critical value limiting the potential gain for selection in breeding [22].

The strong relationships found between the timing of shoot elongation and frequency of stem defects and height growth for the N × N families (Figure 2a,b) correspond to what was found in one of the short-term trials [10]. Once again, it shows the importance of delayed bud-flush to avoid spring frost damages [2,7].

The field test sites cover a wide range of environmental conditions in the Nordic countries where Norway spruce is being planted. Nine of the sites are located at approximately the same latitude (60° N), being situated in the humid coastal climate of West Norway, in East Norway, in Central Sweden, on the Åland Islands and in interior Finland, and one is southern (Hørsholm, Denmark, latitude 56° N) and one northern (Lappkuliden, Sweden, latitude 64° N). The overall results show that quite large geographic zones can be used in the breeding of Norway spruce in the Nordic countries, but also that G × E interactions may depend on local conditions and may be largely unpredictable, as recently demonstrated [2].

The regression analyses identified families and clones that respond differently when the conditions for growth change. Some genotypes have a superior performance at sites with high productivity and less than average performance at poorer sites, others have relatively better growth at the last type of sites. The regression coefficients characterize these differences in response, which are clearly genetically controlled. In addition, the coefficient of determination (R^2) also has a genetic component. Genotypes with a late initiation of shoot elongation generally had the best growth at the productive sites with a steep regression line. Therefore, there are positive relationships between height, Day1, and the regression coefficient, particularly evident in the N × N family group, being more variable in phenology. This shows that the regression coefficient, as a stability parameter, to a large extent is related to these traits.

In forest tree breeding, we want to select materials with high performance across the whole landscape that are able utilize productive sites and at the same time not perform poorly at sites with a lower production level. Therefore, genotypes with a relatively high and stable production at low-yielding sites as well as having capacity to respond with a high production under favourable conditions are sought. We would argue that when applying the regression coefficient method such as used here, candidate genotypes will be characterized by a coefficient close to or slightly above 1. At the same time, they need to have a strong average breeding value estimated across representative sites within the defined breeding zone. However, genotypes might also show strong breeding values even if the performance is not stable, so an index comprising both the coefficient and breeding value might be needed in combination for effective selection in a breeding program. They cannot be completely identified by the regression techniques applied here, and a more complicated curvilinear regression analysis could be used to identify such responses.

A positive relationship was observed between the timing of shoot elongation and the regression coefficient (and mean height increment) for the N × N families. Thus, a selection based on a late bud flush would most likely be favourable for both improving height growth and lower frequency of stem defects.

Author Contributions: Initial project administration and funding acquisition; data collection; data analyses; draft writing: T.S.; further project administration; funding acquisition; review and editing: A.S. All authors have read and agreed to the published version of the manuscript.

Funding: This work was financially supported by The Nordic Forest Research Co-operation Committee (grant number SNS-26 and SNS-60), the Research Council of Norway (grant number 216050), the project B4EST of the Agriculture and the European Union's Horizon 2020 Research and Innovation Programme under grant agreement No 773383, The Norwegian Forest Seed Station, and the Norwegian Institute of Bioeconomy Research (NIBIO).

Institutional Review Board Statement: Not applicable.

Informed Consent Statement: Not applicable.

Data Availability Statement: The data presented in this study are available on request from the corresponding author.

Acknowledgments: Many scientists and technicians have contributed to this study, particularly Jon Dietrichson, Reidar Tutturen, and Øystein Johnsen, at the former Norwegian Forest Research Institute; Gösta Eriksson and Dag Lindgren, at the Swedish University of Agricultural Sciences; Veikko Koski and Lena Yrjänä, at the former Finnish Research Institute; and Hubert Wellendorf and Viggo Jensen, University of Copenhagen. We thank them all for their contributions.

Conflicts of Interest: The authors declare no conflict of interest.

References

1. Becker, H.; Leon, J. Stability analysis in plant breeding. *Plant Breed.* **1988**, *101*, 1–23. [CrossRef]
2. Berlin, M.; Jansson, G.; Högberg, K.A. Genotype by environment interaction in the southern Swedish breeding population of *Picea abies* using new climatic indices. *Scand. J. For. Res.* **2015**, *30*, 112–121. [CrossRef]

3. Persson, B.; Persson, A. *Survival and Quality of Norway Spruce (Picea abies (L.) Karst.) Provenances at the Three Swedish Sites of the IUFRO 1964/68 Picea abies Provenance Experiment*; 1992 Report 29; Dept for Yield Research, SLU: Garpenberg, Sweden, 1992.
4. Skrøppa, T. A critical evaluation of methods available to estimate the genotype × environment interactions. *Studia For. Suec.* **1984**, *166*, 3–14.
5. Hannerz, T.; Sonesson, J.; Ekberg, I. Genetic correlations between growth and growth rhythm observed in short-tern test and performance in long-term field trials of Norway spruce. *Can. J. For. Res.* **1999**, *29*, 768–778. [CrossRef]
6. Benzer, B.; Foster, G.S.; Hellberg, A.R.; Podzorski, A.C. Genotype × environment interaction in Norway spruce involving three levels of genetic control: Seed source, clone mixture and clone. *Can. J. For. Res.* **1988**, *18*, 1172–1181. [CrossRef]
7. Karlsson, B.; Wellendorf, H.; Roulund, H.; Werner, M. Genotype x trial interaction and stability across sites in 11 combined provenance and clone experiments with *Picea abies* in Denmark and Sweden. *Can. J. For. Res.* **2001**, *31*, 1826–1836. [CrossRef]
8. Chen, Z.Q.; Karlsson, B.; Wu, H.X. Patterns of additive genotype-by-environment interaction in tree height of Norway spruce in southern and central Sweden. *Tree Genet. Genomes* **2017**, 13–25. [CrossRef]
9. Berlin, M.; Jansson, G.; Högberg, K.A.; Helmersson, A. Analysis of non-additive genetic effects in Norway spruce. *Tree Genet. Genomes* **2019**, *15*, 42. [CrossRef]
10. Skrøppa, T.; Steffenrem, A. Selection in a provenance trial of Norway spruce. (*Picea abies L. Karst*) produced a land race with desirable properties. *Scand. J. For. Res.* **2016**, *31*, 439–449. [CrossRef]
11. Langlet, O. Mellaneuropeiska granprovenienser i svenske skogsbruk. [Central European provenances in Swedish forestry]. *Kungl. Skogs och Lantbr. Tidskrift.* **1960**, *99*, 259–329.
12. Solvin, T.M.; Steffenrem, A. Modelling the epigenetic response of increased temperature during reproduction on Norway spruce phenology. *Scand. J. For. Res.* **2019**, *34*, 83–93. [CrossRef]
13. Bärring, L.; Berlin, M.; Andersson Gull, B. Tailored climate indices for climate-proofing operational forestry applications in Sweden and Finland. *Int. J. Climatol.* **2017**, 123–142. [CrossRef]
14. Krutzsch, P. *Norway Spruce Development of Buds*; IUFRO S2.02.11; The Royal College of Forestry: Stockholm, Sweden, 1973.
15. Lynch, M.; Walsh, B. *Genetics and Analysis of Quantitative Traits*; Sinauer Associates, Inc.: Sunderland, MA, USA, 1998.
16. Tveite, B. *Bonitetskurver for gran. [Site-index curves for Norway spruce]*; Meddelelser fra Norsk Institutt for Skogforskning: Ås, Norway, 1977.
17. SAS Institute. *SAS/STAT User's Guide*; Version 9; SAS Institute: Cary, CA, USA, 2003.
18. Mullin, T.J.; Morgenstern, E.K.; Park, Y.S.; Fowler, D.P. Genetic parameters from a clonally replicated test of black spruce (*Picea mariana*). *Can. J. For. Res.* **1992**, *22*, 24–36. [CrossRef]
19. Russel, J.; Costa e Silva, J.; Baltunis, B.S. Multisite genetic parameter estimates from a *Callitropsis kookiness* diallel study with clonally replicated progeny. *Can. J. For. Res.* **2015**, *45*, 689–697. [CrossRef]
20. Fottland, H.; Skrøppa, T. The IUFRO 1964/68 provenance experiment with Norway spruce (*Picea abies*) in Norway: Variation in mortality and height growth. *Meld. Skogforsk* **1989**, *43*, 1–30.
21. Falconer, D.S.; Mackey, T. *Introduction to Quantitative Genetics*, 4th ed.; Pearson Education Ltd.: Harlow, UK, 1996.
22. Shelbourne, C.J.A. Genotype-environment interaction: Its study and its implications in forest tree improvement. *Proc. IUFRO Genet. SABRAO Joint Seem Tokyo* **1972**, *B-1*, 1–28.

forests MDPI

Article

Predicting Suitable Habitats of *Camptotheca acuminata* Considering Both Climatic and Soil Variables

Lei Feng [1,2], Jiejie Sun [1,3], Yuanbao Shi [1,2], Guibin Wang [1,2,*] and Tongli Wang [4,*]

1 Co-Innovation Centre for Sustainable Forestry in Southern China, Nanjing Forestry University, Nanjing 210037, China; ChinaFengLei@163.com (L.F.); Chinasunjiejie@126.com (J.S.); ybao_shi@163.com (Y.S.)
2 College of Forestry, Nanjing Forestry University, Nanjing 210037, China
3 College of Biology and the Environment, Nanjing Forestry University, Nanjing 210037, China
4 Department of Forest and Conservation Sciences, Faculty of Forestry, University of British Columbia, Vancouver, BC V6T 1Z4, Canada
* Correspondence: gbwang@njfu.com.cn (G.W.); tongli.wang@ubc.ca (T.W.); Tel.:+86-1377-052-1738 (G.W.); +1-604-822-1845 (T.W.)

Received: 19 June 2020; Accepted: 14 August 2020; Published: 17 August 2020

Abstract: *Camptotheca acuminata* is considered a natural medicinal plant with antitumor activity. The assessment of climate change impact on its suitable habitats is important for cultivation and conservation. In this study, we applied a novel approach to build ecological niche models with both climate and soil variables while the confounding effects between the variables in the two categories were avoided. We found that the degree-days below zero and mean annual precipitation were the most important climatic factors, while the basic soil saturation, soil gravel volume percentage, and clay content were the main soil factors, determining the suitable habitats of *C. acuminata*. We found that suitable habitats of this species would moderately increase in future climates under both the RCP4.5 and RCP8.5 climate change scenarios for the 2020s, 2050s, and 2080s. However, substantial shifts among levels of habitat suitability were projected. The dual high-suitable habitats would expand, which would be favorable for commercial plantations. Our findings contribute to a better understanding of the impact of climate change on this species and provide a scientific basis for the cultivation and conservation purposes.

Keywords: *Camptotheca acuminata*; MaxEnt; climateAP; suitable habitat; climate change; soil

1. Introduction

Camptotheca acuminata is a plant species of the genus *Camptotheca* in the family of *Nyssaceae*, with nicknames Eclipta, water chestnut, and water tung tree, native to southern China and Tibet. It is an important multipurpose tree species with medicinal and ornamental values and was approved as a national second-level key protected wild plant by the State Council of China in 1999. *C. acuminata* is a fast-growing tree species in a temperate climate, but it is not cold-tolerant and requires moisture and fertile soil. Camptothecin (CPT) extracted from the fruit, roots, bark, branches, leaves, and other tissues are the sole inhibitors of Topoisomerase I so far. It has broad-spectrum anticancer activity and can be employed in the acute treatment of gastric cancer, rectal cancer [1], chronic myeloid leukemia, lymphatic leukemia, chorionic epithelial cancer, and lymphosarcoma. It has been considered as the second-most important antitumor medicinal plant in the world after Taxus [2]. The demand for *C. acuminata* has been growing worldwide in recent years. The total sale of *C. acuminata* derivatives has exceeded one billion dollars since 2010, becoming a major clinical agent [3]. In addition, *C. acuminata* plays a significant role in green ornamental and ecological protection [4]. Despite its importance, the impact of climate change on this species has not been assessed.

Unprecedented human-caused climate change has already caused species range shifts and local extirpation [5] and is predicted to have greater future impacts [6]. Thus, the original habitats of some plant species may no longer be appropriate for their growth and survival in future climates, resulting in existing populations with compromised health and productivity. Climate change has culminated in a steady northward shift of forests in the north [5], while forests in the south may be degraded and transformed into shrubs or grasslands in the Northern hemisphere. Changes in species distribution and abundance will depend on the degree of climate change, the intensity of local adaptation, and the potential for migration [7].

Niche-based ecological models have been widely used to predict suitable climatic habitats and to assess the potential impacts of climate change [8,9]. The widely used modeling algorithms include Garp, MaxEnt, Bioclim, randomForest, and Domain. Among them, the MaxEnt is probably the most often used one for being stable and predictable [10], and it can make a better response analysis to species on various environmental gradients [11], including identification of suitable habitats [12], risk assessments related to invasive species [13], potential impacts of climate change [14], design of protected areas, and protection of threatened species protection [15]. In addition, it requires presence data only, which makes the data collection and modeling process easier than others.

Most of the niche-based ecological models are built using only climate variables as predictors [16–18], as climate is the primary factor regulating geographic distributions of plant species [19]. However, the distribution of plant species is also related to soil type [20]. Failure to incorporate soil into the model may overestimate the future habitat adaptability of many plant species [21]. Therefore, some studies included both climate and soil variables as predictors in building niche-based models [22,23], but there are issues to do so due to collinearity [24] and beyond. Soil variables are often correlated with climate variables, as climate is a major soil-forming factor and has a significant influence on the properties of soil. Thus, in building a niche-based model with both climate and soil variables, some variance explained by soil variables in the model could have been explained by climate variables if soil variables were not included and vise versa. As soil conditions are relatively stable in comparison to a rapidly changing climate, and there are no future projections for soil, the explanatory power of soil variables in the model is wasted in future projections. On the other hand, if the opposite happens, the soil effect would be masked by climate variables. However, to our knowledge, this issue has not been addressed yet.

So far, studies on *C. acuminata* are focused on its antitumor mechanism [2]. The distribution of *C. acuminata* has not been reported in detail. The study of the climatic habitats of this species and the potential impact of climate change on its geographical distribution remains blank. This limits the cultivation and genetic conservation of this species. In this study, we used the MaxEnt model to predict the suitable habitats of *C. acuminata* in the current and two climate-change scenarios for three future periods (the 2020s, 2050s, and 2080s). In order to consider the effects of soil on the habitat prediction and to avoid mixed modeling of climate and soil variables, we built niche models with climate variables and soil variables separately and used a climatic niche model to predict suitable climatic habitats and applied a soil niche model to predict suitable soil habitats of the species. We then applied a novel approach to filter the climatic habitats by the soil habitats to incorporate the soil effect into our habitat predictions. To our best knowledge, this was the first attempt to consider both climatic and soil effects with such a two-step approach. Our results would provide a scientific basis for the development of adaptive strategies for this species under climate change, including in situ conservation, reforestation, and resource utilization.

2. Materials and Methods

2.1. Species Occurrence Data

The data of natural distribution sites of *C. acuminata* were obtained from the China Digital Herbarium (http://www.cvh.ac.cn/), a teaching specimen resource-sharing platform (http://mnh.scu.edu.cn/main.aspx), the Global Biodiversity Information Database (GBIF, https://www.gbif.org/), and literature collections. To avoid multiple recording points from the same grid and the impact of sampling bias [25], the original data were filtered geographically to remove some blurred geographical distribution points. By selecting data points with precise coordinates (i.e., latitude and longitude) on Google Earth satellite maps, we finally obtained 351 data points of *C. acuminata* in China (Figure 1) reflecting naturally occurring populations.

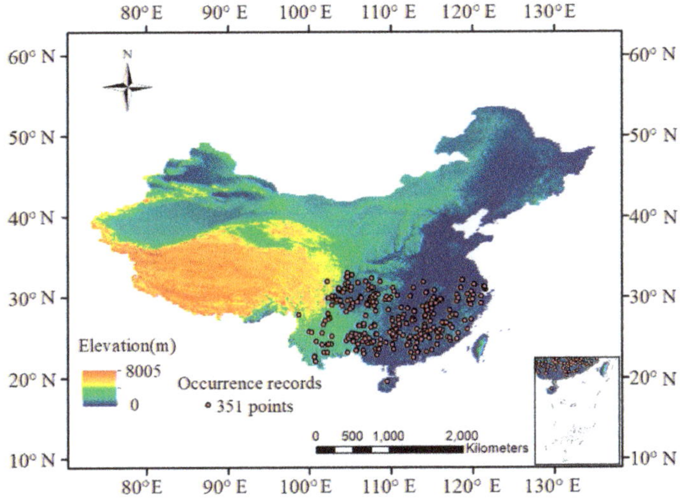

Figure 1. Distribution points of *Camptotheca acuminata*.

2.2. Environment Data

The environmental variables selected in this study include 16 climate variables (Table 1) and 30 soil variables (Table 2). The climate data were generated by ClimateAP (http://ClimateAP.net) [26]. The ClimateAP software generates scale-free climate data for historical and future periods for specific locations based on longitude, latitude, and altitude, instead of grid averages from other climate models. For building the climatic niche model, we used ClimateAP to generate climate variables for the 351 data points for the normal reference period, 1961–1990. For predicting the geographical distributions of the climate habitats for the current and future periods, we also generated grid climate data at 4 × 4 km for the reference period (also called current) (1961–1990), the 2020s (2011–2040), 2050s (2041–70), and 2080s (2071–2100). The future climate data were from the general circulation models (GCMs) of the Coupled Model Intercomparison Project (CMIP5) included in the Intergovernmental Panel on Climate Change (IPCC) Fifth Assessment Report (IPCC 2014). We used the 15-GCMs ensembles of the two climate change scenarios RCP4.5 and RCP8.5 included in ClimateAP for this study. These two scenarios are the most widely used ones and represent intermediate and business-as-usual scenarios, respectively. Soil variables were derived from 30 basic soil indicators (HWSD, Table 3) obtained from the World Soil Database (http://www.iiasa.ac.at/web/home/research/researchPrograms/water/HWSD.html), which contains a key soil attribute raster data layer at a spatial resolution of 30 arc seconds [27].

Table 1. List of the climate variables. Variables in bold are selected for model building.

Code	Description	Code	Description
MAT	Mean annual temperature (°C)	DD > 5	Degree-days above 5 °C, growing degree-days
MWMT	Mean warmest month temperature (°C)	**DD < 0**	**Degree-days below 0 °C, chilling degree-days**
MCMT	Mean coldest month temperature (°C)	NFFD	Number of frost-free days
TD	**Temperature difference between MWMT and MCMT, or continentality (°C)**	**PAS**	**Precipitation as snow (mm)**
MAP	**Mean annual precipitation (mm)**	EMT	Extreme minimum temperature over 30 years
EXT	Extreme maximum temperature over 30 years	Eref	Hargreaves reference evaporation
AHM	Annual heat:moisture index (MAT+10)/(MAP/1000))	**CMD**	**Hargreaves climatic moisture deficit**
DD < 18	Degree-days below 18 °C	**DD > 18**	**Degree-days above 18 °C**

Table 2. List of the soil variables. Variables in bold are selected for model building.

Code	Description	Code	Description
T-GRAVEL	**Topsoil Gravel Content**	**S-GRAVEL**	**Subsoil Gravel Content**
T-SAND	Topsoil Sand Fraction	S-SAND	Subsoil Sand Fraction
T-SILT	**Topsoil Silt Fraction**	S_SILT	Subsoil Silt Fraction
T-CLAY	Topsoil Clay Fraction	**S-CLAY**	**Subsoil Clay Fraction**
T_REF_BULK_DENSITY	**Topsoil bulk density**	S_REF_BULK_DENSITY	Subsoil bulk density
T-OC	Topsoil Organic Carbon	S-OC	Subsoil Organic Carbon
T-PH-H2O	Topsoil pH (H 2 O)	S-PH-H2O	Subsoil pH (H 2 O)
T-CEC-CLAY	**Topsoil CEC (clay)**	**S-CEC-CLAY**	**Subsoil CEC (clay)**
T-CEC-SOIL	Topsoil CEC (soil)	S-CEC-SOIL	Subsoil CEC (soil)
T-BS	**Topsoil Base Saturation**	S-BS	Subsoil Base Saturation
T-TEB	**Topsoil TEB**	S-TEB	Subsoil TEB
T-CACO3	Topsoil Calcium Carbonate	**S-CACO3**	**Subsoil Calcium Carbonate**
T-CASO4	Topsoil Gypsum	**S-CASO4**	**Subsoil Gypsum**
T-ESP	Topsoil Sodicity (ESP)	**S-ESP**	**Subsoil Sodicity (ESP)**
T-ECE	Topsoil Salinity (Elco)	S-ECE	Subsoil Salinity (Elco)

In order to avoid overfitting and inaccurate models caused by the strong correlation between environmental variables in this study, we used the MaxEnt 3.4.1k version (http://www.cs.princeton.edu/schapire/maxent) and ArcGIS software to screen ecological factors and to select the climatic factors with the largest contribution to the distribution of *C. acuminata*. Firstly, following Koeling [28], we used all the 16 climate variables to prebuild the MaxEnt model for three times in succession and discarded the climate variables with no contribution to the models. Secondly, we use ArcGIS to calculate the Pearson correlation coefficient between each pair of the climatic variables. If two climatic variables were strongly correlated at r > 0.8, only one of the two variables with a higher contribution rate was selected to avoid the multicollinearity of environmental variables from violating statistical assumptions [29]. We applied the same procedures for the 30 soil variables. Through the two-step screening process, we selected 6 climate variables (Table 1, in bold) and 12 soil variables (Table 2, in bold) for model building.

2.3. Model Development

We used the MaxEnt algorithm to build a climatic niche model with the 6 selected climate variables and a soil niche model with the 12 selected soil variables, respectively. For comparisons, we also built a combined model with the 6 climate variables and 12 soil variables. We randomly selected 25% of the data points for model validation and used the remaining 75% data points to build the model. The Jackknife method was used to detect the importance of the variables [29]. To consider the uncertainty introduced by splitting the training and validation sets, 10 models were built by 10 repeated runs for cross-validation.

Table 3. Contributions of the environmental variables to the models.

Model	Variable	Unit	Contribution (%)
Climatic	DD < 0	°C	80.7
	MAP	mm	6.3
	TD	°C	4.3
	CMD	mm	3.8
	PAS	mm	3.1
	DD > 18	°C	1.9
Soil	T-BS	%	31.8
	S-GRAVEL	%	22.7
	S-CLAY	%	18.4
	S-CASO4	%	7.1
	T-CEC-CLAY		5.8
	S-ESP		2.9
	S-CEC-CLAY		2.9
	T-TEB		2.7
	T-GRAVEL	%	1.9
	T-REF-BULK		1.7
	T-SILT	%	1.2
	S-CACO3	%	0.9
Climatic + Soil	DD < 0	°C	79.6
	MAP	mm	4.4
	T-BS	%	3.2
	TD	°C	2.8
	CMD	mm	2.7
	DD > 18	°C	1.5
	S-CASO4	%	1.5
	T-GRAVEL	%	1.3
	PAS	mm	0.6
	T-CEC-CLAY		0.5
	S-ESP		0.4
	S-CLAY	%	0.4
	T-REF-BULK		0.4
	S-GRAVEL	%	0.2
	T-SILT	%	0.2
	S-CEC-CLAY		0.1
	S-CACO3	%	0.1
	T-TEB		0.1

The outputs of the MaxEnt models for the reference and future periods were the "cumulative" probability for each pixel on the scale of 0–100% [30]. Based on the suitable habitat evaluation index method of Chun Yan Lu et al. [31], we divided the study area of *C. acuminata* into 4 types of habitats, namely unsuitable ($p < 0.2$), low-suitable ($0.2 \leq p < 0.4$), moderate-suitable ($0.4 \leq p < 0.6$), and high suitable ($0.6 \leq p \leq 1$) for both climatic and soil models, respectively. We collectively called the last three suitable types as "suitable habitats".

2.4. Model Evaluation

We utilized the receiver operating characteristic curve (ROC) built in the MaxEnt software to check the accuracy of the simulation results. The Area Under the Curve (AUC) value ranges from 0 and 1. The closer the AUC value is to 1, the greater the distance from the random distribution, the stronger the correlation between environmental variables and the geographical distribution of predicted species [32]. We used the general evaluation criteria of the ROC as follows: AUC value < 0.6 is the model failure, 0.6 ≤ AUC value of < 0.7 is considered to be a poor model and is barely acceptable,

0.7 ≤ AUC value < 0.8 is considered to be a general model, 0.8 ≤ AUC value < 0.9 is considered to be a good model, and 0.9 ≤ AUC is considered to be a good model [33].

2.5. Model Predictions and Assessments of Climate Change Impacts

Finally, we filtered the four types of climatic suitable habitats by soil suitable habitats, which was the combination of all the three suitable types. The filtered habitats were considered suitable for both climatic and soil conditions and simply called low-, moderate-, and high-suitable habitats, respectively. Changes in the area of suitable habitats were divided into three categories: expansion, contraction, and stable, following Zhang et al. [34]. We generated additional projections combining the high-suitable climatic habitats with the high-suitable soil habitats for establishing commercial plantation purposes. We called this combination the "dual high-suitable habitats".

To assess the impact of climate changes on the suitable habitats of this species, we evaluated changes in spatial distributions, areas, and the centroid of the distributions of both the climatic and the filtered habitats under the six future climate scenarios relative to the reference period. In addition, we also assessed the shift of centroids of both the climatic and the dual high-suitable habitats in future climate scenarios.

3. Results

3.1. Model Performance and Contributing Variables

The AUC values for the model cross-validation training data and test data were 0.932 and 0.931 for the climatic model, 0.922 and 0.907 for the soil model, and 0.926 and 0.917 for the combined model, respectively. The ROC curves of all the three models were far away from the random distribution (0.5), indicating a high accuracy of the two models. Interestingly, the combined model was not the highest in model prediction accuracy, only slightly higher than the soil model and lower than the climatic model.

For the climatic model, variables with the largest contribution rate to the model was DD < 0 (80.7%), followed by MAP (6.3%) and TD (4.3%), which accumulatively interpreted 91.3% of the model (Table 3). Of these three variables, two were temperature-related variables, and one was a precipitation variable (Table 3). The remaining three variables contributed less than 10% to the model. For the soil model, the top three contributing variables were T-BS (31.8%), S-GRAVEL (22.7%), and S-CLAY (18.4%), collectively accounting for 72.9% of the total contribution of all the soil variables. Other soil factors demonstrated a limited impact on suitable habitats. For the combined model, climate variables contributed 91.6% in total, suggesting apparent domination over the soil variables (i.e., 8.4%), although the contribution of each climate variable was also reduced relative to that in the climatic model.

Response curves (Figure 2) of the three most important climate variables showed that their suitable ranges for *C. acuminata* were 0–25 °C for DD < 0, 1100–2000 mm for MAP, and 14–22 °C for TD, respectively. The suitable ranges for the three most significant soil variables were 18–50% for T-BS, 5–30% for S-GRAVEL, and 42–67% for S-CLAY, respectively (Figure 2).

3.2. Predicted Suitable Habitats for the Current

The four types of climatic suitability of *C. acuminata* habitats for the current climate were mapped over all of China (Figure 3a and Table 4). The high-suitable habitats were concentrated between 25–33° N in latitude, accounting for 2.7% of the country's land area. The medium-suitable habitats were distributed around the high-suitable habitats, accounting for 8.3%, and the low-suitable habitats were distributed further away from the high-suitable habitats, accounted for 7.2%. The unsuitable habitat accounted for the remaining 81.8%.

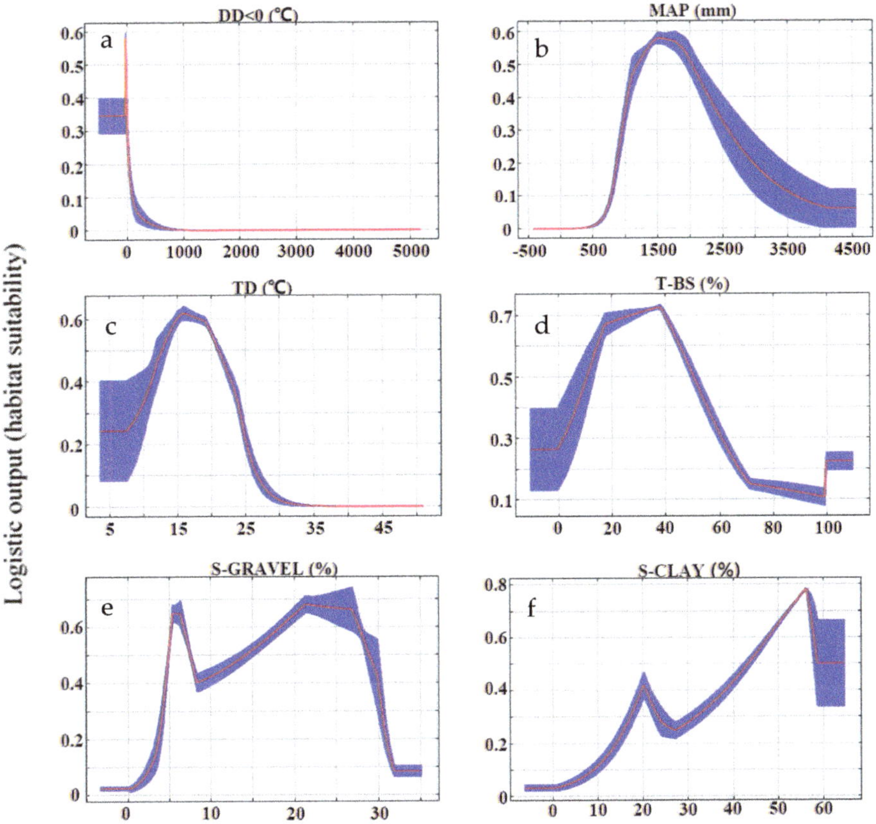

Figure 2. Response curves of the three important climate variables (**a**–**c**) and the three important soil variables (**d**–**f**) in the MaxEnt climatic and soil models, respectively. The maximum entropy logistic output (also known as habitat suitability) is represented by the vertical Y-axis and the variable by the horizontal X-axis. When the logical output value is greater than 0.5, the probability of species presence under this condition is higher than that under a "typical" condition, which indicates that the condition is suitable for the species. The red curves shown are the averages over 10 replicate runs; blue margins show ± 1 standard deviation (SD) calculated over 10 replicates.

Table 4. Distribution areas of the current (1960–1990) habitat suitability of *Camptotheca acuminata* in China.

Classes	High-Suitable		Medium-Suitable		Low-Suitable		Unsuitable	
	Area	Proportion	Area	Proportion	Area	Proportion	Area	Proportion
	km^2	%	km^2	%	km^2	%	km^2	%
Climatic habitats	260,119	2.7	799,626	8.3	693,652	7.2	7,880,658	81.8
Soil habitats	568,409	5.9	973,039	10.1	404,630	4.2	7,687,977	79.8
Climatic habitats filtered by soil habitats	246,400	2.6	733,200	7.6	463,000	4.8	8,179,400	85
Dual high-suitable habitats	83,600	0.87						

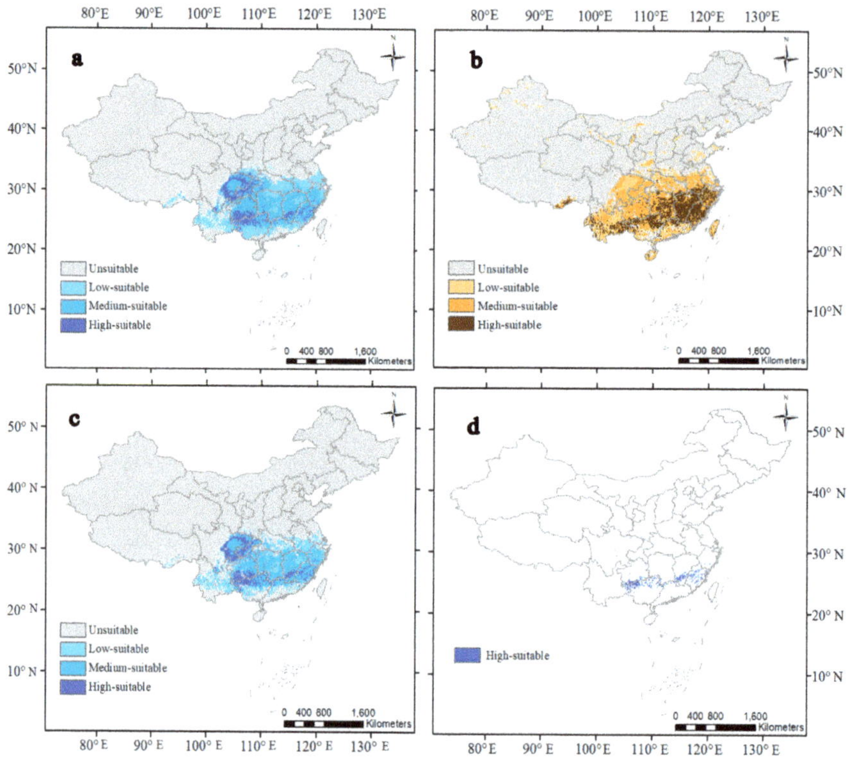

Figure 3. Distributions of the current (1960–1990) suitable habitats of *C. acuminata* in China: (**a**) climatic habitats, (**b**) soil habitats, (**c**) climatic habitats filtered by the soil habitats, and (**d**) dual high-suitable habitats.

For the soil suitable habitats of *C. acuminata* (Figure 3b), the high-suitable habitat were mainly distributed between 18–35° N in latitude, accounting for 5.9% of the total land area. The medium-suitable and low-suitable habitats accounted for 10.1% and 4.2%, respectively. The rest of the country (79.8%) was unsuitable for this species in terms of soil conditions (Table 4). Overall, the distribution of soil-suitable habitats for this species was slightly broader than that for climatic ones. However, there were also some small areas showing nonoverlapping between the two habitats.

After the climatic habitats filtered by the soil habitats (Figure 3c), the areas of the high-suitable habitats accounted for 2.6% of the country. The medium-suitable and low-suitable habitats accounted for 7.8% and 4.8%, respectively. Unsuitable habitats accounted for 85% (Table 4). The difference between the filtered and nonfiltered climatic habitats was increased from high-suitable to low-suitable habitats, suggesting that most of the nonoverlapping areas occurred in the low-suitable habitats of the two categories, as expected. The dual high-suitable habitats (Figure 3d) accounted for 0.87% of the country, which was about a third of the high-suitable habitats (2.6%) (Table 4).

3.3. Projected Changes in Suitable Habitats for Future Periods

Compared with the current distribution (Figure 3a), the climatic suitable habitats show an increasing trend in the future climates by the end of the century under the two climate change scenarios (7–25%, Table 5). However, the increases were less than 11% in most cases, except for a jump for RCP8.5 in the 2080s (25.3%). No substantial geographic shift was projected.

Table 5. Area changes in soil habitat-filtered climatic habitats of *C. acuminata* relative to the current by the 2020s, 2050s, and 2080s under RCP4.5 and RCP8.5, respectively.

Time	Current Area (km^2)	Change by 2020s (%)		Change by 2050s (%)		Change by 2080s (%)	
Emissions Scenarios		**RCP4.5**	**RCP8.5**	**RCP4.5**	**RCP8.5**	**RCP4.5**	**RCP8.5**
Climatic habitats	1,758,000	+7.2	+11.3	+7.7	+10.1	+9.5	+25.3
High-suitable habitats	246,400	−2.8	−3.2	−4.2	−4.9	−28.7	−24.6
Medium-suitable habitats	733,200	+14.8	+18.6	+17.4	+23.1	35.2	+59.8
Low-suitable habitats	463,000	−4.1	−10	−8.6	−19.9	−20.6	−29.6
Total suitable habitats	1,442,600	+5.7	+5.7	+5.4	+4.5	+6.4	+16.7
Dual high-suitable habitats	83,600	+98	+105.3	+100.5	+66.1	−12.7	−20.5

After being filtered by soil habitats (Figure 3c), the high- and low-suitable habitats show a decreasing trend, while the moderate-suitable habitats show an increasing trend in the future climates. The area of the high-suitable habitats would decrease by 2.8% and 3.2% by the 2020s, 4.2% and 4.9% by the 2050s, and 28.7% and 24.6% by the 2080s under the RCP4.5 and RCP8.5 scenarios, respectively (Figure 4 and Table 5). The decline of the low-suitable habitats was at a similar magnitude, as for the high-suitable habitats. However, the area of the moderate-suitable habitats would increase by 14.8% and 18.6% by the 2020s, 17.4% and 23.1% by the 2050s, and 35.2% and 59.8% by the 2080s under the two scenarios, respectively. The total suitable habitats showed smaller increases than climate habitats alone. The centroid of the high-suitable habitats was projected to move northeastward from the current location in East Guizhou to West Hunan by the 2020s and keep moving to Northeast Hunan under the future climates.

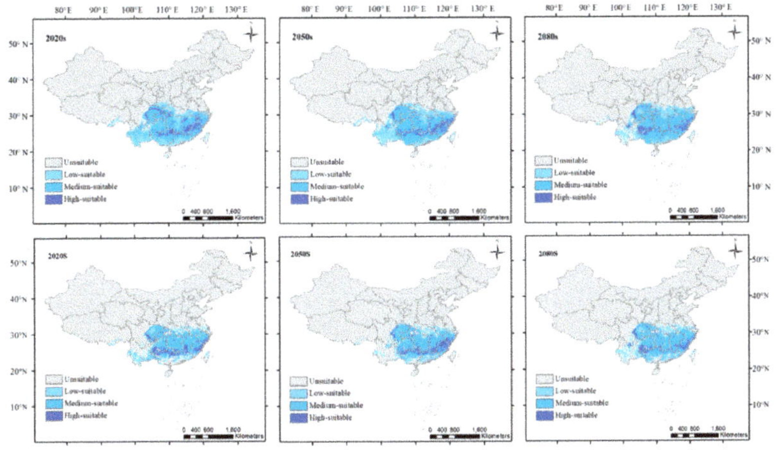

Figure 4. Suitable habitats of *C. acuminata* in China for climatic (1st row), soil-filtered climatic (2nd row), and dual high-suitable habitats (3rd row) in the 2020s, 2050s, and 2080s under RCP4.5, respectively.

However, future projections of the dual high-suitable habitats showed a different pattern. Compared to the current distribution (Figure 3d), their size was projected to increase by 98.0%

and 105.3% by the 2020s and 100.5% and 66.1% by the 2050s under RCP4.5 and RCP8.5, respectively (Figure 5 and Table 5). However, their size was projected to decrease by 12.7% and 20.5% by the 2080s under the two scenarios, respectively. The centroid of the dual high-suitable habitats was projected to move from the current location in South Hunan (Figure 6) northeastward. By the 2080s, the centroid would shift to Jiangxi, with RCP8.5 moving more to the north. Compared with the centroid of high-suitable habitats, the current location and the shifting path were located at much lower latitudes for the centroid of the dual high-suitable habitats.

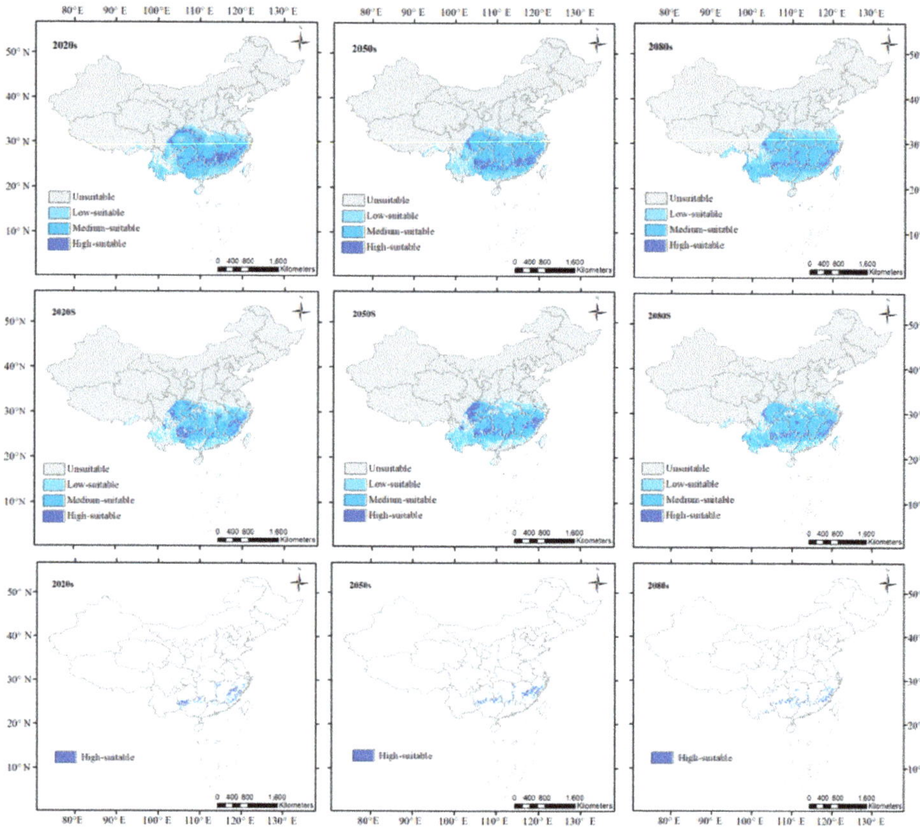

Figure 5. Suitable habitats of *C. acuminata* in China for climatic (1st row), soil-filtered climatic (2nd row), and dual high-suitable habitats (3rd row) in the 2020s, 2050s, and 2080s under RCP8.5, respectively.

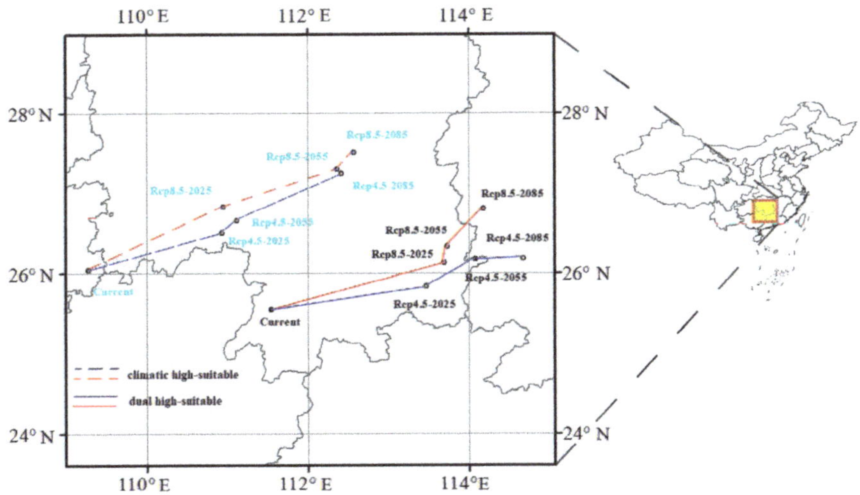

Figure 6. Shifts of centroids of the high-suitable and the dual high-suitable habitats of *C. acuminata*. The dotted line indicates the shifting path of high-suitable habitats. The solid line indicates the shifting path of dual high-suitable habitats.

4. Discussion

Rapid climate changes have affected forest ecosystems and species habitats [35]. Thus, the distribution of suitable habitats of *C. acuminata* may also have been affected. Climatic niche-based ecological models have been widely used to predict the spatial distribution of the suitable habitats of species and project their future shifts under various climate change scenarios [16,36]. Such information can be used to guide management forest resources for adaptation to climate change [36,37]. Although most of the niche-based models involve only climatic variables [16], soil properties are often recommended to be considered, especially for practical applications [38]. In this study, we applied a novel two-step approach to incorporate the soil variables into the future projections without compromising the contribution of climate or soil variables. Thus, our projections of suitable habitats for *C. acuminata* may provide a more credible basis for forest conservation and adaptation to climate change.

4.1. Predicted Suitable Habitats Using Both Climatic and Soil Variables

Using only climate variables is the mainstream in building niche-based models [16–18]. Within a suitable climate niche, the distribution of a species may be restrained by soil conditions. Thus, a number of studies included both climate and soil variables in the same models [20,21]. However, the inclusion of soil variables in a niche-based model may compromise the role of climatic variables or soil variables due to correlations between the variables in the two categories [39]. As a result, it can either compromise the credibility of future predictions as soil variables are treated as constants or mask the soil effect by climate variables. In our case, climate variables (91.6% contribution to the model) clearly masked the contribution from soil variables (8.4%) in the combined model. However, the contribution of each climate variable was also reduced to some extent relative to that in the climatic model. These results suggest that the combined model basically does not represent the effect of soil, while the climatic effect is also slightly compromised. We avoided this problem by using a novel two-step approach. We first predicted the total areas with suitable climate conditions (the climatic niche) for the species—within which, suitable soil conditions may not be available in some areas. Our second step was to exclude such areas where soil conditions were not suitable. This approach can better reflect the true nature of climate and soil variables in determining the distribution of the species than that including both climate and soil variables into the same model. The accuracy of both climate and soil models for *C. acuminata* were high, with the AUC values greater than 0.9. The spatial predictions of suitable habitats for this species by the two models were in high agreement, especially for the high-suitable habitat areas. The restraint of soil conditions on climate habitats exhibited mainly in low-suitable habitat areas, which was expected.

From the perspective of climate conditions, we found that three climate variables had major contributions to the climatic niche model for *C. acuminata*. Two of them were temperature-related variables (DD < 0 and TD) and one precipitation variable (MAP). However, the contribution of DD < 0 was very dominant (80%) over the other two variables, suggesting the suitable climate habitats of this species being mainly restrained by low temperatures, with DD < 0 not greater than 25 °C (Figure 2). The suitable range of MAP, the second-most important climate variable, was between 1100 mm and 2000 mm, suggesting that this species also requires relatively high precipitation. For soil variables, we found that T-BS (32%), S-GRAVEL (23%), and S-CLAY (18%) were important factors for predicting the distribution of *C. acuminata*. T-BS is related to topsoil effective nutrient contents; thus, it is unsurprisingly the most important soil factor in our soil niche model. The content of gravel in the underlying soil (S-GRAVEL) affects the respiration and water absorption of the roots of *C. acuminata*, and the clay content provides higher soil organic carbon; the nutrient-holding and water-holding capacities of organic components contributes significantly to the growth of *C. acuminata*. As soil conditions are strongly affected by climate in the long term, climate change will eventually impact the soil conditions, although this process would be much slower than the rate of climate change. So far, no prediction is available for future soil conditions, and thus, soil variables can only be treated as constants in projecting species suitable habitats for future periods.

4.2. Impact of Climate Change

Our future projections show that the potentially climatic suitable habitats of *C. acuminata* would slightly expand under future climate scenarios. The magnitude of the northward expansion was much smaller than the projections for tree species in North America [40,41]. However, it is comparable to the future projections for the species distributed in the same region of *C. acuminata*, including Chinese fir and Masson pine [42]. The small impact of climate change on the habitats of *C. acuminata* can also be explained by the two most important contributing climate factors, DD < 0 and MAP. Although increasing temperatures due to climate change would expand the climate habitats of this species northward based on DD < 0, the change in MAP was projected to be small in future climates and not able to change the low precipitation status north of China, thus limiting the northward expansion of this species. Projected drought increases in North China, particularly in Dry Belt, may impose a serious challenge to forests, as well as the benefits of forests to mitigate climate change [43].

After being filtered by soil habitats, the area of the total suitable habitats still showed some increase but at a smaller magnitude than the climate habitats alone. This was expected, as the soil habitats were assumed static, and the shift of climatic habitats would cause some breakdown of the matches between climatic and soil habitats. This was evident for the high- and low-suitable habitats, and some of them were probably shifted to moderate-suitable habitats, as the expansion of this type was substantial. Eastern Sichuan Province, which was in the high-suitable habitats of *C. acuminata* at present, would be transformed into low-suitable habitats by the end of this century. More suitable habitats for *C. acuminata* would occur in Jiangsu, Anhui, and Henan in the future. On the other hand, some areas in Yunnan, Guangdong, and Guangxi Provinces would become unsuitable for *C. acuminata* in future climates.

It is worthy of note that our future projections are merely to represent suitable climate and soil conditions for *C. acuminata* to grow rather than projections of the species distribution in the future. Future species distributions would also be affected by many nonclimate and nonsoil factors, such as biotic interactions, seed dispersal, etc. [44], as well human interventions and disturbances, which are substantial or even greater than all biological factors in some regions.

4.3. Implications for Commercial Forest Management

Forest managers are interested in having their commercial plantations established in the most favorable environmental conditions. To address this, we generated projections for the dual high-suitable habitats, which was the combination of the high-suitable climatic habitats and the high-suitable soil habitats. Interestingly, the projected dual high-suitable habitats showed substantial increases in the 2020s and 2050s, while the high-suitable habitats were projected to decrease in the same periods. These inconsistent shifts were due to the matching between the climate and soil layers. The dual high-suitable habitats were only about 30% of the high-suitable habitats, and its near-future projections were better-matched with the soil high-suitable habitats than the high-suitable habitats. The shifting paths of the centroids of the dual high-suitable habitats were also located at much lower latitudes than the suitable-habitats, although their shifting directions were in parallel. The dual high-suitable habitats of *C. acuminata* were currently distributed along the latitude 28° N in China, including the junctions of Guizhou and Guangxi; Hunan and Guangdong; and Fujian, Jiangxi, and Zhejiang. In the future, these would move to near 30° N, including Hunan, Jiangxi, and Zhejiang. The current projections for the dual high-suitable habitats would support forest managers planning their plantations for commercial use to meet the growing market demand.

4.4. Implications for Genetic Conservation

The conservation of forest genetic resources has become a major concern due to climate change [45,46]. Fragmentations due to human activities have resulted in a lower level of gene flow and a higher level of genetic differentiation among populations of *C. acuminata* than that of other species in the region [47]. Thus, protections of local populations appear important. Projections of the shift in suitable habitats can help to develop conservation strategies [44]. We recommend that in situ conservation should be the main conservation measure for *C. acuminata*, as most of its current suitable habitats were projected to remain in future climates. In situ conservation allows the populations to evolve to adapt to changing environments. The shifts among the three levels of suitable habitats in a changing climate revealed in this study may impose selection pressure to promote evolutionary processes for adaptation. Our predictions show that *C. acuminata* has a high risk of habitat loss in the low latitudes under future climates. To prevent the loss of genetic resources in those areas, ex situ conservation through seed storage and assisted migration of the local genetic resources to potentially suitable locations are recommended in these areas.

5. Conclusions

We built niche-based models considering both climate and soil variables and mapped spatial distributions of their suitable habitats for the economically and medicinally important forest tree species *C. acuminata*. We applied a novel approach to integrate the predictions generated by the two models for the current and future climates under different climate-change scenarios. Our future projections indicate that climate change would have a relatively small impact on suitable habitats of this species as a whole, as a low temperature is the dominant limiting factor but would cause shifts among different levels of suitability. The dual high-suitable habitats showed a trend of increase up to the 2050s, which would be favorable for commercial plantations. Our results may provide guidance for the silviculture and conservation planning of this species.

Author Contributions: G.W. and T.W. conceived and designed the experiments; L.F., J.S., and Y.S. performed the experiments, analyzed the data, and wrote the manuscript; and T.W. revised the manuscript. All authors have read and agreed to the published version of the manuscript.

Funding: This research was funded by the National Key Research and Development Program of China (2017YFD0600700) and the Agricultural Science and Technology Independent Innovation Funds of Jiangsu Province (CX (16) 1005).

Conflicts of Interest: The authors declare no conflict of interest.

References

1. Hsiang, Y.H.; Hertzberg, R.; Hecht, S.; Liu, L.F. Camptothecin induces protein-linked DNA breaks via mammalian DNA topoisomerase I. *J. Biol. Chem.* **1985**, *260*, 14873–14878.
2. Zhang, Y.; Jiang, K.; Qing, D.; Bing, H.; Jiang, J.; Wang, S.; Yan, C. Accumulation of camptothecin and 10-hydroxycamptothecin and the transcriptional expression of camptothecin biosynthetic genes in Camptotheca acuminata cambial meristematic and dedifferentiated cells. *RSC Adv.* **2017**, *7*, 12185–12193. [CrossRef]
3. Lorence, A.; Nessler, C.L. Camptothecin, over four decades of surprising findings. *Phytochemistry* **2004**, *65*, 2735–2749. [CrossRef] [PubMed]
4. Zhao, D.; Zhang, X.; Wang, R.; Liu, D.; Sun, J.; Tao, J. Herbaceous peony tryptophan decarboxylase confers drought and salt stresses tolerance. *Environ. Exp. Bot.* **2019**, *162*, 345–356. [CrossRef]
5. Chen, I.C.; Hill, J.K.; Ohlemüller, R.; Roy, D.B.; Thomas, C.D. Rapid Range Shifts of Species Associated with High Levels of Climate Warming. *Science* **2011**, *333*, 1024–1026. [CrossRef] [PubMed]
6. Thomas, C.D.; Cameron, A.; Green, R.E.; Bakkenes, M.; Beaumont, L.J.; Collingham, Y.C.; Erasmus, B.F.; De Siqueira, M.F.; Grainger, A.; Hannah, L. Extinction risk from climate change. *Nature* **2004**, *427*, 145–148. [CrossRef] [PubMed]
7. Peterson, M.L.; Doak, D.F.; Morris, W.F. Incorporating local adaptation into forecasts of species' distribution and abundance under climate change. *Glob. Chang. Biol.* **2019**, *25*, 775–793. [CrossRef]
8. Murray, N.J.; Keith, D.A.; Bland, L.M.; Ferrari, R.; Lyons, M.B.; Lucas, R.; Pettorelli, N.; Nicholson, E. The role of satellite remote sensing in structured ecosystem risk assessments. *Sci. Total Environ.* **2018**, *619*, 249–257. [CrossRef]
9. Pearson, R.G.; Raxworthy, C.J.; Nakamura, M.; Peterson, A.T. Original Article: Predicting species distributions from small numbers of occurrence records: A test case using cryptic geckos in Madagascar. *J. Biogeogr.* **2007**, *34*, 102–117. [CrossRef]
10. Li, G.; Du, S.; Wen, Z. Mapping the climatic suitable habitat of oriental arborvitae (Platycladus orientalis) for introduction and cultivation at a global scale. *Sci. Rep.* **2016**, *6*, 30009. [CrossRef]
11. Guevara, L.; Gerstner, B.E.; Kass, J.M.; Anderson, R.P. Toward ecologically realistic predictions of species distributions: A cross-time example from tropical montane cloud forests. *Glob. Chang. Biol.* **2017**, *24*, 1511–1522. [CrossRef] [PubMed]
12. Esfanjani, J.; Ghorbani, A.; Chahouki, M.A.Z. MaxEnt Modeling for Predicting Impacts of Environmental Factors on the Potential Distribution of Artemisia aucheri and Bromus tomentellus-Festuca ovina in Iran. *Pol. J. Environ. Stud.* **2018**, *27*, 1041–1047. [CrossRef]
13. Mungi, N.A.; Coops, N.C.; Ramesh, K.; Rawat, G.S. How global climate change and regional disturbance can expand the invasion risk? Case study of Lantana camara invasion in the Himalaya. *Biol. Invasions* **2018**, *20*, 1849–1863. [CrossRef]
14. Linhoss, A.C.; Underwood, W.V. Modeling Salt Panne Land-Cover Suitability under Sea-Level Rise. *J. Coast. Res.* **2016**, *32*, 1116–1125. [CrossRef]
15. Ashoori, A.; Kafash, A.; Varasteh Moradi, H.; Yousefi, M.; Kamyab, H.; Behdarvand, N.; Mohammadi, S. Habitat modeling of the common pheasant Phasianus colchicus (Galliformes: Phasianidae) in a highly modified landscape: Application of species distribution models in the study of a poorly documented bird in Iran. *Eur. Zool. J.* **2018**, *85*, 372–380. [CrossRef]
16. Araujo, M.B.; Peterson, A.T. Uses and misuses of bioclimatic envelope modeling. *Ecology* **2012**, *93*, 1527–1539. [CrossRef]
17. Keith, D.A.; Akcakaya, H.R.; Thuiller, W.; Midgley, G.F.; Pearson, R.G.; Phillips, S.J.; Regan, H.M.; Araujo, M.B.; Rebelo, T.G. Predicting extinction risks under climate change: Coupling stochastic population models with dynamic bioclimatic habitat models. *Biol. Lett.* **2008**, *4*, 560–563. [CrossRef]
18. Anderson, B.J.; Akcakaya, H.R.; Araujo, M.B.; Fordham, D.A.; Martinez-Meyer, E.; Thuiller, W.; Brook, B.W. Dynamics of range margins for metapopulations under climate change. *Proc. R. Soc. B Biol. Sci.* **2009**, *276*, 1415–1420. [CrossRef]
19. Woodward, F.I.; Williams, B.G. Climate and plant distribution at global and local scales. *Vegetation* **1987**, *69*, 189–197. [CrossRef]

20. Volis, S.; Zhang, Y.H.; Deng, T.; Dorman, M.; Blecher, M.; Abbott, R.J. Divergence and reproductive isolation between two closely related allopatric Iris species. *Biol. J. Linn. Soc.* **2019**, *127*, 377–389. [CrossRef]
21. Zuquim, G.; Costa, F.R.C.; Tuomisto, H.; Moulatlet, G.M.; Figueiredo, F.O.G. The importance of soils in predicting the future of plant habitat suitability in a tropical forest. *Plant Soil* **2019**, *450*, 151–170. [CrossRef]
22. Figueiredo, F.O.G.; Zuquim, G.; Tuomisto, H.; Moulatlet, G.M.; Balslev, H.; Costa, F.R.C. Beyond climate control on species range: The importance of soil data to predict distribution of Amazonian plant species. *J. Biogeogr.* **2018**, *45*, 190–200. [CrossRef]
23. Syphard, A.D.; Franklin, J. Differences in spatial predictions among species distribution modeling methods vary with species traits and environmental predictors. *Ecography* **2009**, *32*, 907–918. [CrossRef]
24. Dormann, C.F.; Elith, J.; Bacher, S.; Buchmann, C.; Carl, G.; Carré, G.; Marquéz, J.R.G.; Gruber, B.; Lafourcade, B.; Leitão, P.J.; et al. Collinearity: A review of methods to deal with it and a simulation study evaluating their performance. *Ecography* **2013**, *36*, 27–46. [CrossRef]
25. Varela, S.; Anderson, R.P.; Garciavaldes, R.; Fernandezgonzalez, F. Environmental filters reduce the effects of sampling bias and improve predictions of ecological niche models. *Ecography* **2014**, *37*, 1084–1091. [CrossRef]
26. Wang, T.; Wang, G.; Innes, J.L.; Seely, B.; Chen, B. ClimateAP: An application for dynamic local downscaling of historical and future climate data in Asia Pacific. *Front. Agric. Sci. Eng.* **2017**, *4*, 448–458. [CrossRef]
27. Fischer, G.; Velthuizen, H.V.; Shah, M.; Nachtergaele, F. *Global AgroEcological Assessment for Agriculture in the 21st Century*; International Institute for Applied Systems Analysis: Laxenburg, Austria, 2002.
28. Koeling, R. Chunking with maximum entropy models. In Proceedings of the Conference on Computational Natural Language Learning, Lisbon, Portugal, 13–14 September 2000; pp. 139–141.
29. Yang, X.Q.; Kushwaha, S.P.S.; Saran, S.; Xu, J.; Roy, P.S. Maxent modeling for predicting the potential distribution of medicinal plant, Justicia adhatoda L. in Lesser Himalayan foothills. *Ecol. Eng.* **2013**, *51*, 83–87. [CrossRef]
30. Guo, J.; Liu, X.P.; Zhang, Q.; Zhang, D.F.; Liu, X. Prediction for the potential distribution area of Codonopsis pilosula at global scale based on Maxent model. *J. Appl. Ecol.* **2017**, *28*, 992–1000.
31. Lu, C.Y.; Gu, W.; Dai, A.H.; Wei, H.Y. Assessing habitat suitability based on geographic information system (GIS) and fuzzy: A case study of Schisandra sphenanthera Rehd. et Wils. in Qinling Mountains, China. *Ecol. Model.* **2012**, *242*, 105–115. [CrossRef]
32. Hanley, J.A.; Mcneil, B.J. The meaning and use of the area under a receiver operating characteristic (ROC) curve. *Radiology* **1982**, *143*, 29–36. [CrossRef]
33. Swets, J.A. Measuring the accuracy of diagnostic systems. *Science* **1988**, *240*, 1285–1293. [CrossRef] [PubMed]
34. Zhang, K.; Yao, L.; Meng, J.; Tao, J. Maxent modeling for predicting the potential geographical distribution of two peony species under climate change. *Sci. Total Environ.* **2018**, *634*, 1326–1334. [CrossRef] [PubMed]
35. Koralewski, T.E.; Wang, H.H.; Grant, W.E.; Byram, T.D. Plants on the move: Assisted migration of forest trees in the face of climate change. *For. Ecol. Manag.* **2015**, *344*, 30–37. [CrossRef]
36. Keenan, R.J. Climate change impacts and adaptation in forest management: A review. *Ann. For. Sci.* **2015**, *72*, 145–167. [CrossRef]
37. Pecchi, M.; Marchi, M.; Burton, V.; Giannetti, F.; Moriondo, M.; Bernetti, I.; Bindi, M.; Chirici, G. Species distribution modelling to support forest management. A literature review. *Ecol. Model.* **2019**, *411*, 108817. [CrossRef]
38. Antúnez, P.; Suárez-Mota, E.M.; Valenzuela-Encinas, C.; Ruiz-Aquino, F. The Potential Distribution of Tree Species in Three Periods of Time under a Climate Change Scenario. *Forests* **2018**, *9*, 628. [CrossRef]
39. Sehler, R.; Li, J.; Reager, J.; Ye, H. Investigating Relationship Between Soil Moisture and Precipitation Globally Using Remote Sensing Observations. *J. Contemp. Water Res. Educ.* **2019**, *168*, 106–118. [CrossRef]
40. Wang, T.; Campbell, E.M.; O'Neill, G.A.; Aitken, S.N. Projecting future distributions of ecosystem climate niches: Uncertainties and management applications. *For. Ecol. Manag.* **2012**, *279*, 128–140. [CrossRef]
41. Rehfeldt, G.E.; Crookston, N.L.; Warwell, M.V.; Evans, J.S. Empirical analyses of plant-climate relationships for the western United States. *Int. J. Plant Sci.* **2006**, *167*, 1123–1150. [CrossRef]
42. Wang, T.; Wang, G.; Innes, J.; Nitschke, C.; Kang, H. Climatic niche models and their consensus projections for future climates for four major forest tree species in the Asia—Pacific region. *For. Ecol. Manag.* **2016**, *360*, 357–366. [CrossRef]
43. Mátyás, C.; Sun, G. Forests in a water limited world under climate change. *Environ. Res. Lett.* **2014**, *9*, 085001. [CrossRef]

44. Aitken, S.N.; Yeaman, S.; Holliday, J.A.; Wang, T.; Curtis-McLane, S. Adaptation, migration or extirpation: Climate change outcomes for tree populations. *Evol. Appl.* **2008**, *1*, 95–111. [CrossRef] [PubMed]
45. Campbell, E.M.; Antos, J.A. Distribution and severity of white pine blister rust and mountain pine beetle on whitebark pine in British Columbia. *Can. J. For. Res.* **2000**, *30*, 1051–1059. [CrossRef]
46. Allendorf, F.W.; Luikart, G.; Aitken, S.N. *Conservation and the Genetics of Populations*, 2nd ed.; Blackwell Publishing: Malden, MA, USA, 2012; p. 624.
47. Wang, Y.; Clack, B.; Li, S. Genetic Diversity in Camptotheca Decaisne. *Pharm. Crop.* **2014**, *5*, 120–134.

forests

MDPI

Article

Surprising Drought Tolerance of Fir (*Abies*) Species between Past Climatic Adaptation and Future Projections Reveals New Chances for Adaptive Forest Management

Csaba Mátyás [1,*], František Beran [2], Jaroslav Dostál [2], Jiří Čáp [2], Martin Fulín [2], Monika Vejpustková [3], Gregor Božič [4], Pál Balázs [1] and Josef Frýdl [2,*]

1 Institute of Environmental and Earth Sciences, Faculty of Forestry, University of Sopron, 9400 Sopron, Hungary; balazs.pal@uni-sopron.hu
2 Department of Biology and Forest Tree Breeding, Forestry and Game Management Research Institute, 252 02 Jíloviště, Czech Republic; beran@vulhm.cz (F.B.); dostal@vulhm.cz (J.D.); cap@vulhm.cz (J.Č.); fulin@vulhm.cz (M.F.)
3 Department of Forest Ecology, Forestry and Game Management Research Institute, 252 02 Jíloviště, Czech Republic; vejpustkova@vulhm.cz
4 Department of Forest Physiology and Genetics, Slovenian Forestry Institute, 1000 Ljubljana, Slovenia; gregor.bozic@gozdis.si
* Correspondence: matyas.csaba@uni-sopron.hu (C.M.); frydl@vulhm.cz (J.F.)

Abstract: *Research Highlights*: Data of advanced-age provenance tests were reanalyzed applying a new approach, to directly estimate the growth of populations at their original sites under individually generated future climates. The results revealed the high resilience potential of fir species. *Background and Objectives*: The growth and survival of silver fir under future climatic scenarios are insufficiently investigated at the xeric limits. The selective signature of past climate determining the current and projected growth was investigated to analyze the prospects of adaptive silviculture and assisted transfer of silver fir populations, and the introduction of non-autochthonous species. *Materials and Methods*: Hargreaves' climatic moisture deficit was selected to model height responses of adult populations. Climatic transfer distance was used to assess the relative drought stress of populations at the test site, relating these to the past conditions to which the populations had adapted. *ClimateEU* and *ClimateWNA* pathway RCP8.5 data served to determine individually past, current, and future moisture deficit conditions. Besides silver fir, other fir species from South Europe and the American Northwest were also tested. *Results*: Drought tolerance profiles explained the responses of transferred provenances and predicted their future performance and survival. Silver fir displayed significant within-species differentiation regarding drought stress response. Applying the assumed drought tolerance limit of 100 mm relative moisture deficit, most of the tested silver fir populations seem to survive their projected climate at their origin until the end of the century. Survival is likely also for transferred Balkan fir species and for grand fir populations, but not for the Mediterranean species. Conclusions: The projections are less dramatic than provided by usual inventory assessments, considering also the resilience of populations. The method fills the existing gap between experimentally determined adaptive response and the predictions needed for management decisions. It also underscores the unique potential of provenance tests.

Keywords: climate change; common garden; provenance test; silver fir; grand fir; Balkan firs; drought stress; resilience; climate transfer distance; adaptation

Citation: Mátyás, C.; Beran, F.; Dostál, J.; Čáp, J.; Fulín, M.; Vejpustková, M.; Božič, G.; Balázs, P.; Frýdl, J. Surprising Drought Tolerance of Fir (*Abies*) Species between Past Climatic Adaptation and Future Projections Reveals New Chances for Adaptive Forest Management. *Forests* **2021**, *12*, 821. https://doi.org/10.3390/f12070821

Academic Editor: Maciej Pach

Received: 11 May 2021
Accepted: 14 June 2021
Published: 22 June 2021

1. Introduction

The rapid change of climatic conditions, which is unprecedented on the scale of the Holocene, across all forest zones, is the greatest challenge to the long-term stability of forest ecosystems and contemporary forest management [1]. The large-scale shift in forest site conditions challenges the ecosystems along with the rear/xeric limits especially [2,3].

Climatic analyses list subcontinental East-Central Europe among future global climatic hotspots [4]. Compared to the 1981–2010 reference period, climate projections indicate that air temperature e.g., in the Czech Republic will continue to rise by an average of about 1 °C over the next 30 years. Differences between the emission scenarios will increase after 2050. Compared to the reference period, the annual mean temperature will increase by 2.0 °C (RCP4.5, —RCP: Representative Concentration Pathway is the greenhouse gas concentration trajectory assumed for various climate change scenarios), respectively, and by more than 4.0 °C (RCP8.5) until the end of the century. In extensive lowlands, such as in Hungary, more than half of the years may be extremely dry in the last quarter of the present century [5,6]. Less precipitation and higher temperatures in summer threaten the stability of forest ecosystems (Figure 1). Droughts lead to growth decline and mortality, particularly among conifers such as silver fir [7,8], and affect the genetic diversity of forest tree populations [9,10].

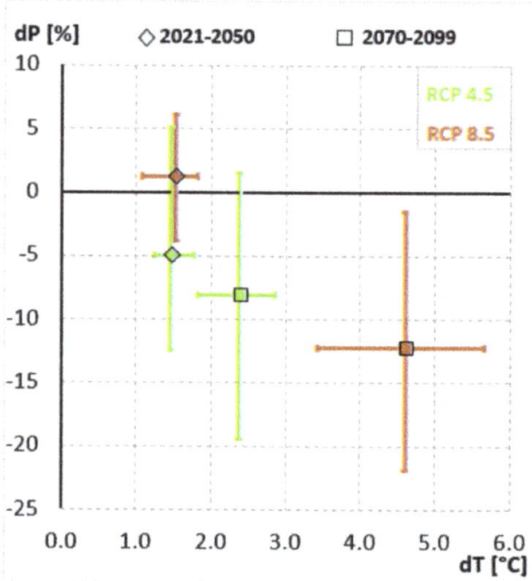

Figure 1. Expected change of temperature (*dT*) and precipitation (*dP*) in the summer quarter (June–July–August) in southwest Hungary at the xeric limits of silver fir based on the RCP4.5 and RCP8.5 scenarios. Reference: mean of 1971 to 2000. Dots: ensemble means (10 simulations) of the projected changes. Error bars: 66% range of the simulations (original design by B. Gálos, method of calculation described in [6]).

The presence of silver fir has significantly declined in Central Europe in the past century, but only partly for climatic reasons, as acid rain and game damage contributed to the damages. After the reduction of air pollution across whole East-Central Europe, silver fir recovered significantly despite recent severe droughts. Regarding the future resilience of silver fir, the assessments of its expected response are incongruent. According to some opinions, the species may benefit from winter warming but may suffer from increasing summer droughts, particularly in poor sites at lower elevations [11,12]. Conversely, more optimistic forecasts emphasize the relatively high resilience of silver fir and predict its survival even under increasing stress [13,14]. Furthermore, numerous studies suggest silver fir as a potential alternative species in ecosystems menaced by increasing droughts [7,15–18]. Introducing preadapted populations and even non-native fir species that possess higher drought tolerance has also garnered increasing attention [19,20]. Assisted migration is an ultimate and still debated option [21], gradually finding approval in East-Central Europe. The unknown long-term adaptive capacity of populations transcending their original climatic niche limits is a reason for concern and challenges the sustainability of forests [3,22].

Provenance trials offer a unique opportunity to explore the inherited potential and limitations of adaptability, allowing for conclusions about drought stress responses of populations exposed to climatic changes mimicked by geographical transfer [23,24]. Test sites close to the xeric (rear) edges of species distribution are particularly promising locations to analyze phenotypic responses to drought stress.

Although questioned by some authors [25,26], the between-population adaptive diversity of silver fir has been proven in provenance tests, but the long-term potential of the species coping with the projected climatic scenarios has not been sufficiently explored. Many of the existing studies were written before gridded climate data became available (e.g., [27–30]). Some results originate from greenhouse/nursery trials [25,31–33] or tests at juvenile age [34]. Since the advent of advanced analysis methods, a plethora of genomic, evolutionary, ecological, phenological, and physiological data improve the awareness of factors determining the adaptation of populations. Concurrent to the adaptive origin of differentiation, interfering historic/evolutionary effects are frequently discussed [35–37]. The assessment of phenological responses in the setting of projected climatic conditions, a precondition for management decisions is, however, still missing.

In this study we concentrate on a single phenotypical trait, height measured at an advanced age, to evaluate long-term adaptive responses of fir species and provenances. We apply a robust climate index to infer drought tolerance profiles for current and future conditions. The aim was to fill the existing gap between experimentally determined adaptive potential and predictions needed for management decisions [6]. The study is based on the reanalysis of results of three older provenance trials of fir species in the Czech Republic [20]. The specific value of the data, measured at an advanced age, justified a closer investigation of the phenotypic response of populations to climatic warming and droughts. The populations in the three trials include provenances of different *Abies* species of regional interest for adaptation/mitigation decisions in forestry practice. Native silver fir was the focus, represented by provenances from the drought-threatened part of the distribution in sub-continental East-Central Europe. The trials also contain other fir species from Southern Europe and from the American Northwest, which could be of potential use for future introductions.

First, we focused on detecting responses to the climatic conditions the populations were exposed to in the trials rather than on the ranking of populations. Height at an advanced age was interpreted as the joint result of the selection driven by past climatic factors at origin and by the response to "current" climate from planting until the last measurement. Geographic provenance data were used exclusively in the context of climatic factors; thus, transfer distances were expressed only in climatic terms. Second, a climate indicator available from harmonized, gridded datasets for selected time periods was chosen to model the observed and predicted responses. Climate projections were derived uniformly for the trial sites and the locations of origin. Third, drought tolerance profiles established in the trials were utilized to estimate the potential of fir species and provenances in the setting of a selected climate scenario for the subcontinental zone in East-Central Europe.

2. Materials and Methods

2.1. Description of the Three Provenance Trials

The Forestry and Game Management Research Institute (FGMRI) established three provenance trials of fir species in the Czech Republic between 1976 and 1984. The trials contain a selection of provenances of Euro-Mediterranean and North American species [38,39]. The Písek trial, organized by the FGMRI, contains native silver fir (*A. alba* Mill.) and other fir species from the Balkan and the Mediterranean (Table 3). As part of a series initiated by IUFRO, an extensive collection of 24 coastal and inland provenances of grand fir (*A. grandis* (Dougl. Ex D. Don) Lindl.) was established in Zbiroh (Table 4). The trial Dražičky, belonging to another international IUFRO series, contains provenances of grand fir and noble fir (*A. procera* Rehder) (Table 5). (Latin names are used in the text only for the less known Balkan and Mediterranean fir species.) Geographic data and maps of the populations' provenance, historic and site details of the trials, as well as all data measured in 2015 are found in Frýdl et al. [20].

Similar to most other provenance tests, the trials are far from being fully representative for any of the species and make only limited comparisons possible. The specific value of these trials is that they were maintained and measured until nearly mature age (32 to 40 years), and, even more importantly, the trials are typically located in climatic environments close to the warm and dry xeric limits of silver fir. (The term xeric limit is preferred as an alternative for rear or trailing limit, to describe the low-elevation, drought-threatened limit of distribution to emphasize the primary role of water scarcity at these borders.) The chosen low-elevation sites were nearly offsite conditions for some *Abies* species and provenances. Such data are very rare and facilitate the "space-for-time" assessment of growth and resilience, even in the absence of mortality observations.

2.2. Source of Data and Method of Analysis

The data of measurements from 2015, which are the basis of the present analysis, have been evaluated by the FGMRI and published by Frýdl et al. [20]. The three trials were investigated individually due to different sets of species and populations. From among the different quantitative traits observed, height consequently yielded the lowest error variation and the best differentiation among populations. In genetic field tests, the high heritability of height has proven its relative independence from environmental and human-caused effects. Therefore, mean height was selected to investigate the sensitivity of provenances to drought.

Phenotypic responses were interpreted as effects of climatic change mimicked by geographic transfer and analyzed by applying the ecodistance concept [24]. "Ecodistance" (here: climatic transfer distance) is defined as the difference between the ecologically relevant variables (in this case, climate) at the test site and the population provenance (origin); i.e., the difference between the conditions undergone by the populations in the experiment and the conditions they had adapted to in the past. Positive differences indicate a transfer to drier and/or warmer sites (=mimicked climatic warming); negative values signify transfer to cooler and/or wetter sites (=mimicked climatic cooling). The 0 value stands for a climate equivalent to that at the site of provenance [40].

Climatic transfer distance is interpreted as the indicator of *relative drought*, individually undergone by populations at the test site, compared to the drought stress they were adapted to in the past. In this sense, the value of relative drought is different for every provenance. To determine climatic transfer distances, the mean of the period from outplanting until the time of measurement was calculated for the "current climate" conditions at the test site. The length of the averaged period was different for each test because the planting years varied, while measurements were made in the same year, in 2015 (Table 1). The reference period for the "past climate at provenance" was set uniformly for 1911–1940. The selected 30-year period covers roughly the middle third of the approximate age (80–100 years) of selected populations from which seeds were harvested. At that age, within-population competition has largely formed the genetic structure of the stand. Climate analyses also indicate that this is the last time period of the Holocene before the onset of global anthropogenic effects [41,42]. Thus, it was taken as the basis for representing past secular adaptation.

Table 1. Basic data of the three trials and their mean height and tree density in 2015. Adapted from [20].

Trial Name	Year of Planting	Altitude (m)	Longitude (Decimal)	Latitude (Decimal)	Age at Measurement (years)	Trial Mean Height (m)	Mean Tree Density (% from Planted)
Písek	1976	395	14.33° E	49.27° N	40	15.49	28.5 [a]
Zbiroh	1980	450–460	13.64° E	49.79° N	36	19.26	50.1
Dražičky	1984	485	14.59° E	49.39° N	32	14.54	55.9

[a] without three Mediterranean populations, showing 100% mortality.

Accordingly, climatic transfer distance, i.e., relative drought, was determined for climatic moisture deficit (*dCMD*) as

$$dCMD = \text{[current value at test site]} - \text{[past value at provenance]}$$

The "current climate" of the Czech test sites was determined using the data of nearby (Czech) meteorological stations. The distances to the nearest meteorological stations were small, and corrections were unnecessary (Table A1). In this study, year-to-year growth (increment) variations or annual weather fluctuations were not the objects of analyses as the focus was on height response attained at the final measurement, determined by long-term means. Besides the annual precipitation and temperature fluctuations, the weather conditions in the course of the trial maintenance have also shown a significant trend of rising temperatures and recurrent drought years. Summer temperature means at the trial sites of Písek and Dražičky increased by 1.3 °C, from 16.6 to 17.9 °C. At the Zbiroh site, the increase reached 1.9 °C. Extremely hot summers were observed at all three trial sites in the years 1992, 1994, 2003, and 2015. Consequently, the climatic moisture deficit has increased during the "current" climate period in all three trials. Nevertheless, the extreme weather conditions did not lead to conspicuous mortality differentiation, with the exception of the Mediterranean fir species in the Písek trial.

Data for past climates were obtained from two databases. For the Euro-Mediterranean locations, the recently published database *ClimateEU* (version 4.80) [43] was utilized. It contains historic climate data for Europe using 15 global climate models, for a representative set of climate variables, for the last 120 years (1901 to 2019) as well as multi-model CMIP5 climate change projections for the 21st century. The similarly developed software package *ClimateWNA* (version July 2020) [44], which also contains historical data and future projections, served to downscale past climate variables for western North American populations of firs in the trials. Scale-free, actualized climate datasets from both databases were downscaled courtesy of T. Wang (UBC Vancouver). During the course of the digital interpolation of past climate data, it became apparent that in numerous cases the original coordinates of sampled forest stands were incorrect. Consequently, the nearest point at proper altitude served to estimate the climate parameters; the actual land cover shown on Google was taken also into consideration. Climate data and executed corrections are indicated in Tables 3–5. Due to the different methods of determining climate parameters, the climate data in this study are not identical to those in Frýdl et al. [20].

Hargreaves' climatic moisture deficit (*CMD*) was selected to calculate the climatic transfer distances. This bioclimatic variable was calibrated originally for estimating potential evaporation relative to precipitation under semi-arid agrarian conditions [45]. Comparisons of results with other, straightforward variables have indicated that *CMD* is also well suited for describing drought stress under forest conditions. The comparison of response regressions using *CMD* values both for annual and summer periods surprisingly indicated that annual *CMD* values yielded higher determination coefficients, so these were selected for the calculation of climatic transfer distances ($dCMD_{\text{ann}}$). Hargreaves' *CMD* means of 30-year periods were available from both databases (*ClimateEU* and *ClimateWNA*). The climatic transfer distances expressed as differences of *CMD* values (*dCMD*) were calculated not only for current vs past climates but also for future climatic changes. The transfer distances of the two basic variables, temperature, and precipitation, are shown in Appendix A to illustrate their contribution to the responses at the test sites (Figures A1–A3).

The statistical analysis of the trials was performed using the *QC.Expert* 3.1 [46] and *NCSS* 10 (version: 10.0.6) programs [20]. Due to the non-normality of the data, the differences among provenances were tested using the Kruskal-Wallis one-way *ANOVA* test. Regressions were calculated between height and climatic transfer distance of provenances// but were presented only if statistically significant and biologically appropriate. These *tolerance profiles* are transfer functions (even if shown only as scatter of data points), comparing the responses of different populations in the same common garden test. They have to be distinguished from reaction norms of *individual* populations expressing their phenotypic plasticity across numerous test sites [40]. Lacking data of similar-age parallel trials, the results did not allow for the calculation and comparison of reaction norms. However, the tolerance profile of species, i.e., the variation of phenotypic response between differently adapted provenances, offers a hint to estimate the species-specific range of climatic resilience and allows certain comparisons between species.

3. Results

Tables 1 and 2 show the main data of the test sites and their mean height and mean density (remaining percentage from planted) measured in 2015. Tree density numbers were the lowest at the driest site Písek, partly due to the higher age of the trial. The density data could not be used for inferring survival differentiation between provenances. The reasons were the absence of extreme events triggering significant mortality and the routine silvicultural tendings, applied to keep relatively even competition conditions in the trials. Partly due to the relatively small plot size for the advanced-age trees, the rather strong differentiation of tree numbers per plot could not be linked to climate factors, with a few exceptions (discussed later).

Table 2. Current climate parameters of the three provenance tests, calculated according to the described protocol.

Trial Name	Annual Mean Temp. (°C)	Mean Temp. in Warmest Quarter (°C)	Annual Mean Precip. (mm)	Mean Precip. in Warmest Quarter (mm)	Mean Annual Moisture Deficit (CMD_{ann})
Písek	8.1	17.2	570	229	213
Zbiroh	8.1	17.0	595	241	179
Dražičky	8.1	17.3	592	229	189

The results of the Kruskal-Wallis one-way *ANOVA* tests rejected the hypothesis of equal mean heights at the $p \leq 0.05$ level for all three experiments. Matrices of significance, based on the obtained statistics are presented in Appendix A in Figures A4–A6. The statistics have proven significant differences mostly at the level of provenance groups. The significance of differences might be underestimated due to the applied one-way *ANOVA* test.

Contemplating climatic transfer distances ($dCMD_{\mathrm{ann}}$), i.e., relative drought values at the trial sites, annual climatic moisture supply was sufficient or even advantageous for some populations, first due to the summer rainfall maxima typical for Central Europe. Table 2 summarizes the main climatic data of the three Czech test sites for the "current climate" period. Data of provenances are presented in Tables 3–5.

Table 3. Basic data of the provenance trial 64, Písek: species and populations represented and basic data used for analysis (sequence and geographic data of provenances are identical to the source publication (Adapted partly from [20]). Climate distances refer to the trial location.

Populations of the Provenance Trial 64 Písek, CZ						Past Climate at Origin (1911–1940)						Climate Distance (D)			Data 2015	
Prov No.	*Abies* Species Name	Provenance Name	Alt. (m)	Long. E	Lat. N	January Mean Temp. (°C)	Annual Mean Temp. (°C)	Mean Temp. Warmest Quarter (°C)	Annual Mean Prec. (mm)	Mean Prec. Warmest Quarter (mm)	*CMD* (mm, Ann.)	Temp. Change Warmest Quarter (°C)	Prec. Change Warmest Quarter (mm)	*dCMD* (Index Change, mm)	Mean Height (m)	Mean Tree Density (% From Planted)
74	*A. alba*	Milevsko, Klučenice CZ	410	14.2°	49.6°	−2	8.4	17.3	492	189	284	−0.1	40.0	−71	17.19	22.0
81	*A. alba*	Vyšší Brod, Vítkův Kámen CZ	900	14.3°	48.6°	−4.2	5.9	14.5	854	296	122	2.7	−67.0	91	15.47	10.0
89 *	*A. cilicica*	Kammouha LIB	1100	36.0°	34.0°	6.3	14.9	22.2	1167	4	763	−5.0	225.0	−550	0.00	0.0
109	*A. cephalonica*	Centr. Peloponnese, Vytina GR	1250	22.1°	37.6°	1.9	10.9	18.7	794	51	573	−1.5	178.0	−360	14.19	29.3
121 *	*A. cilicica*	Djebel el Chouk, Lattakia SYR	1300	36.0°	35.8°	2.2	12.6	20	1039	24	525	−2.8	205.0	−312	0.00	0.0
130	*A. alba*	Nasavrky, Podhůra CZ	370	15.8°	49.8°	−2.7	8.2	16.8	487	196	253	0.4	33.0	−40	16.89	32.7
132 *	*A. alba*	Rilskije gory, Borovec BG	1200	23.6°	42.2°	−3	7.3	15.7	515	137	362	1.5	92.0	−149	16.85	31.3
135 *	*A. pinsapo*	Malaga, La Yunquera ES	878	4.7° W	36.7°	6.8	14.7	22.2	633	18	711	−5.0	211.0	−498	0.00	0.0
136	*A. cephalonica*	Peloponnese, Vytina GR	1010	22.2°	37.7°	3.4	12.2	20	753	48	626	−2.8	181.0	−413	14.34	34.7
137 *	*A. borisii-regis*	Mt. Pindos, Pertuli GR	1200	21.3°	39.8°	−0.3	9.6	18.2	727	88	503	−1.0	141.0	−290	14.59	35.3
223 *	*A. alba*	Sanski Most BH	1050	16.6°	44.6°	−3.5	6.9	15.3	943	226	170	1.9	3.0	43	15.61	28.0
228 *	*A. alba*	Regello, Vallombrosa IT	1010	11.5°	43.7°	0.2	9.5	17.5	967	163	180	−0.3	66.0	33	14.82	30.7
S 2 *	*A. alba*	Banská Bystrica, Radvaň SK	780	19.0°	48.7°	−5.1	6.4	15.1	759	258	152	2.1	−29.0	61	14.92	31.3
		Data and current climate of the test site **	395	14.3°	49.2°	−1.5	8.1	17.2	570	229	213				*15.49*	*28.5*

Climate data for Kammouha, LB were estimated from Worldclim database, 1961–1990; * Corrected locations ** CZ met data (original or corrected). Last row: numbers in italics indicate averages excluding the populations with 0.0 values.

Table 4. Basic data of the provenance trial 213 Zbiroh: location and climate of grand fir populations used for analysis (sequence and geographic data of provenances are identical to the source publication (Adapted partly from [20]). Climate distances refer to the trial location.

Populations of the Provenance Trial 213 Zbiroh, CZ						Past Climate at Origin (1911–1940)						Climate Distance (D)			Data 2015	
Prov No.	*Abies* Species Name	Provenance Name	Alt. (m)	Long. W	Lat. N	January Mean Temp. (°C)	Annual Mean Temp. (°C)	Mean Temp. Warmest Quarter (°C)	Annual Mean Prec. (mm)	Mean Prec. Warmest Quarter (mm)	*CMD* (mm, ann.)	Temp. Change Warmest Quarter (°C)	Prec. Change Warmest Quarter (mm)	*dCMD* (Index Change, mm)	Mean Height (m)	Mean Tree Density (% from Planted)
12040	*A. grandis*	Salmon River CAN/BC *	50	125.8°	50.3°	0.6	8.4	15.6	1444	150	211	1.4	91.0	−32	21.81	44.0
12041	*A. grandis*	Oyster Bay CAN/BC	5	125.2°	49.9°	1.2	9.1	16.3	1137	121	286	0.7	120.0	−107	20.90	70.0
12042	*A. grandis*	Buckley Bay CAN/BC	45	124.9°	49.5°	0	8.8	16.3	1391	87	319	0.7	154.0	−140	19.32	62.0
12043	*A. grandis*	Sproat Lake CAN/BC	25	125.0°	49.3°	0.8	9.4	17	1827	117	293	0.0	124.0	−114	20.05	58.0
12044	*A. grandis*	Kay Road CAN/BC	50	124.3°	49.3°	2	9.4	16.5	863	85	342	0.5	156.0	−163	20.02	72.0
12045	*A. grandis*	Yellow Point CAN/BC	30	123.8°	49.1°	2.7	9.7	16.6	916	81	335	0.4	160.0	−156	20.03	57.0
12046	*A. grandis*	Mount Provost CAN/BC	75	123.8°	48.8°	1.8	9.5	16.7	1199	78	353	0.3	163.0	−174	21.38	63.0
12047	*A. grandis*	Sooke CAN/BC	20	123.8°	48.4°	2.9	9.6	14.9	1111	72	296	2.1	169.0	−117	20.05	60.0
12002	*A. grandis*	Tulalip USA/WA	30	122.3°	48.1°	3.4	9.5	14.8	651	53	369	2.2	188.0	−190	22.29	48.0
12003	*A. grandis*	Indian Creek USA/WA	140	123.6°	48.1°	2.1	9	14.7	1011	64	334	2.3	177.0	−155	21.48	57.0
12004	*A. grandis*	Gardiner USA/WA	30	122.9°	48.1°	2.9	9.6	15.2	497	61	401	1.8	180.0	−222	21.72	56.0
12006	*A. grandis*	Eagle Creek—low USA/WA *	760	120.6°	47.7°	−4.2	7.6	17.8	831	56	531	-0.8	185.0	−352	19.05	46.0
12008	*A. grandis*	Jack Creek USA/WA *	825	120.8°	47.3°	−3.7	6.6	15.9	877	56	487	1.1	185.0	−308	17.30	43.0
12011	*A. grandis*	Clear Lake USA/WA *	945	121.3°	46.6°	−3.8	5.4	14	948	69	394	3.0	172.0	−215	17.89	48.0
12013	*A. grandis*	Cooper Spur USA/OR	1040	121.7°	45.5°	−1.9	6.7	14.8	2097	115	232	2.2	126.0	-53	18.91	36.0

Table 4. *Cont.*

Populations of the Provenance Trial 213 Zbiroh, CZ						Past Climate at Origin (1911–1940)							Climate Distance (D)			Data 2015	
Prov No.	*Abies* Species Name	Provenance Name	Alt. (m)	Long. W	Lat. N	January Mean Temp. (°C)	Annual Mean Temp. (°C)	Mean Temp. Warmest Quarter (°C)	Annual Mean Prec. (mm)	Mean Prec. Warmest Quarter (mm)	*CMD* (mm, ann.)	Temp. Change Warmest Quarter (°C)	Prec. Change Warmest Quarter (mm)	*dCMD* (Index Change, mm)	Mean Height (m)	Mean Tree Density (% from Planted)
12015	*A. grandis*	Sisi Butte USA/OR *	975	121.8°	44.9°	−0.8	7.4	15.2	1595	98	318	1.8	143.0	−139	18.19	36.0
12016	*A. grandis*	Santiam Summit USA/OR	1400	121.9°	44.4°	−2.2	5.5	13.1	2033	127	266	3.9	114.0	−87	16.22	34.0
12019	*A. grandis*	Roaring River USA/OR *	1310	122.0°	43.5°	−0.7	6.8	14.3	1733	129	286	2.7	112.0	−107	17.88	24.0
12020	*A. grandis*	Crescent Creek USA/OR *	1375	121.9°	43.5°	−2.7	5.7	13.6	782	56	524	3.4	185.0	−345	17.90	24.0
12025	*A. grandis*	Buckskin Creek USA/ID *	1220	116.2°	48.0°	−5.4	4.8	14.4	1087	129	304	2.6	112.0	−125	14.55	56.0
12026	*A. grandis*	Plummer Hill USA/ID	850	116.9°	47.3°	−3.2	7.6	17.3	642	80	511	-0.3	161.0	−332	16.96	20.0
12031	*A. grandis*	Bertha Hill USA/ID *	1430	115.8°	46.8°	−4.4	5.7	15.5	1341	120	307	1.5	121.0	−128	18.17	63.0
12038	*A. grandis*	Clearwater USA/ID *	760	115.4°	46.6°	−3.5	7.9	18.3	965	114	479	−1.3	127.0	−300	21.48	72.0
12037	*A. grandis*	Stanley Creek USA/MT *	800	115.9°	48.3°	−5.4	6.1	16.1	698	97	494	0.9	144.0	−315	18.58	53.0
Data and current climate of the test site **			456	13.64° E	49.79°	−1.4	8.1	17.0	595	241	179				19.26	50.1

* Corrected locations, ** CZ met data (original or corrected).

Table 5. Basic data of the provenance trial 219 Dražičky: location and climate of species and populations used for analysis (sequence and geographic data of provenances are identical to the source publication (Adapted partly from [20]). Climate distances refer to the trial location.

Populations of the Provenance trial 219 Dražičky, CZ						Past Climate at Origin (1911–1940)							Climate Distance (D)			Data 2015	
Prov No.	*Abies* Species Name	Provenance Origin, Name	Alt. (m)	Long. W	Lat. N	January Mean Temp. (°C)	Annual Mean Temp. (°C)	Mean Temp. Warmest Quarter (°C)	Annual Mean Prec. (mm)	Mean Prec. Warmest Quarter (mm)	*CMD* (mm, ann.)	Temp. Change Warmest Quarter (°C)	Prec. Change Warmest Quarter (mm)	*dCMD* (Index Change, mm)	Mean Height (m)	Mean Tree Density (% from Planted)
CZ 0	*A. alba*	Adršpach CZ	620	16.1°E	50.6°	−5.1	6.2	14.6	590	231	137	2.7	−2.0	52.0	6.22	32.0
12001	*A. grandis*	Buck Creek (Skamania) USA/WA	400	121.4°	48.3°	−0.6	8.5	16.5	2291	174	199	0.8	55.0	−10.0	20.65	74.0
12002	*A. grandis*	Tulalip (Ellensburg) USA/WA	30	122.3°	48.1°	3.4	10.2	16.1	909	111	296	1.2	118.0	−107.0	19.75	66.0
13004 *	*A. procera*	Mary's Peak USA/OR	1065	123.6°	44.5°	2.8	9.6	16.1	3060	89	269	1.2	140.0	−80.0	12.41	44.0
13006 *	*A. procera*	Snow Peak USA/OR	1060	122.6°	44.6°	1.5	7.9	14.3	2532	173	177	3.0	56.0	12.0	13.68	42.0
13011 *	*A. procera*	Larch Mtn. USA/OR	975	122.1°	45.5°	−0.3	7.2	14.2	2969	209	157	3.1	20.0	32.0	14.42	68.0
13014 *	*A. procera*	Red Mtn. USA/WA	1220	121.8°	45.9°	−2.4	5.4	13	2489	150	171	4.3	79.0	18.0	14.53	63.0
13018	*A. procera*	McKinley Lake USA/WA	900	122.1°	46.6°	−0.4	7	14.2	2175	184	162	3.1	45.0	27.0	15.16	64.0
13021 *	*A. procera*	Stevens Pass USA/WA	1000	121.1°	47.7°	−3.8	5.1	13.2	2075	159	185	4.1	70.0	4.0	14.00	50.0
Data and current climate of the test site **			488	14.59° E	49.39°	−1.8	8.1	17.3	592	229	189				14.54	55.9

* Corrected locations, ** CZ met data (original or corrected).

3.1. Comparison of Euro-Mediterranean Fir Species and Provenances in Trial 64, Písek

The trial contains seven populations of silver fir and six other fir populations from Southern Europe and the Near East (Table 3). The silver fir (*Abies alba* L.) provenances represent a geographically and climatically differentiated set of locations, mostly from the drought-threatened subcontinental part of the species distribution, which enabled the modelling of the intra-specific effect of growing drought stress across climates. The species were separately analyzed.

Five out of the seven silver fir populations in the Písek test (Vyšší Brod/CZ, Sanski Most/BH, Banská Bystrica/SK, Borovec/BG and the Sub-Mediterranean Regello/IT) are higher-altitude provenances (>700 m a.s.l.), with lower annual temperatures and higher precipitation than at Písek, with the exception of the extremely dry location at Borovec/BG. The two Czech low-elevation sources (<500 m a.s.l., Milevsko and Nasavrky) were transferred to a climate similar to their origin (Table 3, Figure 2, Figure A1a,b). The Italian provenance is the only population outside the subcontinental climate zone (Figures 2 and 3).

Higher altitude provenances are adapted to less moisture deficit stress, appearing mainly during the summer months. Their transfer to the test site Písek caused increased exposure to relative (provenance-related) drought, causing weaker growth response. The low-elevation populations were more stressed at their origin, where the water supply deficit is also significant in the spring and autumn months. Their relative drought stress at Písek was less; thus, their height growth response was better. The response of the high-elevation Bulgarian population from Borovec, Rila Mts. (1200 m a.s.l.) is specific; at its provenance annual precipitation is only 515 mm, and the annual *CMD* amounts to 362 mm. It is adapted to high moisture deficit also outside the summer season. Consequently, its response was similar to low-elevation populations, indicating high drought tolerance (Table 3, Figure 2). The Sub-Mediterranean population Regello (IT) displayed a weak drought tolerance close to the other high-elevation populations; its annual *CMD* at origin is just half of the Bulgarian provenance (180 mm, Table 3).

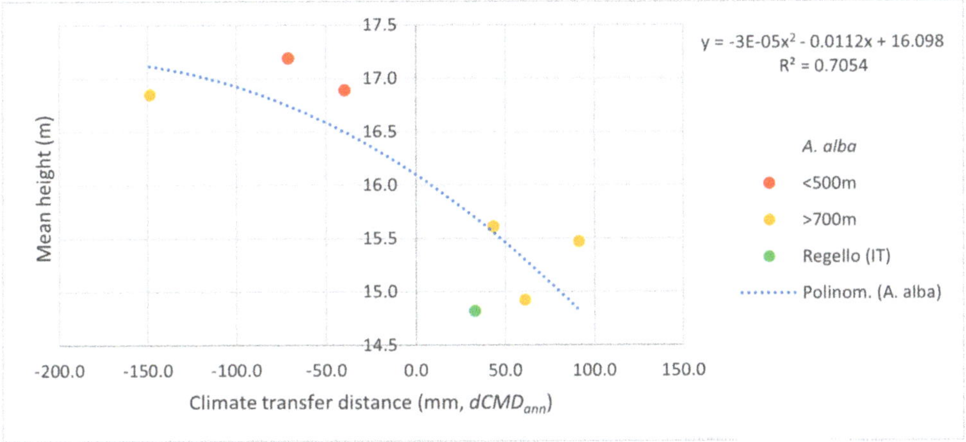

Figure 2. Mean height response and unilateral transfer equation of silver fir provenances to climate transfer distance (current vs past) expressed in annual moisture deficit units (mm precipitation, $dCMD_{ann}$) in the trial Písek. The regression is significant at $p \leq 0.05$.

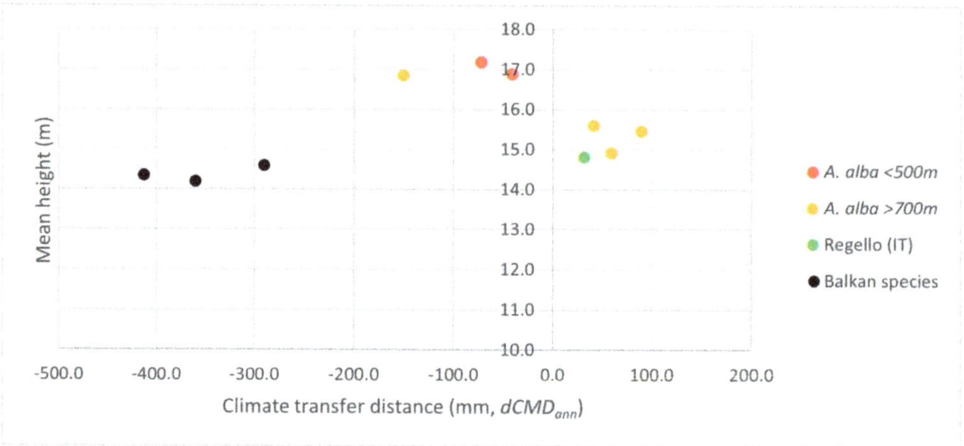

Figure 3. Mean height response of all provenances to climate transfer distance (current vs past) expressed in annual climatic moisture deficit (mm precipitation $dCMD_{ann}$) in the trial Písek. The provenances *A. cilicica* and *A. pinsapo* are not shown, having suffered total mortality (see also Figure A1a,b for distances expressed in temperature and precipitation changes).

Thus, a connection between height growth, respectively, drought tolerance at an older age, and the climate at origin to which the populations were adapted was detected. The annual *CMD* values have shown a better correlation with height response than the summer values. The calculated regression is exponential and attains a significant R^2 value (0.705, $p \leq 0.05$). The equation is unilateral, i.e., it shows the change of growth response towards only one limit of tolerance, here towards the xeric limit [40]. The regression mean height vs. $dCMD_{ann}$ indicates the best height response for provenances with origins close to the test site, but the example of Borovec (BG) shows that transferred populations may attain similar performance. The result implies a considerable climate sensitivity of silver fir, which should be treated with caution as the result is based on only seven populations. The Kruskal-Wallis test matrix (Figure A4) supports the credibility of the result. The test demonstrated significant differences in height growth of silver fir by elevation and region; high elevation populations appeared set apart from the low-elevation Czech populations Milevsko and Nasavrky (74, 130), indicating effective local adaptation by altitude.

The comparison of all provenances in the Písek test shows a completely different picture (Figure 3). The three provenances from the Balkan species (*A. borisii-regis*, *A. cephalonica*, 109, 136, and 137) enjoyed a considerably cooler/wetter summer climate than at their origin. They responded with slower growth and shared this response with two silver fir populations transferred from locations further to the south than Písek; these are the high-elevation Slovak (S2 Radvaň) and Italian (228 Regello) populations. The Kruskal-Wallis test confirmed their height growth to be significantly weaker, but there was no significant difference between them (Figures 3 and A4). The climate tolerance profile of Balkan populations displays high moisture surplus (negative $dCMD_{ann}$) values and does not show any trend linked to changing surplus moisture supply at Písek.

It is remarkable that *A. cilicica* and *A. pinsapo* populations from geographically distant and climatically extreme locations did not survive the climate of the test site (Table 3). The total mortality of these populations is most likely due to winter and late frosts, which were considerably harsher than at their original sites [38]. The January mean temperature at the origin of these populations is over 0 °C (2.2–6.8 °C), while the summer mean precipitation is extremely low; e.g., for *A. cilicica* from Kammouha, Lebanon, it amounts to only 4 mm (Table 3). The calculated high "surplus moisture" ($dCMD_{ann}$ in Table 3) played no role in preventing their mortality.

For the projection of future performance of silver fir provenances in the digital space of climate models, we used the unilateral transfer equation of silver fir, calculated for the "current climate" as a basis. Extending the equation for the periods 2041–2070 and 2071–2100, estimations of height responses are achieved for the second half of the 21st century. For the sake of comparison, the future height data are shown uniformly for the age in the year 2015, i.e., in the sense of a site index at the age of 40 years, i.e., in 2015. The multi-model ensemble climate data in *ClimateEU* under the high-emission RCP8.5 and the less stringent mitigation RCP4.5 pathways were used both for the original locations of provenance and for the test site in Písek. The mid-century projections (2041–2070) did not show climatic changes substantial enough to warrant them being analyzed separately. Furthermore, only the results of the pathway RCP8.5 for the period 2071–2100 are displayed (Figure 4, Table A3). The pessimistic high-emission pathway was preferred as it was considered more realistic, considering also unknown physical and biological risks expected to affect the resilience of populations in the future.

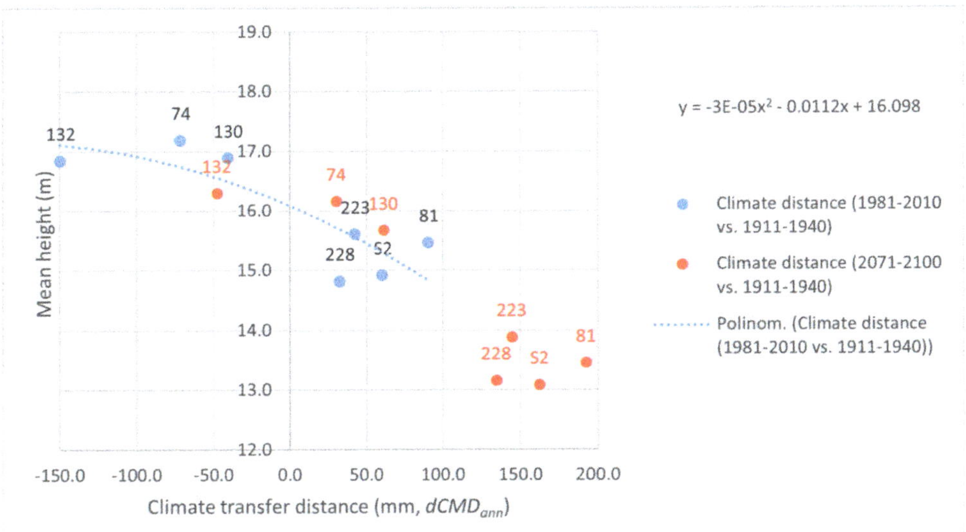

Figure 4. Current and the future climatic transfer distance and mean height response of silver fir provenances in the Písektrial. Their performance is compared under two transfer situations, current heights at the test site (data measured in 2015, blue dots) and projected heights (climate period 2071–2100, derived from *ClimateEU*, pathway RCP8.5, red dots). The transfer equation for current vs past climate was extrapolated for future data. The horizontal axis shows climate transfer distances (precipitation mm deficit). For the sake of comparison, current and future heights are both presented as the site index value for the age of 40 years, i.e., in 2015.

The projected response of the silver fir populations to the change of climate in the period 2071–2100 was calculated with the transfer equation gained for the current vs. past climates, assuming its validity for the whole century (Figure 4). The virtual transfer changed the climate distance by approximately +100 mm moisture deficit increase for all populations. Height response shifts of provenances from current to future climates are basically of the same magnitude, but follow the exponential character of the equation; the height of the provenances that already had moisture deficits in the current climate experienced stronger declines under future conditions. The populations with extreme moisture deficit attained the lowest projected heights, and the responses reveal the most vulnerable populations. These are all high-altitude populations, adapted to milder drought stress of shorter duration: the Czech Vyšší Brod (81), the Slovak Radvaň (S2), further the

Italian Regello (228) and the Bosnian Sanski Most (223). The high-altitude provenance Borovec (132) is adapted to a high moisture deficit at its origin and may attain a top position in Písek under the future climate (Table A3). No transfer equation is available for the Balkan fir species due to their currently sufficient moisture supply (Figure 3). Assuming a drought stress change of similar magnitude, i.e., a transfer shift ($dCMD_{ann}$) of approximately 100 mm, all populations remain in the sufficient moisture supply zone in the period 2071–2100. This predicts a low drought stress exposure in the future for the high-altitude Balkan populations. The future performance of distant Euro-Mediterranean provenances cannot be estimated due to a lack of data in parallel test locations in milder climates.

A further step is to estimate the population performance at their locations of origin, which is also calculated with the help of the equation for "current" vs. past data for silver fir, here using the climate projections for the individual sites of origin, run for the 2071–2100 period, using the database *ClimateEU*, pathway RCP8.5. While the climate projection for the future was the same for all populations in the Písek test, here every provenance has a different climate projection according to its original geographic location. (The differences in local site potential between the locations of provenance are not considered.) Comparing Figures 4 and 5, the scatter of projected points appears less drastic for the sites of origin (Figure 5), and their rank is different due to the change of the reference location.

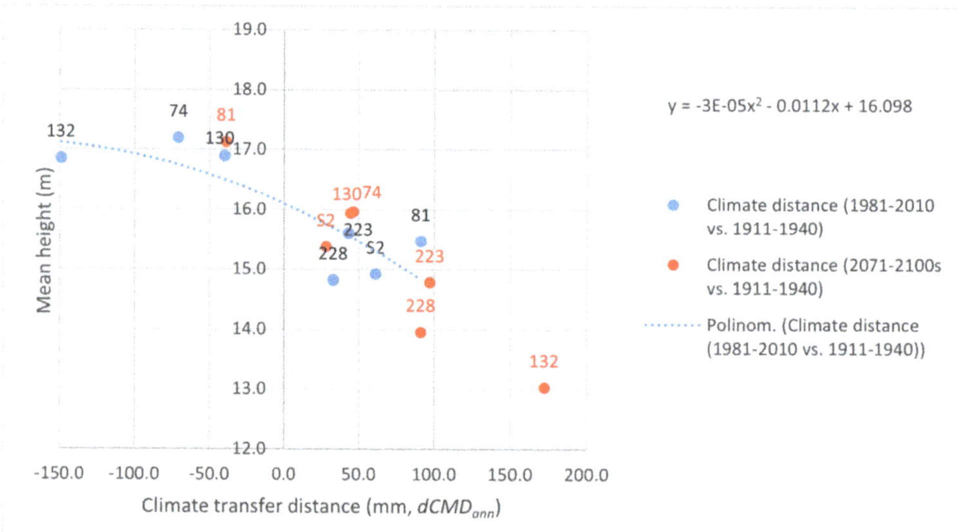

Figure 5. Comparison of current height at Písek and estimated future mean height response to climatic change of silver fir populations at their original location of provenance. Their performance is compared under two climatic transfer situations, current heights at the test site (data measured in 2015, blue dots) and projected heights at origin in the climate period 2071–2100 (derived from *ClimateEU*, pathway RCP8.5, red dots). The transfer equation for current vs past climate was extrapolated for future data. The horizontal axis shows climate distances in *dCMD* units (mm precipitation deficit). For the sake of comparison, both current and future heights appear as the site index value for the age of 40 years, i.e., in 2015.

The projections reveal the most vulnerable populations at their origins at the end of the century. The change of the position of provenance 81 (Vyšší Brod CZ) is notable, as it shows best the differentiation caused by individually projected climates. The high-elevation population has responded with weak height in Písek, and its survival at the test site was the lowest. The projected climatic position at the original high-elevation site is, however, very suitable: −39 mm moisture deficit (i.e., moisture surplus), and the expected response is the best-projected height: 17.13 m (Table A4). The silver fir provenance with the highest drought tolerance, the Bulgarian Borovec (132), will suffer from an extreme moisture deficit increase of +172 mm at its origin. Its projected height response is the lowest with 13.0 m. While it may survive in Písek, it will most likely go extinct in Bulgarian Borovec. Thus, the population will be more threatened by droughts at its original location than in the low-elevation site Písek towards the end of the century. Other populations with extreme moisture deficit are the Italian Regello and the Bosnian Sanski Most; both were among the weak performers also in the Písek trial. The two Czech low-elevation provenances of Nasavrky and Milevsko, as well as the Slovak Radvaň, will maintain their medium positions at their original location. In the case of the Balkan fir species, projections were calculated by the same equation for silver fir. Although their predicted relative drought change is above the assumed survival limit of 100mm (for details, see the discussion), their current $dCMD_{ann}$ values are so low that they will remain in the high moisture surplus zone at their origin, similar to the Písek trial location (Figure 3, Table A4).

3.2. Comparison of Grand Fir Provenances in Trial 213, Zbiroh

The trial contains 24 provenances of grand fir, among them 8 from Vancouver Island, Canada, three from the coastal belt north of Seattle, WA, and 16 sources from the Inland NW of the USA, representing higher altitude stands from the Cascades and the inland range in Idaho and Montana (Table 4). The shortcoming of this rather detailed collection of provenances is it contains no control populations of native silver fir in the trial; nonetheless, the within-species mean heights are worth comparing, in view of their possible use in East-Central Europe.

The Canadian sources all originate from low elevations of the eastern coast of Vancouver Island and receive high annual rainfall above 1000 mm, but low summer precipitation (94–229 mm). The three populations from coastal Washington (USA) enjoy somewhat less annual precipitation but also lower summer rainfall. The Cascade sources have similar rainfall ranges, while inland provenances in Idaho and Montana receive less annual precipitation (684–933 mm). Temperature conditions across the huge range from mild-coastal to continental-inland climates are relatively comparable in summer with means of 12.6–17.4 °C, while January mean temperatures show larger differences between coastal and inland locations (between 3.8 and −6.0 °C). All climate data related to the past reference period 1911–1940 (Table 4).

The Kruskal-Wallis test results for mean heights quite clearly differentiate two main groups within the grand fir provenances. The first main group contains coastal populations from Vancouver Island (reg. numbers 12040 to 12047) and Puget Sound, Washington, USA (12002 to 12004). The latter populations display the best growth from all tested provenances. The second main group originates partly from higher elevations (>700 m) of the Washington and Oregon Cascades (12006 to 12020), with significantly weaker growth, while the populations from inland Idaho and Montana (12026 to 12031) form an overlapping heterogeneous group, representing highly variable ecological conditions and discontinuous ranges in the Rocky Mts. A significant outlier is Clearwater, Idaho (12038), a top grower (Figure A5).

In the trial of Zbiroh, "current" temperature conditions at the trial site (annual mean of 8.1 °C and summer mean 17.0 °C) are within the data range of the sources from the American Northwest. The amount and distribution of rainfall is, however, notably different. "Current" annual precipitation amounts to only 595 mm in Zbiroh, but over one-third of these falls in the summer quarter (241 mm), producing a summer rainfall peak.

It is very illustrative to compare the climate charts of the weather stations at Zbiroh and Everett, a coastal station in the state of Washington, USA. The data for the charts were derived from [47], drawn by the *ClimateChart* application (*climatecharts.net*) accessed on 10 November 2020. Figures 6 and 7 display the diametrical difference of the climate of Central-Southeast Europe versus the mild, oceanic "Sub-Mediterranean type" climate along the inland coast of Washington State (Puget Sound). At the test site, most grand fir provenances experienced a drastic decrease in their annual precipitation with the exception of inland sources, where the decrease was less (Table 4). The mean summer rainfall at Zbiroh, however, was higher. These differences caused all provenances to experience a wetter and warmer summer climate at the test site than at origin (Figure 8, Figure A2a,b). Adapted to relatively dry summers, grand fir populations did not suffer from summer drought stress at Zbiroh for obvious reasons (Figure 8).

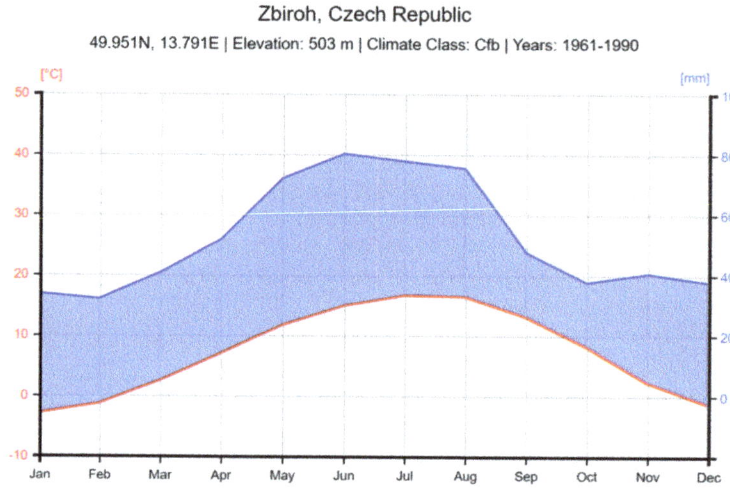

Figure 6. Climate chart of the trial Zbiroh (mean annual data for 1961–1990: temperature 7.2 °C; precipitation: 586 mm). Note the quantity and seasonal distribution of precipitation with a summer maximum.

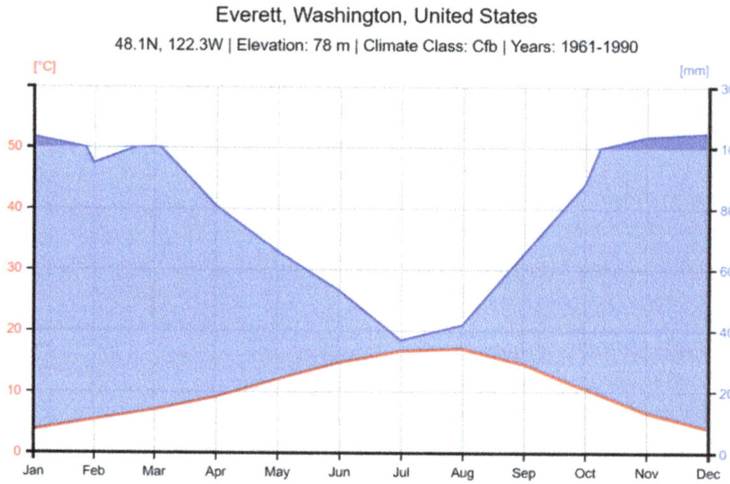

Figure 7. Climate chart of the weather station Everett, Washington (mean annual data for 1961–1990: temperature 10.3 °C; precipitation: 950 mm). Note the "Sub-Mediterranean type" distribution of rainfall with minimum precipitation in July.

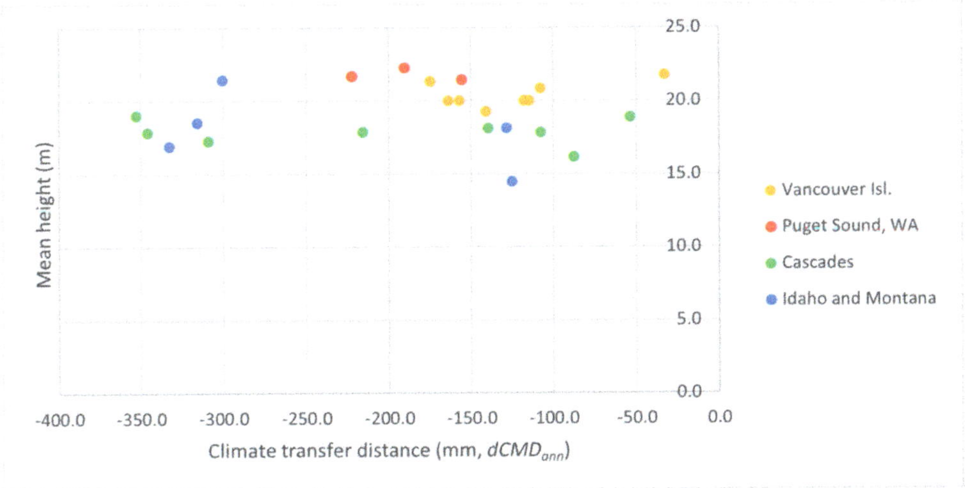

Figure 8. Mean height response of grand fir provenances to climate transfer distance (current vs past) expressed in annual climatic moisture deficit (mm precipitation deficit $dCMD_{ann}$), in the trial Zbiroh. Colors indicate origin of populations (see also Figure A2a,b for climatic distances, expressed in temperature and precipitation changes). Note the differentiation by provenance groups.

Thus, the high resilience potential and better growth of grand fir appears to be linked to their adaptation to the high summer aridity at their origin. This result obtained under subcontinental climate conditions contrasts the opinion that at the interspecific level, the precipitation in the wettest month or in the wettest quarter at the location of origin is the best predictor for summer drought response [35].

The distribution of height data across the ecological space of tested populations indicates no correlation with the amount of moisture surplus in any of the provenance groups. Neither the drastic increase of summer temperature (up to 3.9 °C, for Santiam Summit USA/OR) nor the increase of summer precipitation (up to 185 mm, for Eagle Creek, WA, USA) have produced a significant trend change of heights (Table 4, Figure A2), which indicates a considerable resilience of the species to the change of moisture supply. A direct comparison with silver fir was not possible in these trials, but the top performance of grand fir provenances in other European trials [48] corroborates the extraordinary growth and resilience of the species. One reason for the superiority of the species across Central Europe is most probably the wetter summer climate when compared to the original sites.

The projection of the phenotypic response of the populations at lower elevations in the Czech Republic (Písek) for the climate period 2071–2100 was also attempted for grand fir. Most populations indicated moisture surplus conditions at the test site; therefore, it may be assumed that grand fir provenances will tolerate the projected future climate, especially at higher elevations in East-Central Europe.

3.3. Comparison of Noble Fir Provenances in the Trial 219, Dražičky

The six tested noble fir provenances originate from ecologically rather similar, medium elevations (900–1065 m a.s.l.) of the Central-Southern Cascades (Table 5). Their climatic differentiation is moderate; the annual and summer mean temperature difference between northern and southern sources is less than 2 °C. Precipitation conditions vary between 1800 and 2800 mm annually, out of which, however, only 90–200 mm falls in the summer quarter, which is less than the summer mean at the trial site (229 mm). One silver fir and two grand fir populations were added for comparison. The main climatic effect of transfer for the two American species was the markedly lower annual precipitation; however, rainfall was somewhat higher in the summer quarter (by 20–140 mm) than at the original locations (Table 5, Figures 7, 9 and A3a,b).

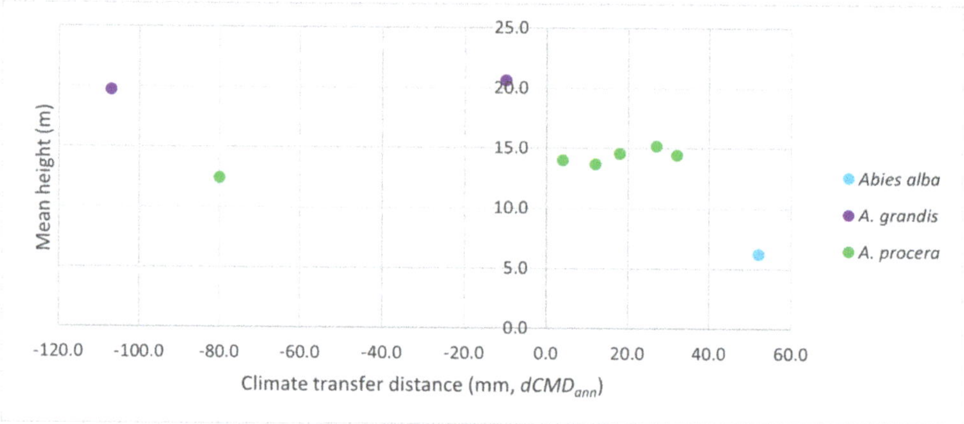

Figure 9. Mean height response of noble fir provenances to climate transfer distance (current vs past) expressed in annual climatic moisture deficit (mm precipitation deficit $dCMD_{ann}$), in the trial Dražičky (see also Figure A3a,b for distances expressed in temperature and precipitation).

Noble fir has shown a low sensitivity to summer aridity conditions, similar to grand fir, and tolerated a drastic decline of annual precipitation (over 1500 mm/year less), while summer precipitation was higher than adapted to at the original location. The Kruskal-Wallis test did not separate the majority of noble fir populations, only the southernmost ones (13004 and 13006, see Table 5 and Figure A6). The provenances have shown minor differentiation of height at age 32, related to climatic transfer distance, which did not change even in the case of the extreme provenance Stevens Pass, WA (13021). This population has shown the highest resilience to changing climatic conditions; it tolerated a five-degree increase of summer and annual temperature with no decline in height. With the exception of one provenance (13004 Mary's Peak, OR), all displayed moderate moisture deficits, with no effect on growth (Table 5). The projection of future climatic moisture deficits (calculated the same way as before) for the climate period 2071–2100 parallels the results of grand fir. Future moisture deficit values may reach 130 mm; thus, some populations may show a significant decline in the last third of this century. Judged on its performance, noble fir does not qualify for introduction.

Buck Creek and Ellensberg, the two low-elevation provenances of grand fir serving for comparison, have shown superior growth responses, similar to the results of the populations in the Zbiroh trial. The weaker performance of the noble fir provenances is statistically confirmed (Figure A6).

Adršpach, the Czech silver fir standard in the trial Dražičky, is certainly not representative of high-altitude Czech provenances and its growth is unsuitable for a realistic comparison. The extremely poor mean height indicates the harsh site conditions at its origin (Table 5 shows for the population the lowest January mean temperature, −5.1 °C). Its slower growth has probably declined further due to the competition of neighboring plots, indicated also by its worst survival among all provenances in the trial.

4. Discussion

4.1. Adaptive Differentiation in Drought Tolerance

The Písek trial shows a significant adaptive differentiation for silver fir indicated by the clear relation between climate transfer distance (relative drought stress) and mean height of provenances. The regression depicts an exponential decline of height growth with increasing relative drought stress, determined by the past drought conditions at the site of origin. The regression is similar to unilateral climate transfer distance regressions calculated in provenance tests of other species (e.g., beech: [49]; sessile oak: [40]). Earlier findings [34,35] indicated that silver fir provenances display high within-species differentiation regarding adaptive stress response when compared to other species are upheld by these results.

In some studies, intraspecific differentiation in silver fir is interpreted concurrently with adaptive selection as a result of tradeoff effects [33,50] or linked to alternate biogeographic histories [8,36,37,51]. Our results cannot ascertain either opinion, supporting that the growth response to warming/drying climate scenarios seems to be determined compellingly by adaptation to past climates. The correlation of responses with climates of the recent past suggests that the differentiation may have evolved in a shorter evolutionary period than usually assumed and might be enhanced by epigenetics. The maintenance of the differentiation has been observed until advanced age. The generally superior growth and drought resilience of grand fir populations found also in other low-elevation tests [52], is attributed to the summer rain peak in Central Europe, which causes less summer aridity stress than at the populations' origins.

The existence of tradeoff effects, i.e., a negative correlation between drought tolerance and inherent height growth, may be assumed for the Balkan fir species (*A. borisii-regis* and *A. cephalonica*). Their weaker growth may be explained by higher drought tolerance at the expense of compromised photosynthetic capacity, especially close to the rear/xeric limits [33,35,50,53].

4.2. Projected Adaptive Response of Silver Fir to Future Climatic Challenges

Until the end of the century, the running of the *ClimateEU* model under pathway RCP8.5 calculates a drastic annual mean temperature increase of ca. 4.5 °C at all three sites, while the increase of precipitation is moderate (Table A2). Consequently, the local climatic moisture deficit may increase, reaching the highest value of 315 mm at Písek (current value: 213 mm). The calculated future climatic transfer distances ($dCMD_{ann}$) and the projected heights of silver fir provenances at Písek and their origins (Figures 4 and 5) may provide estimations for their survival and the chances for transfer (assisted migration) for silver fir.

Whether the climate projected for the period 2071–2100 surpasses the limit of drought tolerance and survival of individual silver fir populations cannot be judged directly from their present growth in the trial Písek due to the lack of reliable mortality data. Bearing in mind the within-species differentiation in drought tolerance, a species-specific uniform tolerance limit for climatic moisture deficit of transferred populations cannot be set. The relatively high values of past moisture deficit indicate that the inherited drought resilience of populations is not fully challenged by the present test conditions.

An indirect approach for estimating the tolerance limit was attempted by selecting sites for comparison, artificially regenerated with silver fir, which were identified as actual climatic limits of the species in the subcontinental climate (Table 6). In the Czech Republic, a specific experiment, planted in 1975 in Jíloviště/Cukrák (FGMRI, trial No. 211) right at the assumed climatic limit of silver fir in the hornbeam-oak mixed forest zone, was selected [54]. Similar sites were identified in Hungary (Sopron 85 K, planted in 1900, Figure 10) and Slovenia (Zg. Cruškovje). The low-elevation Slovenian site is comparatively mild, it turned out to be of natural origin at the low-elevation SE limit of the species. The artificially established Czech and Hungarian locations show moisture supply conditions at the dry, subcontinental limits of the species (Table 6). Their "current" conditions indicate annual moisture deficits of 296 and 276 mm, respectively, projected to increase to 389 and 393 mm by the end of the century. These values qualify the current annual moisture deficit of the test site Písek as quite distant from the xeric limit conditions for artificially regenerated silver fir.

Table 6. Past, "current" and projected future climate data for selected locations in the Czech Republic, Hungary and Slovenia considered as xeric/rear limits of silver fir distribution at low elevations in Central-Eastern Europe (source of climate data: *ClimateEU*, projected pathway RCP8.5; for comparison with test site conditions see also Table A2).

Location, Elevation (m), Geogr. Coordinates (Decimal Degrees)	Climate Period	Annual Mean Temperature (°C)	Anual Mean Precipitation (mm)	Annual *CMD* (mm)
Jíloviště CZ, Trial No.211 310 m Lat. 49.944, Long. 14.356	1911–1940 1981–2010 2070–2100	9 9.2 13.5	416 477 522	331 296 389
Sopron HU, comp.85 K. 340 m Lat. 47.667, Long. 16.563	1911–1940 1981–2010 2070–2100	9.3 9.5 14.1	539 623 661	312 276 393
Zgornje Gruskovje SI, 325 m Lat. 46.272, Long. 15.860	1911–1940 1981–2010 2070–2100	9.8 10 14.6	831 965 978	202 172 271

Figure 10. Vigorous natural regeneration of silver fir in a regerenation cut, in the mixed sessile oak/hornbeam forest zone in Sopron, Hungary, at the xeric limits of artificially re-introduced silver fir (elevation 340 m, current climatic moisture deficit: 276 mm, further details in the text and Table 6; photo: P. Balázs).

Primarily the level of relative drought ($dCWD_{ann}$) is assumed to decide on population survival. A basis for the rough estimation of the relative drought tolerance limit might be the difference between the "artificial" (Jíloviště) and the "natural" (Písek) moisture deficit values, 296 vs. 213 mm, respectively (see Tables 6 and A2). The difference of 83 mm comes close to the observed maximum of relative moisture deficits in Písek. The highest value of 91 mm relative deficit was determined for the Czech provenance 81, Vyšší Brod. Its apparently low survival (10%, Table 3) may indicate declining adaptation to changed conditions. Hereafter, an arbitrary value of 100 mm relative moisture deficit is assumed for a limit of relative drought tolerance for silver fir at subcontinental zonal sites. (Zonal refers here to rain-fed sites with sufficient water holding capacity of the soil.)

The relative moisture deficit differences ($dCMD_{ann}$) projected for the transferred populations in the period 2071–2100 in Písek are shown in Figure 4 and Table A3. It is evident that high-elevation provenances (81, 223, 228 and S2) are exposed to relative drought values above the set limit. While these populations survived and performed sufficiently well in the current climate, this could change by the end of the century, and they may go extinct. On the contrary, the high-elevation Bulgarian provenance (132, Borovec), adapted to extreme drought stress, may survive at the low-elevation Písek site (see also the results chapter). Hence, it is expected that low-elevation populations may survive the climate of Písek until the end of the century, but some of the high-elevation populations will be lost. Their transfer to a low-elevation site as an "adaptive measure" would be pointless and illogical.

Of course, the survival of populations *at their original locations* and the prospects for assisted transfer is of primary interest for the projected future. Figure 5 shows the expected relative drought values for the end of the century at the original locations, i.e., at their provenance. It is self-evident that the survival of populations is more probable at their original sites than at Písek. Only the extremely high relative drought value of Bulgarian provenance 132, Borovec, indicates probable local extinction at the end of the century. Other, partly low-elevation populations will be stressed, but will probably survive. Favorable climates are projected for two provenances from high elevations (74 Vyšší Brod CZ and S2 Radvaň SK); the site conditions there may even improve due to projected higher rainfall. Calculating the relative drought values based on their current climate, the future survival of populations introduced to sites outside the natural range limits of silver fir (such as Jíloviště and Sopron, Table 6) appears questionable.

The individual differentiation between provenances due to diverging climate projections is apparent when comparing projected survivals at the test site and their origin. The population 132, Borovec BG is an extreme example, which may go locally extinct at its origin, but survives in the climate of the low-elevation site Písek. Conversely, the Czech and Slovak high-elevation provenances will survive at their origin but may be lost in the trial Písek. The location-bound, individual projections may provide concrete guidance for selecting reproductive material for future climates and for identifying climatically threatened valuable populations for evacuation.

Some of the conclusions contradict generally accepted opinions. The advantage of local provenances compared to non-locals is generally derived from transfer functions based on common garden results. However, if response norms are available for individual populations, the assumed superiority of artificially regenerated (transferred) local populations may often be questioned [40,55,56]. Here, this was also found for silver fir. In view of the speed of climatic changes, the superiority of local provenances cannot be taken for granted for the future; growth and survival will depend on projected local climate conditions. Similarly, the experience that southern and eastern silver fir provenances respond with the best growth and resilience, as found already in the first provenance test [27,30] and confirmed in later trials as well [8,29,57], is not generally valid. Further, the negative correlation between drought tolerance and inherent height growth has led to the conclusion that superior growth is generally linked with lower drought tolerance. Although this trade-off effect exists across a wide range of temperate and boreal tree species [14,40,56], our results did not ascertain this for silver fir, and a recent, very detailed experimental series of a silver fir [57] gave the same result. The projections also reveal that transfers may safeguard the survival of otherwise threatened populations. These conclusions support that generalized opinions on the expected performance of populations should be refined by a detailed inspection of past and future climatic projections. The inherent adaptive capacity, i.e., the resilience of the populations, may surpass the often implicitly assumed "perfect" adaptation under undisturbed, natural conditions. To project future species distributions without taking into consideration the adaptive capacity of populations may therefore underestimate the climatic resilience of species [3,14].

4.3. Limitations of the Study Results

None of the investigated provenance trials is perfect regarding the representativeness of the set of provenances, and the design of the experiment. Similarly, the applied datasets for projecting past and future climates represent the present level of knowledge. Hence, the results reflect the limitations of the investigated material and the applied methods. The results are based on the input provided by the selected climate variable and the climate projections, which cannot forecast unforeseen geophysical, biological (and socio-economic) events. Therefore, it remains a model based on real-time data that anticipates validation via coming episodes of terrestrial evolution.

The confidence of evaluating responses of species and populations was determined by the current condition of the tests. Owing to their advanced age, the plot sizes and the remaining number of trees per plot were far from optimal, which made statistical evaluations nearly impracticable (see details of variance statistics in Figures A7–A9). An analysis of survival/mortality was not performed, due to missing records on intermediate thinnings. Lacking reliable data on current mortality was also the reason why the climate extremes during the long maintenance period of the tests were not analyzed; these details would not have contributed much to the planned aim of the paper, which was the transparent presentation of the new results that these aged experiments were able to provide. The manifold methodical and practical limitations of the three trials were taken as a challenge to extort useful conclusions from the dataset. The authors are confident that the results justify the compromises, which had to be made.

5. Conclusions

The projections of the study may affect the future management of silver fir in East-Central Europe. Careful management supporting the successive adaptation of silver fir may be assumed as sufficient until the end of the century, especially in less-threatened higher elevations. High-elevation populations may even enjoy more suitable growth conditions. In other instances, particularly close to the xeric limits where enhancement of ecosystem resilience is required, artificial regeneration and assisted migration are options to preserve or restore silver fir occurrences. The introduction of provenances of the closely related Balkan species *A. cephalonica* and *A. borisii-regis* is a realistic choice at exposed lower elevations. The geographically and climatically distant Mediterranean fir species cannot be alternatives for silver fir [20,35]. The breeding for resilience proposed by some authors is only a theoretical possibility, considering the time necessitated.

The cautious use of grand fir in mountain forests is also conceivable (e.g., in Poland: [58]). Regarding health risks, investigation in the trials Dražičky and Zbiroh were inconclusive; however, both fungus and insect damage was reported in other grand fir stands in the Czech Republic, apparently triggered by recent extreme droughts [20]. Bearing in mind the technical and economic efforts needed, as well as plant health risks of non-native plant material [59], the introduction of grand fir will probably remain on the level of color splashes.

When choosing reproductive material for regeneration and transfer, resilience and tolerance to extremes must have priority over growth and yield traits. Selecting pre-adapted populations for transfer, the genetic risks using material of isolated, small occurrences should be considered, as well as the fact that transferred reproductive material may be at a competitive disadvantage in the initial years compared to local populations. To reduce risks of transfers and introductions, the gathering of available information, the field monitoring of extreme events [35,60,61], and antagonist organisms of any kind should not be ignored.

Methodically, the study presents a new approach to interpret experimental data in the context of projected climate scenarios. The introduction of relative drought, based on climatic moisture deficit change explains current responses of transferred tree populations and may predict their future performance and survival. Our results underscore the potential of provenance tests to provide practice-oriented information in order to mitigate the effects of a rapidly changing climatic environment. They may provide less dramatic and more detailed projections of future performance, counting with the resilience of populations, which cannot be forecasted by inventory-type, in-situ field analyses (e.g., [62]). The presented method, as well as the lengthy time needed to obtain reliable results from new experiments, may promote the reassessment of existing provenance tests and the reanalysis of their data supported by improved global and regional climate databases.

Author Contributions: Conceptualization, C.M. and J.F.; formal analysis, J.D., P.B. and M.V.; investigation, F.B., J.Č. and M.F.; writing—original draft preparation, C.M.; writing—review & editing, C.M., J.F. and G.B. visualization, P.B.; project administration, J.F., C.M.; funding acquisition, J.F., C.M. All authors have read and agreed to the published version of the manuscript.

Funding: Research work in the Czech Republic was supported by the fund MZE-RO0118, provided by the Ministry of Agriculture.

Institutional Review Board Statement: Not applicable.

Informed Consent Statement: Not applicable.

Data Availability Statement: Not applicable.

Acknowledgments: Authors are grateful to all members of FGMRI Department of Biology and Forest Tree Breeding research team, involved in field works and data processing activities connected with the provenance trials. Csaba Mátyás was supported in Hungary by the joint national-EU research project "VKSZ_12-1-2013-0034-Agrárklima.2". Gregor Božič was supported by the Slovenian Research Agency (P4-0107) and the national research projects V4-1818, V4-1819 co-financed by the Ministry of Agriculture, Forestry and Food. Special thanks are expressed to Tongli Wang (UBC Vancouver, Canada) for downscaling climate datasets for locations analyzed in the study, to Raphael Klumpp (Fed. Forest and Landscape Research Inst., Vienna, Austria) providing data on the provenance test in Knödelhütte, and to Anikó Kóczán-Horváth (University Sopron, Hungary) for statistical analyses. The careful, thorough linguistic correction of the manuscript by Frank Berger (University Sopron, Hungary) and the helpful comments of anonymous reviewers are gratefully acknowledged.

Conflicts of Interest: The authors declare no conflict of interest. The funders had no role in the design of the study; in the collection, analyses, or interpretation of data; in the writing of the manuscript, or in the decision to publish the results.

Appendix A.

Appendix A.1. Additional Climate Data of Test Sites and Provenances

Table A1. Origin of climate data used for test sites.

Test Site Name	Meteorological Station Name	Altitude of Station	Distance from Site (km)	Altitude of Site (m a.s.l.)
Písek	Vráž	433	19	395
Zbiroh	Zbiroh	480	11	450–460
Dražičky	Tábor	461	7	485

Table A2. Past, "current" and projected future climate data for the three trial sites in the Czech Republic (source of climate data: *ClimateEU*; projected pathway RCP8.5).

Location Elevation (m) Coordinates	Climate Period	Mean Temperature (°C)		Mean Precipitation (mm)		*CMD* (mm)	
		Annual	Summer	Annual	Summer	Annual	Summer
Písek. CZ Trial No.64	1911–1940	8.7	17.6	548	205	273	150
395 m	1981–2010 *	8.1 *	17.2 *	570 *	229 *	213	137
Lat. 49.27	2040–2070	11.4	20.9	709	246	234	154
Long. 14.33	2070–2100	13.2	23.2	691	222	315	211

Table A2. *Cont.*

Location Elevation (m) Coordinates	Climate Period	Mean Temperature (°C)		Mean Precipitation (mm)		*CMD* (mm)	
		Annual	Summer	Annual	Summer	Annual	Summer
Zbiroh. CZ Trial No.213	1911–1940	7.7	16.4	533	191	234	134
456 m	1981–2010*	8.1 *	17 *	595 *	241 *	179	120
Lat. 49.79	2040–2070	10.4	19.7	689	230	199	136
Long. 13.64	2070–2100	12.1	22	673	209	274	189
Dražičky. CZ Trial No.219	1911–1940	8.1	16.9	539	206	250	137
488 m	1981–2010 *	8.1 *	17.3 *	592 *	229 *	189	122
Lat. 49.39	2040–2070	10.9	20.3	696	247	209	140
Long. 14.59	2070–2100	12.7	22.6	678	224	288	196

* Respectively the current testing period, CZ met. Data.

Table A3. Climate parameters and heights of silver and Balkan firs in past, current and projected climates and heights in the Písek trial (see also Figure 4).

Populations of the Provenance Trial 64 Písek, CZ				Past Climate at Origin (1911–1940)			Current Climate	Future Climate (2071–2100)			
Nr.	*Abies* Species Name	Provenance Name	Alt. (m)	Ann. Mean Temp. (°C)	Ann. Mean Prec. (mm)	Ann. *CMD* (mm)	Mean Height (Year 2015) (m)	Temp. Annual Change (°C)	Prec. Annual Change (mm)	Ann. *dCMD* (mm)	Proj. Mean Height (m) ***
74	*A. alba*	Milevsko, Klučenice CZ	410	8.4	492	284	17.19	4.8	199	31	16.17
81	*A. alba*	Vyšší Brod, Vítkův Kámen CZ	900	5.9	854	122	15.47	7.3	−163	193	13.46
109	*A. cephalonica*	Centr. Peloponnese, Vytina GR	1250	10.9	794	573	14.19	2.3	−103	−258	14.94
130	*A. alba*	Nasavrky, Podhůra CZ	370	8.2	487	253	16.89	5.0	204	62	15.68
132	*A. alba*	Rilskije gory, Borovec BG *	1200	7.3	515	362	16.85	5.9	176	−47	16.31
136	*A. cephalonica*	Peloponnese, Vytina GR	1010	12.2	753	626	14.34	1.0	−62	−311	15.41
137	*A. borisii-regis*	Mt. Pindos, Pertuli GR *	1200	9.6	727	503	14.59	3.6	−36	−188	14.91
223	*A. alba*	Sanski Most BH *	1050	6.9	943	170	15.61	6.3	−252	145	13.89
228	*A. alba*	Regello, Vallombrosa IT *	1010	9.5	967	180	14.82	3.7	−276	135	13.16
S 2	*A. alba*	Banská Bystrica, Radvaň SK *	780	6.4	759	152	14.92	6.8	−68	163	13.09
Data and current climate of the test site **			395	8.1	570	213	15.49				
Future climate of the test site (2071–2100, RCP 8.5)			395	13.2	691	315					

* Corrected locations; ** CZ met data (except *CMD* value of the test site); *** Projected heights of Balkan fir species are estimated using the equation for silver fir. Source of climate data: *ClimateEU* v4.80. Projected heights are calculated also for the age of 40 years, for the sake of comparison. Current climate data are for all provenances the data of the Písek site (Table 3).

Table A4. Climate parameters of silver and Balkan firs in the past, current heights (in Písek) and projected climatic changes (related to the past) and projected (Proj.) heights at their original location of provenance (see also Figure 5).

Populations of the Provenance Trial 64 Písek, CZ				Past Climate at Origin (1911–1940)			Current Climate	Future Climate (2071–2100)			
Nr.	*Abies* Species Name	Provenance Name	Alt. (m)	Ann. Mean Temp. (°C)	Ann. Mean Prec. (mm)	Ann. *CMD* (mm)	Mean Height (Year 2015) (m)	Temp. Annual Change (°C)	Prec. Annual Change (mm)	Ann. *dCMD* (mm)	Proj. Mean Height (m) ***
74	*A. alba*	Milevsko, Klučenice CZ	410	8.4	492	284	17.19	4.5	127	46	15.97
81	*A. alba*	Vyšší Brod, Vítkův Kámen CZ	900	5.9	854	122	15.47	4.5	226	-39	17.13
109	*A. cephalonica*	Centr. Peloponnese, Vytina GR	1250	10.9	794	573	14.19	4.2	−71	131	12.06
130	*A. alba*	Nasavrky, Podhura CZ	370	8.2	487	253	16.89	4.5	123	44	15.94
132 *	*A. alba*	Rilskije gory, Borovec BG	1200	7.3	515	362	16.85	5.0	20	172	13.03
136	*A. cephalonica*	Peloponnese, Vytina GR	1010	12.2	753	626	14.34	4.2	−65	135	12.77
137 *	*A. borisii-regis*	Mt. Pindos, Pertuli GR	1200	9.6	727	503	14.59	4.8	−31	162	11.26
223 *	*A. alba*	Sanski Most BH	1050	6.9	943	170	15.61	4.7	130	97	14.78
228 *	*A. alba*	Regello, Vallombrosa IT	1010	9.5	967	180	14.82	4.5	109	91	13.95
S 2 *	*A. alba*	Banská Bystrica, Radvaň SK	780	6.4	759	152	14.92	4.8	180	28	15.38
	Data and current climate of the test site **		395	8.1	570	213	15.49				

* Corrected locations; ** CZ met data (except CMD value of the test site; *** Projected heights of Balkan fir species are estimated using the equation for silver fir. Source of climate data: ClimateEU v4.80. Projected heights are calculated for the age in the year 2015, for the sake of comparison. Current climate data are for all provenances the data of the Písek site (Table 3).

Appendix A.2. Separate Presentation of Height Responses to Temperature and Precipitation Changes in the Summer (Warmest) Quarter, Caused by Transfer of Populations to the Trials

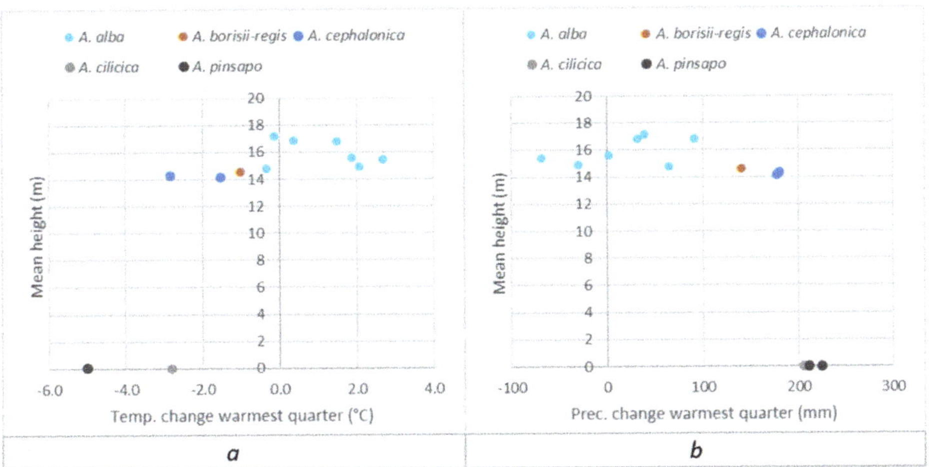

Figure A1. Mean summer temperature (**a**) and precipitation (**b**) change versus mean height of silver fir (blue), *A. cephalonica* (red) and *A. borisii-regis* (deep blue) provenances at the age of 40 years, in the trial 64, Písek. The Mediterranean provenances of *A. cilicica* and *A. pinsapo* (gray) suffered total mortality, depicted at 0 height.

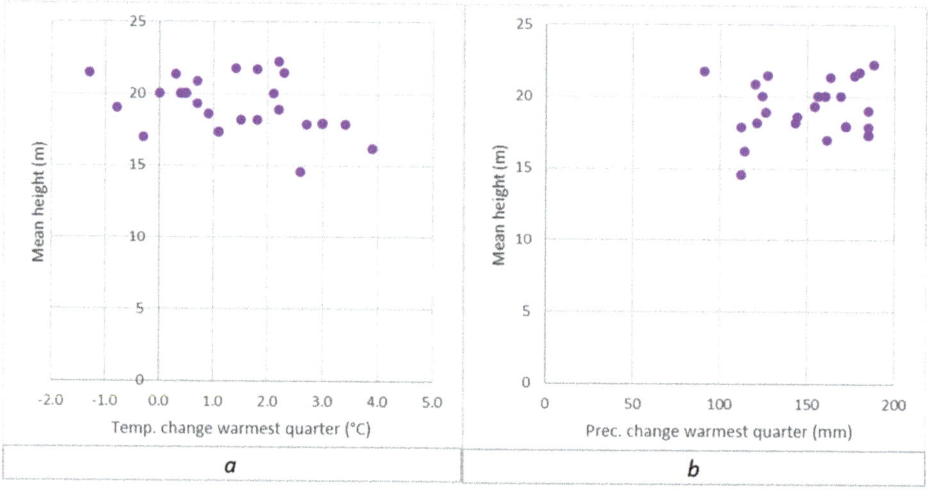

Figure A2. Mean summer temperature (**a**) and precipitation (**b**) change versus mean height of grand fir provenances at the age of 36 years, in the trial 213, Zbiroh. Note that nearly all populations were transferred into a warmer summer climate with higher rainfall.

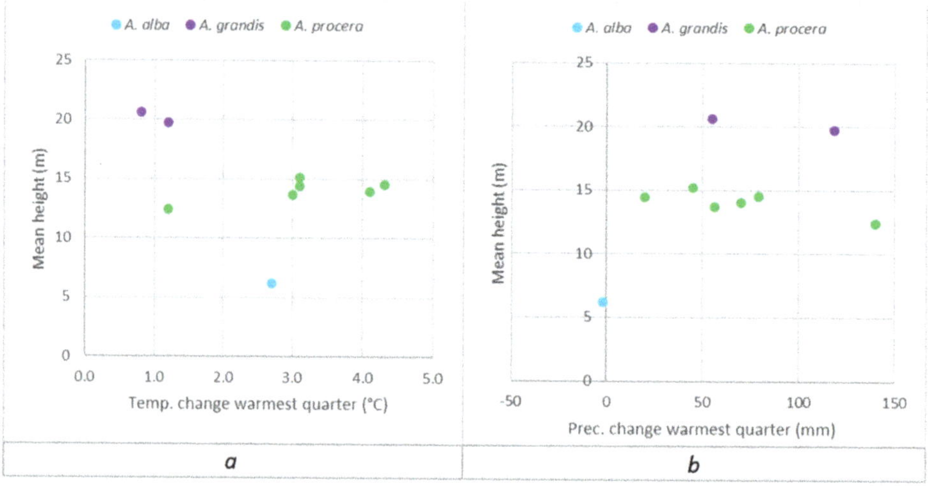

Figure A3. Mean summer temperature (**a**) and precipitation (**b**) change versus mean height of provenances of three species at the age of 32 years, in the trial 219, Dražičky. Note the differences between grand fir, noble fir and the autochthonous standard Adršpach from the Czech Sudeten Mts.

Appendix A.3. Matrices of Significance Calculated for the Three Provenance Tests, Based on Statistics in [20]

The Kruskal-Wallis one-way ANOVA results indicate significant differences of mean heights between pairs of populations. The populations are sorted by the average heights of groups in a quadrat (respectively, by geographic position in case of Zbiroh). Populations with top height are placed to the top left corner. As the height differences within the separated groups were found to be insignificant, the sequence of individual provenances within the group is not necessarily by their mean height. Filled circles indicate significant differences, colored fields signify differentiated groups. The quadrats of nonsignificantly different populations may overlap.

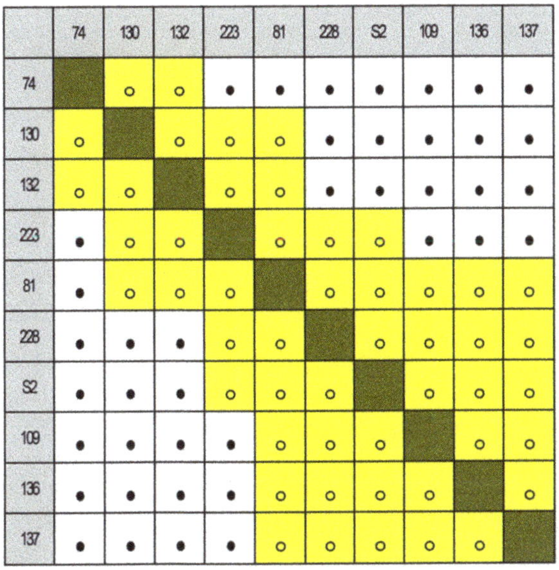

Figure A4. Matrix of significance of population mean height differences of the trial Písek, sorted by the group average of mean heights (best in top left).

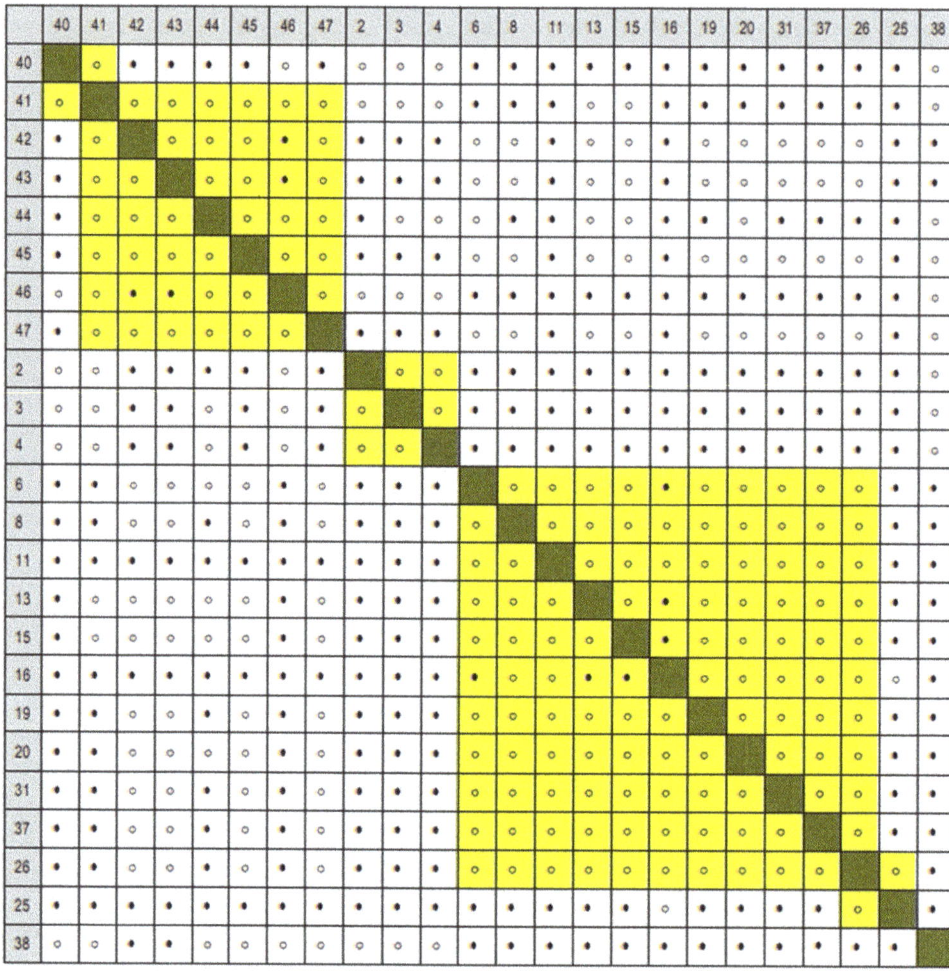

Figure A5. Matrix of significance of population mean height differences of the trial Zbiroh, sorted by their *geographic* position from NW to SE (the first three digits of the reg. numbers are omitted).

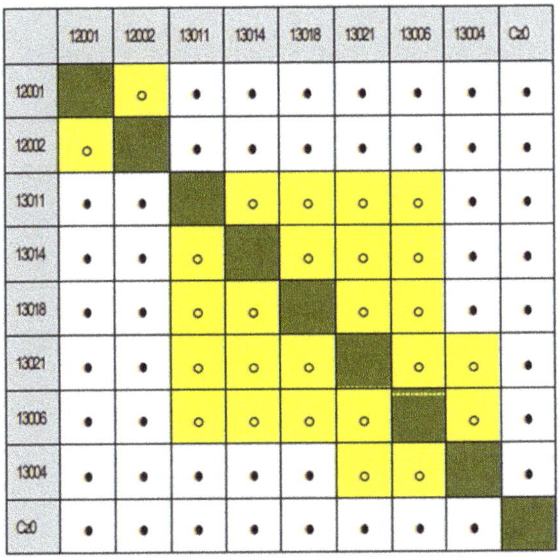

Figure A6. Matrix of significance of population mean height differences of the trial Dražičky, ordered by the group average of mean heights (best in top left).

Appendix A.4. Variance of Populations Sorted by the Sequence of Significance Statistics (The Sequence of Provenances follow the Respective Matrices of Significance)

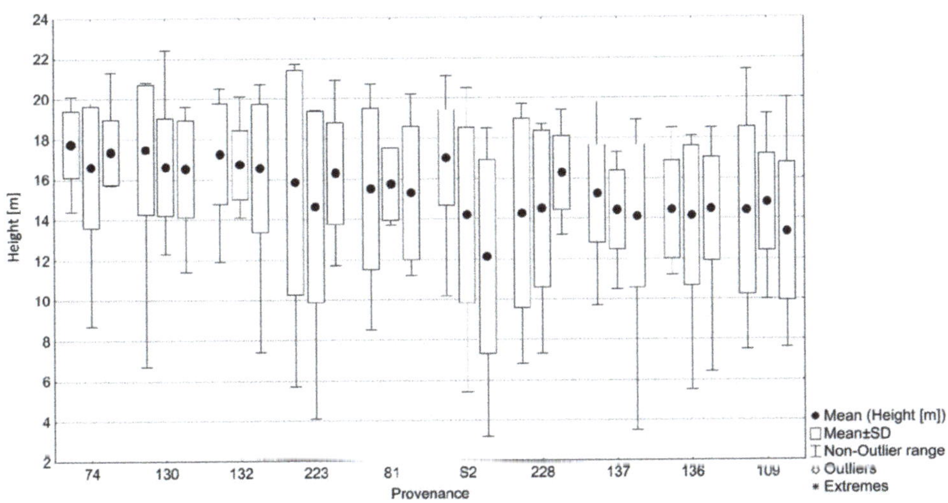

Figure A7. Box plot of height variation statistics by repetitions and by provenances in the trial Písek. The provenances are sorted by groups of mean heights (compare Figure A4).

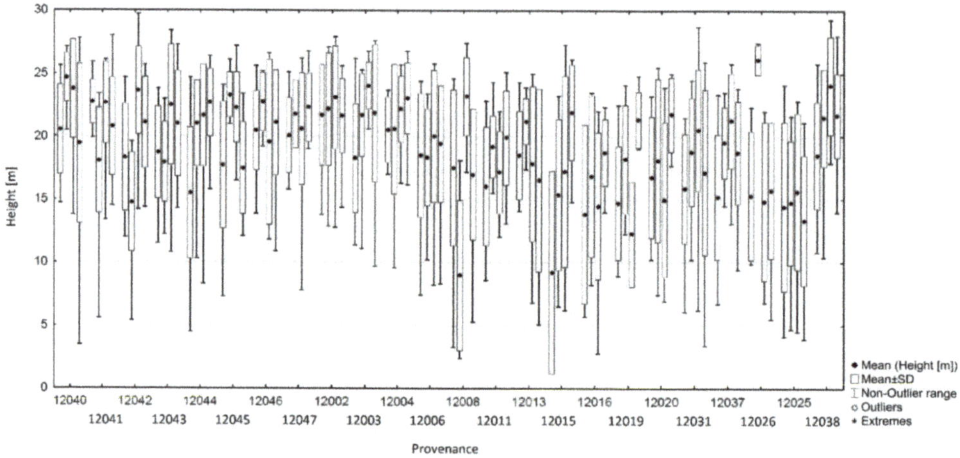

Figure A8. Box plot of height variation statistics by replications and by provenance in the trial Zbiroh. The provenances are sorted by *geographic* position (compare Figure A5).

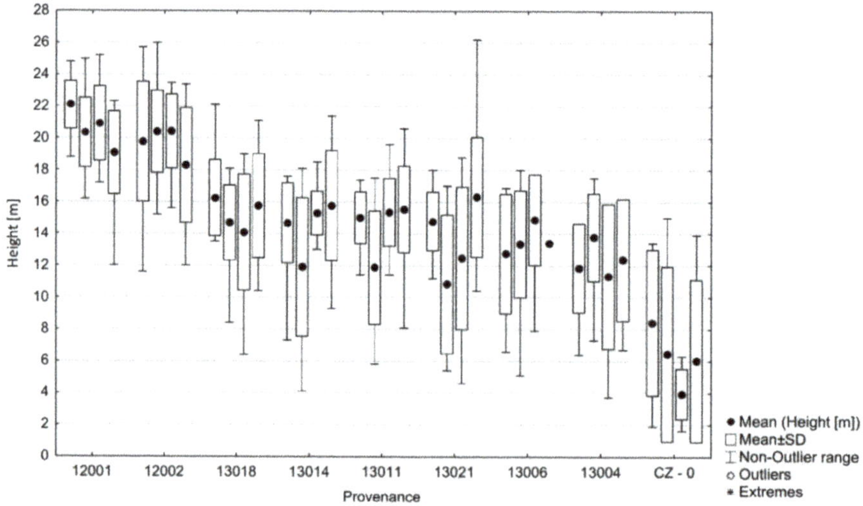

Figure A9. Box plot of height variation statistics by replications and by provenance in the trial Dražičky. The provenances are sorted by groups of mean heights (compare Figure A6).

References

1. Allen, C.D.; Breshears, D.D.; McDowell, N.G. On underestimation of global vulnerability to tree mortality and forest die-off from hotter drought in the Anthropocene. *Ecosphere* **2015**, *6*, 129. [CrossRef]
2. Mátyás, C. Forecasts needed for retreating forests (Opinion). *Nat. Intern. Weekl. J. Sci.* **2010**, *464*, 1271.
3. Mátyás, C.; Berki, I.; Bidló, A.; Csóka, G.; Czimber, K.; Führer, E.; Gálos, B.; Gribovszki, Z.; Illés, G.; Hirka, A.; et al. Sustainability of forest cover under climate change on the temperate-continental xeric limits. *Forests* **2018**, *9*, 489. [CrossRef]
4. Spinoni, J.; Barbosa, P.; Bucchignani, E.; Cassano, J.; Cavazos, T.; Christensen, J.H.; Dosio, A. Future global meteorological drought hotspots: A study based on CORDEX data. *J. Climate.* **2019.** [CrossRef]
5. Gálos, B.; Führer, E.; Czimber, K.; Gulyás, K.; Bidló, A.; Hänsler, A.; Jacob, D.; Mátyás, C. Climatic threats determining future adaptive forest management—A case study of Zala County. *Időjárás* **2015**, *119*, 425–441.
6. Czimber, K.; Gálos, B. A new decision support system to analyse the impacts of climate change on the Hungarian forestry and agricultural sectors. *Scand. J. For. Res.* **2016**, 664–673. [CrossRef]

7. Gazol, A.; Camarero, J.J.; Gutiérrez, E.; Popa, I.; Andreu-Hayles, L.; Motta, R.C.; Carrer, M. Distinct effects of climate warming on populations of silver fir (*Abies alba*) across Europe. *J. Biogeogr.* **2015**, *42*, 1150–1162. [CrossRef]
8. Matías, L.; Gonzalez-Díaz, P.; Quero, J.L.; Camarero, J.J.; Lloret, F.; Jump, A.S. Role of geographical provenance in the response of silver fir seedlings to experimental warming and drought. *Tree Physiol.* **2016**, *36*, 1236–1246. [CrossRef]
9. Brázdil, R.; Trnka, M. Sucho v českých zemích: Minulost, současnost, budoucnost. In *Drought in the Czech Lands: Past, Present, Future*, 1st ed.; Research Centre for Global Change, Academy of Science of the Czech Republic: Brno, Czech Republic; UNI Press: Turnov, Czech Republic, 2015. (In Czech)
10. Šindelář, J.; Frýdl, J. K problematice genetických změn v populacích lesních dřevin v důsledku narušování lesů škodlivými vlivy prostředí. Problems of genetic changes in forest tree populations due to disturbances by harmful environmental influences. *Zprávy Lesn. Výzk.* **2008**, *53*, 116–119. (In Czech)
11. Lebourgeois, F.; Rathgeber, C.B.K.; Ulrich, E. Sensitivity of French temperate coniferous forests to climate variability and extreme events (*Abies alba, Picea abies* and *Pinus sylvestris*). *J. Veget. Sci.* **2010**, *21*, 364–376. [CrossRef]
12. George, J.-P.; Grabner, M.; Campelo, F.; Karanitsch-Ackerl, S.; Mayer, K.; Klumpp, R.T.; Schüler, S. Intra-specific variation in growth and wood density traits under water-limited conditions: Long-term-, short-term-, and sudden responses of four conifer tree species. *Sci. Total Environ.* **2019**, *660*, 631–643. [CrossRef] [PubMed]
13. Vitasse, Y.; Bottero, A.; Rebetez, M.; Conedera, M.; Augustin, S.; Brang, P.; Tinner, W. What is the potential of silver fir to thrive under warmer and drier climate? *Eur. J. Forest Res.* **2019**. [CrossRef]
14. Benito-Garzón, M.; Robson, T.M.; Hampe, A. ΔTraitSDM: Species distribution models that account for local adaptation and phenotypic plasticity. *New Phytol.* **2019**, *222*, 1757–1765. [CrossRef] [PubMed]
15. Bouriaud, O.; Popa, I. Comparative dendroclimatic study of Scots pine, Norway spruce, and silver fir in the Vrancea Range, Eastern Carpathian Mountains. *Trees* **2008**, *23*, 95–106. [CrossRef]
16. Tinner, W.; Colombaroli, D.; Heiri, O.; Henne, P.D.; Steinacher, M.; Untenecker, J.; Conedera, M. The past ecology of *Abies alba* provides new perspectives on future responses of silver fir forests to global warming. *Ecol. Monogr.* **2013**, *83*, 419–439. [CrossRef]
17. Zang, C.; Hartl-Meier, C.; Dittmar, C.; Rothe, A.; Menzel, A. Patterns of drought tolerance in major European temperate forest trees: Climatic drivers and levels of variability. *Global Chang. Biol.* **2014**, *20*, 3767–3779. [CrossRef]
18. Vitali, V.; Büntgen, U.; Bauhus, J. Silver fir and Douglas fir are more tolerant to extreme droughts than Norway spruce in south-western Germany. *Glob. Chang. Biol.* **2017**, *23*, 5108–5119. [CrossRef] [PubMed]
19. Kapeller, S.; Schüler, S.; Huber, G.; Božič, G.; Wohlgemuth, T.; Klumpp, R. Provenance Trials in Alpine Range—Review and Perspectives for Applications in Climate Change. In *Management Strategies to Adapt Alpine Space Forests to Climate Change Risks*; Cerbu, G., Ed.; IntechOpen: London, UK, 2013; ISBN 978-953-51-1194-8. [CrossRef]
20. Frýdl, J.; Dostál, J.; Beran, F.; Čáp, J.; Fulín, M.; Frampton, J.; Božič, G.; Mátyás, C. Exotic Abies Species in Czech Provenance Trials: Assessment after Four Decades. *Acta Silv. Lign. Hung.* **2018**, *14*, 1, 9–34. [CrossRef]
21. Sáenz-Romero, C.; Mendoza-Maya, E.; Gómez-Pineda, E.; Blanco-García, A.; Endara-Agramont, A.R.; Lindig-Cisneros, R.; Vargas-Hernández, J.J. Recent evidence of Mexican temperate forest decline and the need for *ex situ* conservation, assisted migration, and translocation of species ensembles as adaptive management to face projected climatic change impacts in a megadiverse country. *Can. J. For. Res.* **2020**, *50*, 843–854. [CrossRef]
22. Booth, T.H. Assessing species climatic requirements beyond the realized niche: Some lessons mainly from tree species distribution modelling. *Clim. Chang.* **2017**, *145*, 259–271. [CrossRef]
23. Mátyás, C.; Yeatman, C.W. Effect of geographical transfer on growth and survival of jack pine (*Pinus banksiana* Lamb.) populations. *Silvae Genet.* **1992**, *41*, 370–376.
24. Mátyás, C. Modelling climate-change effects with provenance test data. *Tree Physiol.* **1994**, *14*, 797–804. [CrossRef] [PubMed]
25. Sagnard, F.; Barberot, C.; Fady, B. Structure of genetic diversity in *Abies alba* Mill. from southwestern Alps: Multivariate analysis of adaptive and non-adaptive traits for conservation in France. *For. Ecol. Manag.* **2002**, *157*, 175–189. [CrossRef]
26. Latreille, A.C.; Pichot, C. Local-scale diversity and adaptation along elevational gradients assessed by reciprocal transplant experiments: Lack of local adaptation in silver fir populations. *Ann. For. Sci.* **2017**, *74*, 77. [CrossRef]
27. Mayer, H.; Reimoser, F.; Kral, F. Ergebnisse des Internationalen Tannenherkunftsversuches Wien 1967—1978. Morphologie und Wuchsverhalten der Provenienzen. *Cbl. Ges. Forstw.* **1982**, *99*, 169–191.
28. Ruetz, W.F.; Dimpflmeier, R.; Kleinschmit, J.; Svolba, J.; Weisgerber, H.; Rau, H.M. The IUFRO *Abies procera* provenance trial in the Federal Republic of Germany: Field results at age 9 and 10 years. In *Proceeding Joint Meeting Western Forest Genetics Assoc. and IUFRO Working Parties S2.02-05, 06, 12 and 14*; Weyerhaeuser Co.: Olympia, WA, USA, 1990.
29. Ruetz, W.; Stimm, B. Der Süddeutsche Weißtannen-Provenienzversuch: IV. Entwicklung der Herkünfte der Aussaat 1982 auf den Versuchsflächen in Bayern bis zum Alter von 12 Jahren. In *Ergebnisse des 7. Weisstannen-Symposiums in Altensteig, Deutschland*; Eder, W., Ed.; IUFR0-Verlag: Mainz, Germany, 1994; pp. 17–29.
30. Klumpp, R.T. Der Tannenprovenienzversuch "Knödelhütte Wien 1967": Ergebnisse im Alter 24. *Mitt. Forsch. Anst. Waldökol. Forstw. Rheinland-Pfalz.* **2002**, *50/03*, 44–49.

31. Gugerli, F.; Frank, A.; Rellstab, C.; Pluess, A.R.; Moser, B.; Ahrend, M.; Sperisen, C.; Wohlgemuth, T.; Heiri, C. Genetische Variation und lokale Anpassung bei Waldbaumarten im Zeichen des Klimawandels. (Genetic variation and local Adaptation in forest tree species under climate change.). In *Wald im Klimawandel. Grundlagen für Adaptationsstrategien. (Forests in Climate Change. Basics of Adaptation Strategies.)*; Pluess, A.R., Augustin, S., Brang, P., Eds.; BAFU, WSL Birmensdorf, Haupt Verl.: Bern, Switzerland, 2016; pp. 93–118.
32. Frank, A.; Sperisen, C.; Howe, G.T.; Brang, P.; Walthert, L.; St. Clair, J.B.; Heiri, C. Distinct genecological patterns in seedlings of Norway spruce and silver fir from a mountainous landscape. *Ecology* **2017**, *98*, 211–227. [CrossRef] [PubMed]
33. Csilléry, K.; Buchmann, N.; Fady, B. Adaptation to drought is coupled with slow growth, but independent from phenology in marginal silver fir (*Abies alba* Mill.) populations. *Evol. Appl.* **2020**, *13*. [CrossRef]
34. Hansen, J.K.; Larsen, J.B. European silver fir (*Abies alba* Mill.) provenances from Calabria, Southern Italy: 15-year results from Danish provenance field trials. *Eur. J. For. Res.* **2004**, *123*, 127–138. [CrossRef]
35. George, J.P.; Schueler, S.; Karanitsch-Ackerl, S.; Mayer, K.; Klumpp, R.T.; Grabner, M. Inter- and intra-specific variation in drought sensitivity in *Abies* spec. and its relation to wood density and growth traits. *Agric. For. Meteorol.* **2015**, *214–215*, 430–443. [CrossRef]
36. Bošel'a, M. Jedl'a biela: Perspektívna drevina v čase meniacej sa klímy? [Silver fir: A species with perspective in case if climate changes?]. In *Zachovanie druhovej rozmanitosti lesov Slovenska v čase globálnych environmentálnych zmien. [Conservation of species diversity of Slovakia's forests in times of global environmental change]*; Mális, F., Ed.; *Bulletin Správy*; Forestry Faculty, TU Zvolen: Zvolen, Slovakia, 2017; pp. 3–9. (In Slovak)
37. Csilléry, K.; Ovaskainen, O.; Sperisen, C.; Buchmann, N.; Widmer, A.; Gugerli, F. Adaptation to local climate in multi-trait space: Evidence from silver fir (*Abies alba* Mill.) populations across a heterogeneous environment. *Heredity* **2020**, *124*, 77–92. [CrossRef] [PubMed]
38. Šindelář, J.; Beran, F. Comparison of some exotic species of *Abies* genus with chosen silver fir provenances on the plots of town Písek. *Comm. Inst. For. Bohem.* **2008**, *24*, 99–113.
39. Šindelář, J.; Beran, F. Exotic species of fir (*Abies* spec. div.) at the age of 30 years in the nature forest region No. 10-Středočeská pahorkatina (Central Bohemian Upland). *Comm. Inst. For. Bohem.* **2008**, *24*, 115–130.
40. Mátyás, C. Adaptive pattern of phenotypic plasticity and inherent growth reveal the potential for assisted transfer in sessile oak (*Quercus petraea* L.). *For. Ecol. Manag.* **2021**, *482*, 118832. [CrossRef]
41. Nielsen, R. *The Little Green Handbook*; Scribe Publications: London, UK, 2006; ISBN 0312425813. Available online: http://home.iprimus.com.au/nielsens/ (accessed on 2 December 2020).
42. Steffen, W.; Broadgate, W.; Deutsch, L.; Gaffney, O.; Ludwig, C. The Trajectory of the Anthropocene: The Great Acceleration. *Anthr. Rev.* **2015**, *2*, 81–98. Available online: http://hdl.handle.net/1885/66463 (accessed on 4 December 2020). [CrossRef]
43. Marchi, M.; Castellanos-acuña, D.; Hamann, A.; Wang, T.; Ray, D.; Menzel, A. ClimateEU, scale-free climate normals, historical time series, and future projections for Europe. *Sci. Data* **2020**, *7*, 428. [CrossRef] [PubMed]
44. Wang, T.; Hamann, A.; Spittlehouse, D.L.; Murdock, T.Q. ClimateWNA-high-resolution spatial climate data for western North America. *J. Appl. Meteorol. Climatol.* **2012**, *51*, 16–29. [CrossRef]
45. Hargreaves, G.H. Defining and using reference evapotranspiration. *J. Irrigat. Drain. Eng.* **1994**, *120*, 6. [CrossRef]
46. QC-Expert 3.1 TriloByte Statistical Software. TriloByte Ltd., Pardubice—Staré Hradiště, Czech Republic. Available online: http://www.trilobyte.cz/Nase-Produkty/QC-Expert.html(accessed on 10 November 2020).
47. CEDA Archive. Climate Research Unit, UK, Time Series Version 3.23, High Resolution Gridded Data. 2018. Available online: http://catalogue.ceda.ac.uk/uuid/5dca9487dc614711a3a933e44a933ad3 (accessed on 10 November 2020).
48. Alizoti, P.; Bastien, J.C.; Chakraborty, D.; Klisz, M.; Kroon, J.; Neophytou, C.; Schueler, S.; van Loo, M.; Westergren, M.; Konnert, M.; et al. Non-native forest tree species in Europe: The question of seed origin in afforestation. *Forestry* **2019**. Available online: http://mc.manuscriptcentral.com/forest (accessed on 10 November 2020).
49. Mátyás, C.; Bozic, G.; Gömöry, D.; Ivankovic, M.; Rasztovits, E. Juvenile growth response of European beech (*Fagus sylvatica* L.) to sudden change of climatic environment in SE European trials. *iForest, J. Biogeosc. For.* **2009**, *2*, 213–220. [CrossRef]
50. Konôpková, A.; Pšidová, E.; Kurjak, D.; Stojnić, S.; Petrík, C.P.; Fleischer, A.P.; Ditmarová, L. Photosynthetic performance of silver fir (*Abies alba*) of different origins under suboptimal growing conditions. *Funct. Plant. Biol.* **2020**, *47*. [CrossRef]
51. Mosca, E.; Cruz, F.; Gómez-Garrido, J.; Bianco, L.; Rellstab, C.; Brodbeck, S.; Neale, D.B. A Reference Genome Sequence for the European Silver Fir (*Abies alba* Mill.): A Community-Generated Genomic Resource. *G3 Genes Genomes Genetics* **2019**, *9*, 2039–2049. [CrossRef]
52. Fulín, M.; Remeš, J.; Tauchman, P. Růst a produkce jedle obrovské (*Abies grandis* Lindl.) ve srovnání s jinými jehličnany. Growth and production of grand fir (*Abies grandis* Lindl.) compared with other conifer species. *Zprávy lesn. výzk.* **2013**, *58*, 186–192. (in Czech).
53. Roschanski, A.M.; Csilléry, K.; Liepelt, S.; Oddou-Muratorio, S.; Ziegenhagen, B.; Huard, F.; Fady, B. Evidence of divergent selection for drought and cold tolerance at landscape and local scales in *Abies alba* Mill. in the French Mediterranean Alps. *Mol. Ecol.* **2016**, *25*, 776–794. [CrossRef] [PubMed]
54. Čáp, J.; Novotný, P.; Frýdl, J. Vyhodnocení provenienční výzkumné plochy s jedlí bělokorou (Abies alba Mill.) č. 57-Lesy Jíloviště, Cukrák ve věku 35 let. Evaluation of a provenance research trial with silver fir at the age of 35 years. *Zprávy Lesn. Výzk.* **2009**, *54*, 33–43. (In Czech)

55. Rehfeldt, G.E.; Leites, L.P.; Joyce, D.G.; Weiskittel, A.R. Role of population genetics in guiding ecological responses to climate. *Glob. Chang. Biol.* **2018**, *24*, 858–868. [CrossRef]
56. Rehfeldt, G.E.; Warwell, M.V.; Monserud, R.A. Species, climatypes, climate change and forest health: A conversion of science to practice for inland northwest (USA) forests. *Forests* **2020**, *11*, 1237. [CrossRef]
57. Mihai, G.; Alexandru, A.M.; Stoica, E.; Birsan, M.V. Intraspecific growth response to drought of *Abies alba* in the Southeastern Carpathians. *Forests* **2021**, *12*, 387. [CrossRef]
58. Wilczyński, S.; Kulej, M. The growth of seven *Abies grandis* provenances in the climatic conditions of the Polish Carpathian Mountains. *Dendrobiology* **2019**, *81*, 1–13. [CrossRef]
59. Grady, K.C.; Kolb, T.E.; Ikeda, D.H.; Whitham, T.G. A bridge too far: Cold and pathogen constraints to assisted migration of riparian forests. *Restor. Ecol.* **2015**, *23*, 811–820. Available online: https://onlinelibrary.wiley.com/doi/abs/10.1111/rec.12245 (accessed on 5 December 2020). [CrossRef]
60. Adams, H.D.; Barron-Gafford, G.A.; Minor, R.L.; Gardea, A.A.; Bentley, L.P.; Law, D.J.; Huxman, T.E. Temperature response surfaces for mortality risk of tree species with future drought. *Environ. Res. Lett.* **2017**, *12*, 115014. [CrossRef]
61. Berki, I.; Rasztovits, E.; Móricz, N.; Kolozs, L. The role of tree mortality in vitality assessment of sessile oak forests. *South-East Eur. For.* **2016**, *7*, 91–97. [CrossRef]
62. Střelcová, K. Stres suchom a lesné porasty. [Drought stress and forest trees]. In *Zachovanie Druhovej Rozmanitosti Lesov Slovenska v čase Globálnych Environmentálnych Zmien. [Conservation of species diversity of Slovakia's forests in times of global environmental change]*; Mális, F., Ed.; *Bulletin Správy*; Forestry Faculty, TU Zvolen: Zvolen, Slovakia, 2017; pp. 59–71. (In Slovak)

forests MDPI

Article

Reciprocal Common Garden Altitudinal Transplants Reveal Potential Negative Impacts of Climate Change on *Abies religiosa* Populations in the Monarch Butterfly Biosphere Reserve Overwintering Sites

Ana Laura Cruzado-Vargas [1], Arnulfo Blanco-García [2], Roberto Lindig-Cisneros [3], Mariela Gómez-Romero [2,4], Leonel Lopez-Toledo [5], Erick de la Barrera [3] and Cuauhtémoc Sáenz-Romero [5,*]

1 Instituto de Investigaciones Agropecuarias y Forestales, Universidad Michoacana de San Nicolás de Hidalgo (UMSNH), Morelia 58330, Michoacán, Mexico; al.cruva@gmail.com
2 Facultad de Biología, Universidad Michoacana de San Nicolás de Hidalgo (UMSNH), Morelia 58030, Michoacán, Mexico; arnulfo.blanco@umich.mx (A.B.-G.); mgomezr@conacyt.mx (M.G.-R.)
3 Instituto de Investigaciones en Ecosistemas y Sustentabilidad, Universidad Nacional Autónoma de México (UNAM), Morelia 58190, Michoacán, Mexico; rlindig@iies.unam.mx (R.L.-C.); delabarrera@unam.mx (E.d.l.B.)
4 Cátedra, Consejo Nacional de Ciencia y Tecnología (CONACyT), Ciudad de México 03940, Mexico
5 Instituto de Investigaciones sobre los Recursos Naturales, Universidad Michoacana de San Nicolás de Hidalgo (UMSNH), Morelia 58330, Michoacán, Mexico; leonellopeztoledo@gmail.com
* Correspondence: csaenzromero@gmail.com; Tel.: +52-443-334-0475 (ext. 118)

Citation: Cruzado-Vargas, A.L.; Blanco-García, A.; Lindig-Cisneros, R.; Gómez-Romero, M.; Lopez-Toledo, L.; de la Barrera, E.; Sáenz-Romero, C. Reciprocal Common Garden Altitudinal Transplants Reveal Potential Negative Impacts of Climate Change on *Abies religiosa* Populations in the Monarch Butterfly Biosphere Reserve Overwintering Sites. *Forests* **2021**, *12*, 69. https://doi.org/10.3390/f12010069

Received: 9 December 2020
Accepted: 4 January 2021
Published: 9 January 2021

Abstract: Research Highlights: Reciprocal altitudinal transplants of *Abies religiosa* seedlings within the Monarch Butterfly Biosphere Reserve (MBBR) allow prediction of the impacts of climatic change, because they grow in sites with a climate that differs from that of their origin. Background and Objectives: Climatic change is generating a mismatch between the sites currently occupied by forest populations and the climate to which they have adapted. This study determined the effect on the survival and growth of *A. religiosa* seedlings of transfer to sites that were warmer or colder than that of the origin of their seeds. Materials and Methods: Eleven provenances of *A. religiosa*, collected along an altitudinal gradient (3000 to 3550 m a.s.l.), were assayed in common gardens in three sites of contrasting altitude: 3400, 3000 and 2600 m a.s.l. The results were evaluated by fitting a response curve with a mixed model. Results: The climate transfer distance for the seasonal balance between the temperature conducive to growth (degree days above 5 °C) and the available precipitation (a ratio expressed as dryness index) dominated the shape of the response function curve. The rainy season (June–October) dryness index transfer distance was critical for survival, while that of the cold and dry season (November–February) was critical for aerial biomass, and the annual index was critical for the increase in basal diameter. The effect of climatic transfer distance is much more negative (triggering about 45% mortality) when transfer is toward warmer and dryer sites (at 400 m lower in altitude, +1.9 °C warmer and 16% less precipitation), than when shifting toward colder and wetter sites (400 m higher in altitude, resulting in 95% survival). Conclusions: The projected higher temperatures and lower precipitation due to climatic change will undoubtedly cause severe mortality in young *A. religiosa* seedlings. A 400 m shift upwards in altitude to compensate for climatic change (assisted migration) appears to be a feasible management action.

Keywords: climate transfer distance; dryness index; climate change; assisted migration; *Abies religiosa*; survival; mortality; biomass; basal diameter; Monarch Butterfly Biosphere Reserve

1. Introduction

In the near future, climatic change will generate warmer climates and displace the climatic zones towards the poles and towards higher altitudes in mountainous regions [1].

Given that, as the result of a long process of evolutionary adaptation, species and their populations are distributed in localities that are generally coupled to a climate suitable for them, climatic change will generate a decoupling between the populations and the climate to which they are adapted [2,3], influencing phenological changes, population and growth dynamics [4] and reproduction [5]. Stress induced by this decoupling favors the frequency and intensity of the development of pests and diseases [6–10] and defoliation through drought stress [11], inducing a severe process of forest decline, which is already underway globally [12], especially at the xeric limit (lower altitudinal or south in the northern hemisphere) [13,14].

Forest tree species that present a wide geographic distribution can present an important genetic differentiation among populations in terms of quantitative characters, such as adaptation to the diversity of environmental conditions [3,15,16]. This differentiation is expressed in different variables such as growth rate, seasonality of elongation of the apical bud and resistance to frost damage [17–20]. Provenance assays allow the expression of these differences by cultivating plants from different geographic origins (provenances) under the same environmental conditions [21]. In general, for forest populations, a pattern of variation of growth has been reported in which the provenances originating from the colder (higher altitude) part of the natural distribution express a lower potential for growth in height, but a greater resistance to frost damage while, in contrast, those from the low part of the altitudinal distribution (warmer) have greater potential for growth but are more susceptible to frost damage [22–24].

In Mexico, one of the most important species for biological conservation is *Abies religiosa* (sacred fir or oyamel), because it hosts the overwintering sites of the monarch butterfly (*Danaus plexippus* L.) for five months (November–March) [25]. The areas of the greatest extension of *A. religiosa* are found within the Trans-Mexican Volcanic Belt, at between 2400 and 3600 m a.s.l. and mainly between 19° and 20° in latitude North [26]. The Monarch Butterfly Biosphere Reserve (MBBR) is found within the Trans-Mexican Volcanic Belt, on the border between the states of Michoacán and Estado de México. The forests of *A. religiosa* can be strongly affected by climatic change—it will induce a decoupling between the populations of *A. religiosa* and the environment suitable for them, such that by the decade 2090, the models project that suitable climatic habitat for *A. religiosa* will have disappeared completely from the MBBR [26]. Previous studies have reported that *A. religiosa* presents a high percentage of non-viable seeds (63–79%) [27,28], in addition to very low recruitment and high mortality of sacred fir seedlings, necessitating the implementation of an active restoration program or other management actions [29]. Additionally, so that the populations of this species can remain coupled to their favorable climatic habitat, their upwards altitudinal displacement of between 300 and 500 m will be necessary by 2060 [30], due to the projected increases in temperature and diminution in precipitation. This could have a considerable effect on the overwintering of the monarch butterfly, in addition to severe environmental impacts. For these reasons, human-assisted displacement of seed sources towards higher altitudes (through reforestation programs) appears to be an option for consideration as a strategy of adaptive management to climatic change. This is because, while the populations of forest species can, and in fact currently are, migrating altitudinally towards higher altitudes through natural means, the speed at which they can accomplish this is lower than that necessary to remain coupled to the climate that favors them [31,32].

The results of two previous field tests indicate that it is feasible to conduct assisted altitudinal migration to above the provenances of *A. religiosa*, translocating the seed sources up to 400 m higher in altitude than their origin, but only when planted beneath the shade protection of local shrubs (such as *Baccharis conferta*) serving as nurse plants [33]. However, to date, no experiments have determined what would happen if young *A. religiosa* seedlings are exposed to greater temperatures and lower precipitation, as projected in scenarios of climatic change [26,34]. There is also evidence that there are serious difficulties in terms of the natural regeneration of young seedlings, because the warm and dry season of March–May is now more dry and hot than usual, causing a severe mortality of seedlings in

April [29]. Although an important genetic differentiation has been demonstrated among populations of *A. religiosa* for resistance to frost damage along altitudinal gradients [20], a significant genetic differentiation has not been demonstrated among populations for resistance to drought, for example, in the differential accumulation of solutes, such as the amino acid proline, that confer physiological resistance to drought [35,36].

Considering the above, the objective of this study was to evaluate the effects of climatic transfer (climatic difference between the site of origin and the site of plantation) on the survival and growth of young *A. religiosa* seedlings, through reciprocal transplants of eleven provenances of *A. religiosa* to common gardens at three contrasting altitudes. The climatic transfer was conducted by simulating three possible scenarios: (i) towards altitudes lower than the seed source origin (simulating a future climate of higher temperature and lower precipitation); (ii) towards higher altitudes (to confirm the viability of assisted migration); and (iii) to a site of intermediate altitude as a reference. The latter two sites were located in the core zone and on the edge of the buffer zone of the MBBR, respectively, in Estado de México, in central-western Mexico. The intention was to generate guidelines for decision-making in terms of the transfer of seed sources in a program of assisted migration, as well as to quantify the impacts of climatic change and the risks of failing to practice adaptive management in the face of climatic change.

2. Materials and Methods

2.1. Study Site

The study was conducted within the Monarch Butterfly Biosphere Reserve (MBBR), which forms part of the Trans-Mexican Volcanic Belt, and in Tlalpujahua de Rayón, Michoacán. The altitudes of the MBBR range from 2200 to 3640 m a.s.l. and the annual average temperatures range from 8 to 22 °C [37]. The vegetation is mainly comprised by forests of fir (*A. religiosa*), pine (*Pinus* spp.), oak (*Quercus* spp.), and cedar (*Cupressus* spp.) [38].

2.1.1. Seed Collection Sites

Cones were collected (in December 2017) from eleven provenances of *A. religiosa*, along an altitudinal transect (3000 to 3550 m a.s.l. in altitude; altitudinal difference of 50 m between sites), within the MBBR (Table 1). At each site, cones were collected at random from ten trees, at a distance of at least 30 m apart [39], in order to reduce the possibility of endogamy.

Table 1. Geographic location and some climatic variables (their acronyms as in Table 2) of *Abies religiosa* seed collection sites within the Monarch Butterfly Biosphere Reserve (MBBR).

Altitude (m a.s.l)	Lat. N	Long. W	MAT (°C)	MAP (mm)	MTCM (°C)	ADI Index	WDSDI Index	CDSDI Index	RSDI Index
3552	19.564	100.229	8.5	1107	6.2	0.034	0.152	0.160	0.029
3491	19.567	100.233	8.8	1094	6.5	0.035	0.161	0.176	0.030
3457	19.571	100.235	9.0	1089	6.7	0.036	0.165	0.184	0.031
3411	19.573	100.237	9.3	1079	7.0	0.038	0.173	0.199	0.032
3364	19.575	100.234	9.5	1065	7.2	0.039	0.183	0.206	0.033
3300	19.579	100.231	9.9	1048	7.5	0.041	0.191	0.222	0.035
3233	19.580	100.224	10.2	1029	7.8	0.043	0.201	0.233	0.036
3210	19.581	100.220	10.3	1022	8.0	0.044	0.204	0.238	0.037
3143	19.581	100.214	10.6	1001	8.3	0.046	0.216	0.250	0.039
3099	19.586	100.214	10.8	987	8.5	0.047	0.226	0.260	0.040
3003	19.595	100.210	11.3	969	8.9	0.050	0.253	0.303	0.041

Sites at altitudes from 3552 to 3300 m a.s.l and at 3099 m a.s.l. were collected at Ejido La Mesa. Sites from 3233 to 3143 m a.s.l. were collected at Ejido Buenavista Casablanca. Site at 3003 m a.s.l. was collected at Ejido Guadalupe Buenavista, Municipality of San José del Rincón, Estado de México.

Table 2. Climatic variables estimated for each provenance (seed origin, taken from [40]) and for each test site (taken from data loggers HOBOs and rainwater capture data).

Code	Unit	Definition
MAT	°C	Mean annual temperature
MAP	mm	Mean annual precipitation
GSP	mm	Growing season precipitation (total precipitation from April–September)
WDSP	mm	Warm and dry season (total precipitation March–May)
RSP	mm	Rainy season (total precipitation June–October)
CDSP	mm	Cold and dry season (total precipitation November–February)
MTCM	°C	Mean temperature in the coldest month
MMIN	°C	Mean minimum temperature in the coldest month
MTWM	°C	Mean temperature in the warmest month
MMAX	°C	Mean maximum temperature in the warmest month
DD5	°C	Degree-days > 5 °C
WDSDD5	°C	Warm and dry season (March–May degree days > 5 °C)
RSDD5	°C	Rainy season (June–October degree days > 5 °C)
CDSDD5	°C	Cold and dry season (November–February degree days > 5 °C)
ADI	index	Annual dryness index ($ADI = \sqrt{DD5}/MAP$)
GSDI	index	Growing season dryness index ($GSDI = \sqrt{GSDD5}/GSP$)
WDSDI	index	Warm and dry season (March–May) dryness index ($WDSDI = \sqrt{WDSDD5}/WDSP$)
RSDI	index	Rainy season (June–October) dryness index ($RSDI = \sqrt{RSDD5}/RSP$)
CDSDI	index	Cold and dry season (November–February) dryness index ($CDSDI = \sqrt{CDSDD5}/CDSP$)

2.1.2. Provenance Test Sites in Common Gardens

Three field provenance tests were established in common gardens (Table 3): (a) high altitude site at 3400 m a.s.l., to confirm the viability of altitudinally upwards assisted migration, within the core zone of the MBBR (at an approximate linear distance of 300 m from a monarch butterfly overwintering site); (b) intermediate altitude site at 3000 m a.s.l., on the edge of the buffer zone of the MBBR; and (c) low altitude site at 2600 m a.s.l., to quantify the effects of a warmer, drier climate than that of the seed origin (considering that the temperature changes by approximately 0.5 °C for each 100 m of altitudinal difference [34]).

Table 3. Location and some climatic variables (their acronyms as in Table 2) of sites of tests of provenances in common gardens.

Site	Altitude (m a.s.l.)	Lat. N	Long. W	MAT (°C)	MAP (mm)	MTCM (°C)	ADI Index	WDSDI Index	CDSDI Index	RSDI Index
Llano Grande [1]	3400	19.57	100.23	9.3	1076	7.0	0.026	0.198	0.041	0.026
La Mesa [2]	3000	19.58	100.18	11.3	951	8.9	0.056	0.537	0.251	0.044
Tlalpujahua [3]	2600	19.80	100.16	12.8	906	9.6	0.078	0.930	0.611	0.057

[1] Core zone of the Monarch Butterfly Biosphere Reserve (MBBR). [2] Ejido La Mesa, Municipality of San José del Rincón, Estado de México (near the buffer zone of the MBBR). [3] Michoacán state.

2.2. Production of Plants in Nursery

Once the seed was obtained from the mother tree, it was stratified at 4 °C for 14 days [39,41]. Each lot of seeds subjected to stratification comprised a similar quantity of seeds originating from each mother tree, so that the progenies were represented equally in the provenances.

Plant production was conducted for 14 months (May 2018–June 2019) in a shade-house (35% shade mesh) in the Instituto de Investigaciones sobre los Recursos Naturales (INIRENA), in Morelia city in Michoacán, Mexico. Plants were germinated in a seedbed and later transplanted into containers of 380 cm^3, with a mixed substrate comprising loamy soil, agrolite, vermiculite and peat moss (volume 60:15:15:10, respectively). Transplantation was conducted with an equal quantity of seeds germinated from each mother tree, so that they were represented equally in the provenances. To prevent differential growth among the provenances in the nursery as a result of possible environmental differences within the shade house, the plants were positioned in a completely random block design (6 blocks, 11 provenances, 54 plants per plot).

2.3. Test of Provenances in Common Gardens in the Field

In each field site (as in Table 3), three raised beds were constructed, with wooden frames of 5 × 1.5 × 0.4 m in length:width:height. The internal base of each raised bed was covered by a metal mesh (aperture size 3.5 cm) to prevent access by gophers. To ensure that the sites differed only in terms of the climate as a result of their altitude, and thus to evaluate the impact of the climatic transfer in each provenance, the raised beds were filled with the same forest soil, transported from a forest site dominated by *A. religiosa* at the intermediate altitude (3000 m a.s.l.), thus avoiding any confounding effect of the local soil. Finally, the raised beds were covered by a shade mesh (35%) at a height of 2.5 m, to simulate the protective effect of nurse plants [33] (Figure 1).

Figure 1. Raised beds of the provenance tests (common gardens) in the site at 2600 m a.s.l., October 2020, 15 months after transplantation. Each raised bed contained two complete randomized blocks.

The seedlings produced in the nurseries were transplanted to the raised beds in July 2019 (beginning of the rainy season), at age 14 months. In each site, six completely random blocks were established (11 provenances per block, five seedlings per provenance per plot in line, at a spacing of 0.2 × 0.2 m and with two blocks per raised bed. At the extremes of the raised bed, a protection strip was planted, comprising a mixture of seedlings of all of the provenances, in order to control the edge effect (Figure 1).

The trees (the seeds of which were collected at different altitudes) represented in the common garden tests by their seedlings are herein termed populations, while the location of origin of a population is called the provenance.

2.4. Measurements

For sixteen months, monthly or bimonthly evaluations of plant survival and height were conducted, with semestral evaluations of the basal diameter. The first evaluations of plant height and basal diameter were conducted six months after plantation, in the expectation that the plants would be well established by that time. The final measurement was taken at the end of the growing season of the following year (November 2020). Plant height was measured with a 1 mm precision up to the tip of the apical bud, supporting the ruler on a bar of aluminum, which was in turn supported on the edge of the wooden frame (to avoid measurement errors due to the irregular surface of the substrate). Basal diameter was measured at the base of the stem with a digital vernier (KNOVA®, Cupertino, CA, USA) at a 0.1 mm precision.

The increase in seedling height and basal diameter was obtained by subtracting the first measurement (December 2019) from the final measurement (November 2020). Survival was estimated as the percentage of live plants per plot. Substrate moisture content was measured with a W.E.T (water content, electrical conductivity and temperature) sensor of type W.E.T-2 (Delta-T-Devices, Cambridge, UK) as well as gravimetrically, by measuring the wet and dry weights of soil samples.

2.5. Estimation of Aerial Biomass

Following the methodology of Martinez-Luna, et al. [42], the aerial biomass of the seedlings in the common garden field assay was estimated using a regression formula from the measurements of total height and basal diameter. The formula was fitted by measuring (through destructive harvest) seedling height, basal diameter and dry weight of the stem, branches, and foliage (dried for 48 h at 65 °C in a Shel Lab drying oven (model FX4) and weighed on a Mettler Toledo analytical balance, model AB204-S). Five seedlings of diverse sizes were selected for this harvest from each of the 11 assayed provenances (harvested from a surplus lot of plants). The regression was fitted using Proc Reg of SAS [43,44], from which the following parameters were obtained:

$$Ba = -1.62490 + (0.03538 \times Ht) + (0.77238 \times Db) \quad (1)$$

where Ba is the aerial biomass (g), Ht is the total height (mm), and Db is the basal diameter (mm 0.1).

2.6. Climatic Data

Climatic data from the provenance of origin of the seeds were obtained for the reference period 1961–1990 through climatic spline models (based on [34]), available at the website [40]. The climate of the provenance is considered the recent historic climate to which the provenances have evolved and adapted [45]. The climate of the field assay sites, which was that experienced by the seedlings while growing in the field common gardens, was estimated using three data loggers HOBOs (temperature/external channel data logger; Onset Computer Corporation®, Bourne, MA, USA) per site, set to record temperature every 30 min. Monthly precipitation was measured by capturing rainwater, averaging the results from two rainwater traps per site.

The climatic variables estimated for both the provenances and the sites are described in Table 2, and are considered of physiological importance for representing the seasonal or annual balance between the available humidity and the temperature in terms of seedling growth [17,46]. The seasonal indices warm and dry season dryness index (WDSDI), rainy season dryness index (RSDI) and cold and dry season dryness index (CDSDI) aim to represent the seasonality of each of the *A. religiosa* study sites and are based on the seasonality observed in [33].

From the climate of the provenances and of the field test sites, the climatic or altitudinal transfer distance (CTD or ATD = climate or altitude of the test site − climate or altitude of the source of the provenance) was estimated in order to represent the impact of growing in a site with a climate (or altitude) that differed from that of the seed origin.

2.7. Statistical Analysis

A curve was estimated for each of the four response variables studied (survival, increase in basal diameter, increase in height, and aerial biomass of *A. religiosa* seedlings), to measure the effect of the climatic or altitudinal transfer (growing in a site with a climate or altitude that differed from that of the seed origin), using a mixed model in Proc Mixed of SAS [43,44], based on refs [16,46].

The fixed effects considered three components of the effect of climate as a selective force that produces a genetic differentiation among populations and/or a response of phenotypic plasticity (also considered as a genotype × environment interaction [15]): (a) the climate of the provenance (C); (b) the climatic transfer distance (D = climate of the test site − climate of the provenance); and (c) the interaction between climate of the provenance and climatic transfer distance (C × D). Random effects provided the variance that was not explained by the fixed effects; i.e., those derived from the experimental design: site, provenance, block and the interaction between the site and the provenance. The climatic transfer distance is a concept originally known as "ecological distance" [47]; more details regarding the reasoning of this model can be found in refs [16,48].

2.7.1. Mixed Model

The following mixed model was fitted, based on refs [16,46]:

$$Y_{ijkl} = \mu + \beta_0 + \beta_1 D_{ij} + \beta_2 D_{ij}^2 + \beta_3 C_j + \beta_4 (D_{ij} \times C_j) + \beta_5 S_i + \beta_6 P_j + \beta_7 B_k(S_i) + \beta_8 (S_i \times P_j) + e_{ijkl} \quad (2)$$

where Y_{ijkl} is the survival or aerial biomass or increase in basal diameter or increase in plant height, corresponding to the *l*th tree for the *j*th provenance in the *k*th block in the *i*th test site, and β_0 is the intercept. D_{ij} is the climatic transfer distance (difference between the value of a climatic variable in the test site and that of the provenance) for the *j*th provenance in the *i*th test site, C_j is the value of the climatic variable in the *j*th provenance, and $D_{ij} \times C_j$ is the interaction between the climatic transfer distance for the *j*th provenance in the *i*th test site and the climatic variable in the *j*th provenance. S_i is the effect of the *i*th test site, P_j is the effect of the *j*th provenance, $B_k(S_i)$ is the effect of the *k*th block nested within the *i*th test site, and e_{ijkl} is the error term.

2.7.2. Selection of Variables to Fit the Best Mixed Model

The climatic variables for use in the mixed model were selected using the following steps, independently for each response variable, based on refs [16,46]:

(1) Five climatic variables were selected that best described the climate of the provenance, estimating the Spearman correlations between the values of the climatic variables and the average value per provenance across sites of the response variables, and selecting those with the highest $|r|$ value.
(2) The climatic variables for the climatic transfer distance were selected by fitting a reduced mixed model, eliminating from the model (Equation (2)) the term of climate of the provenance (C_j) and its respective interaction ($D_{ij} \times C_j$). The five climatic variables selected were those for which the model obtained the lowest (and thus the best) value of the Akaike information criterion (AIC) and, in addition, that necessarily presented the estimated parameter of the quadratic term both negative and significant, in order to ensure that it was biologically sound [15,16].
(3) Subsequently, 5 × 5 = 25 full "competing" models were run, which included all of the possible combinations of the five variables of the climate of provenance and the five variables of climatic transfer distance preselected in the previous two steps. The best model was selected based on the AIC value.

In total, 100 models (25 "competing" models for each of the four response variables) were run, in order to determine the best model for understanding the response of survival, increase in basal diameter, increase in plant height and aerial biomass.

Finally, from the regression parameters estimated for the fixed effects, quadratic curves of response to the climatic transfer were estimated for each response variable, in order to visualize the effect on survival and growth of growing in a site with a climate that differs to that of the origin of the populations.

3. Results

3.1. Climatic Variables That Best Explained the Climatic Transfer Distance Response Function

The survival response function was best explained (lowest AIC value = 1651.1) by the quadratic term of the rainy season dryness index transfer distance (RSDI_td^2; $p = 0.0112$), although the linear term was not significant ($p = 0.3542$). The other fixed effect terms, the climate of the provenance term selected (warm and dry season dryness index, WDSDI; $p = 0.0941$) and the interaction between climatic transfer distance and the climate of the provenance (RSDI_td × WDSDI; $p = 0.1273$), were also not significant. Regarding the random effect terms, only the block nested within the site was significant (p = 0.0252), contributing 21% of the total variance, where 100% is the sum of the contribution of the random terms only (Table 4).

Table 4. Analysis of the mixed model for seedling survival, biomass and increase in basal diameter. Akaike information criterion (AIC), estimated parameters (coefficients of regression for fixed effects and variances estimated for random effects), contribution to the total variance (of the random effects) and significance (p).

Parameter or Source of Variation	Survival			Biomass			Increase in Basal Diameter		
Fixed Effects	**Estimate**		p	**Estimate**		p	**Estimate**		p
Akaike Information Criterion	1651.1		—	2397.2		—	1917.3		—
Intercept	119.7		0.0115	2.4		0.0478	0.081		0.8824
Climate at seed source									
Warm and dry season dryness index (WDSDI)	−111.3		0.0941	6.8		0.0198	—		—
Mean temperature in the coldest month (MTCM)	—		—	—		—	0.19		0.0036
Climate transfer distance									
Rainy season dryness index (RSDI)	1426.8		0.3542	—		—	—		—
Cold and dry season dryness index (CDSDI)	—		—	8.7		0.0002	—		—
Annual dryness index (ADI)	—		—	—		—	143.5		<0.0001
(Climate transfer distance)^2									
RSDI^2	−55,955		0.0112	—		—	—		—
CDSDI^2	—		—	−10.6		<0.0001	—		—
ADI^2	—		—	—		—	−1291.2		<0.0001
Interaction climate seed source × Climate transfer distance	−10,202		0.1273	−20.6		0.0395	−14.7		<0.0001
Random Effects	**Variance**	**% ***	p	**Variance**	**% ***	p	**Variance**	**% ***	p
Site	0	0	1	0	0	1	0	0	1
Population	1.5	0.34	0.4412	0.042	3.10	0.1217	0.010	1.4	0.1364
Block (Site)	91.2	21.02	0.0252	0.097	7.19	0.0196	0.108	15.04	0.0095
Site × Population	0	0	1	0.019	1.43	0.2036	0	0	1
Error	341.0	78.64	<0.0001	1.195	88.28	<0.0001	0.598	83.55	<0.0001

* Contribution to total variance expressed as a percentage, where 100% is the sum of the contribution to the total variance of the random terms only.

The biomass response function was best explained (lowest AIC value = 2397.2) by the significant linear and quadratic term of the cold and dry season dryness index transfer distance (CDSDI_td, p = 0.0002; CDSDI_td^2, $p < 0.0001$; respectively). The climate of the provenance term selected (warm and dry season dryness index, WDSDI) was also significant (p = 0.0198), as was the interaction CDSDI_td × CDSDI (p = 0.0395). The random effect term that contributed most to the total variance (among the random effects) was block nested within site (p = 0.0196), contributing 7% of the total variance (Table 4).

The increase in basal diameter response function was best explained (lowest AIC value = 1917.3) by the significant linear and quadratic term of the annual dryness index (ADI) transfer distance ($p < 0.0001$). The climate of the provenance term selected (mean temperature of the coldest month, MTCM) was also significant (p = 0.0036), as was the interaction ADI_td × MTCM ($p < 0.0001$). The random effect term that most contributed to the total variance was block nested within site (p = 0.0095), contributing 15% of the total variance (Table 4).

The increase in seedling height response function was best explained (lowest AIC value = 5122.4) by RSDI, although, as with transfer distance, it was not significant (linear: p = 0.23; quadratic: p = 0.3517), as well as the climate of the provenance term selected (MMIN, p = 0.1605). Again, only block nested within site was significant (p = 0.0096) among the random terms, contributing 11% of the total variance (results not shown in Table 4 for simplicity).

In all of the four response variables analyzed, site was not significant, and the error term contributed the most among all the random terms (from 79% for survival to 88% for aerial biomass; Table 4).

3.2. Curves of Response to the Climatic and Altitudinal Transfer Distance

The predicted response curve for survival indicated that when the seed sources were moved toward warmer and dryer sites (toward lower altitude, right side of the curve in Figure 2a), there is a severe induced mortality. In sharp contrast, when the shift is conducted toward a colder and moister site (to higher altitudes, sites at 3400 m a.s.l.), survival is maintained mostly above 95%. This is even more evident when fitting a simple quadratic regression of survival against altitudinal transfer distance (Figure 3a), where exceeding the 400 m shift to lower altitudes (grossly equivalent to +2 °C, based on a lapse rate of 0.5 °C per 100 m [34]) causes significantly increased mortality. It should also be noted in Figure 2a that the maximum value of the response curve is displaced slightly towards the left side (colder and moister sites), either for the average species curve or for the extreme lower altitudinal population curve. This suggests that the optimum of survival for the species in general, and for the extreme lower population, would be a colder site than that occupied at present.

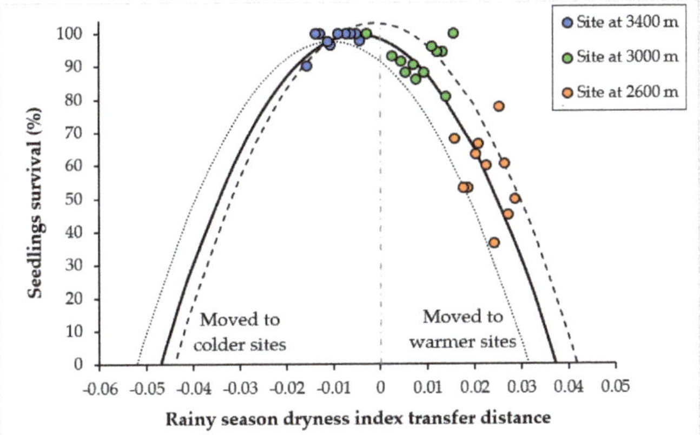

(a)

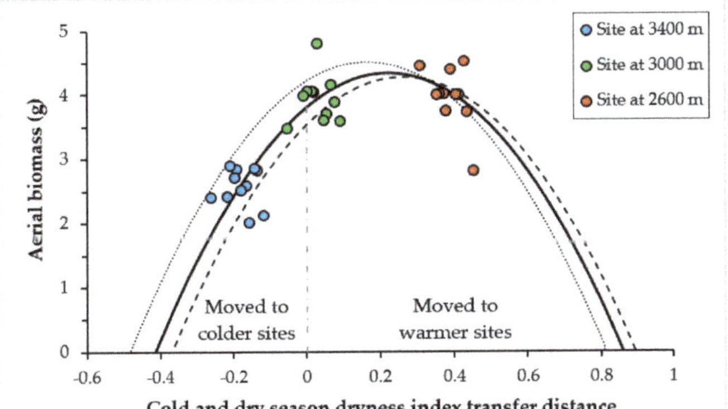

(b)

Figure 2. *Cont.*

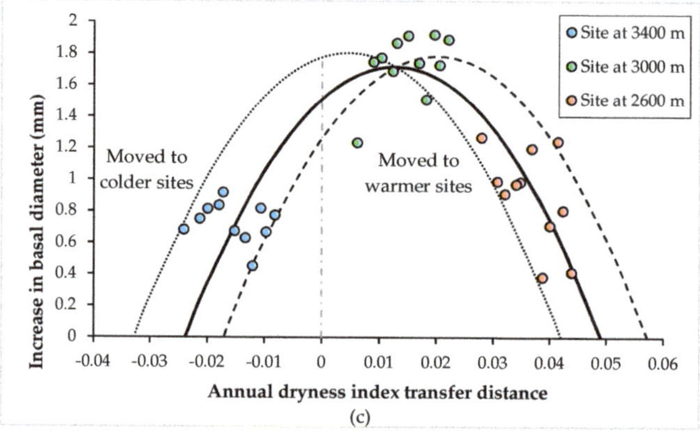

Figure 2. Seedling response to climatic transfer distance for: (**a**) seedling survival; (**b**) aerial biomass (dry weight); and (**c**) increase in basal diameter. Predicted response curves (constructed with the estimated parameters according to the fixed terms of the mixed model, as in Table 4), are indicated for the species (solid line) and for the populations of the coldest site (dashed line) and the warmest site (dotted line). Symbols are observed averages per population per site (the color distinguishes the particular test site). Respective climatic transfer distance on the x-axis signifies the transfer to drier and/or warmer sites if the values are positive, and transfer to cooler and/or moister sites if they are negative. Zero signifies a climate similar to that of the site of provenance.

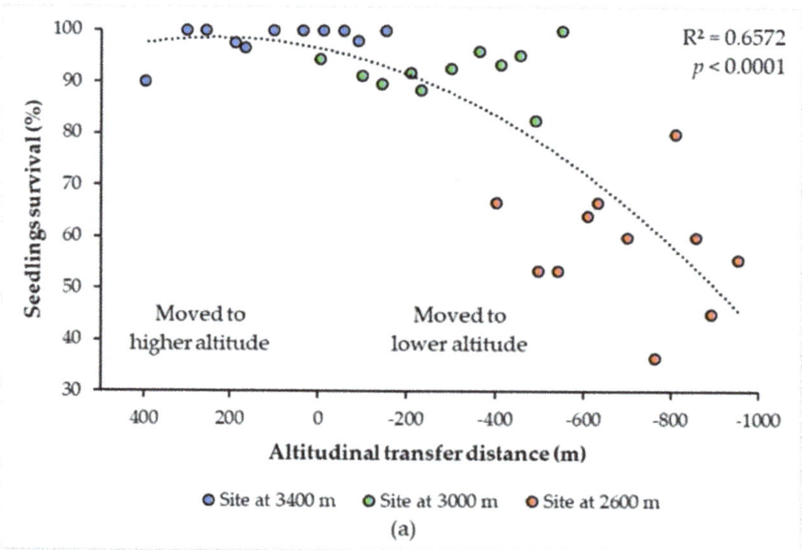

Figure 3. *Cont.*

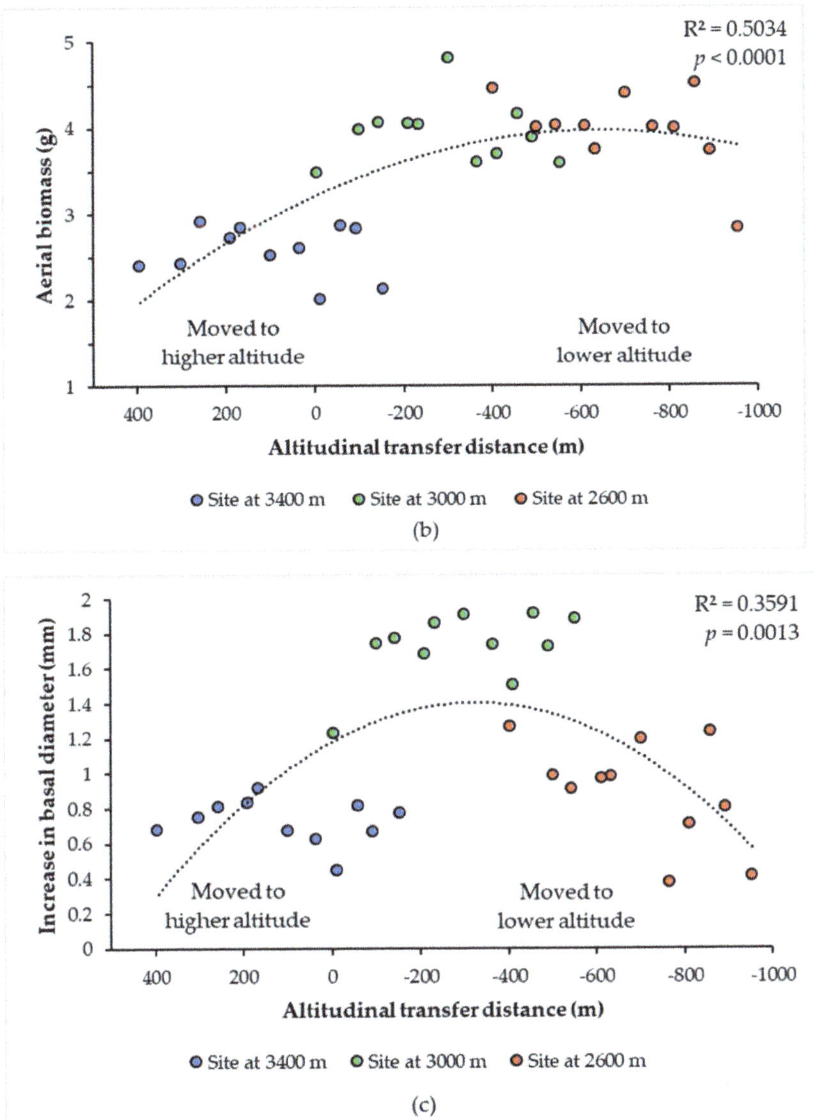

Figure 3. Estimated response curve (based on a quadratic regression) of the plants to the altitudinal transfer for: (**a**) survival; (**b**) aerial biomass (in dry weight); and (**c**) increase in basal diameter. Symbols indicate the average observed values per provenance per site. Symbol color denotes the particular assay site: blue, 3500 m a.s.l.; green, 3000 m a.s.l.; orange, 2600 m a.s.l. Positive values on the *x*-axis indicate movement towards sites of higher altitude, which are colder and moister, while negative values indicate movement towards sites of lower altitude, which are warmer and drier.

The predicted response curve for aerial biomass indicated that, when seed sources are moved toward warmer and dryer sites (to a lower altitude, right side of curve the in Figure 2b), there is little negative effect on the growth expressed as aerial biomass: the observed values fluctuated around the maximum predicted value of the curve (about 4.2 g of aerial biomass in dry weight). In contrast, when seed sources are shifted toward colder and moister sites, there is an important loss of accumulated aerial biomass. This becomes more evident when plotting the average biomass per population and per site against the altitudinal transfer distance (Figure 3b).

The predicted response curve for the increase in basal diameter indicated that the greatest growth occurs at a site that is warmer and dryer than that occupied by the populations at present, given that the curves are displaced to the right (Figure 2c). It is also clear that when they are moved toward colder and moister sites (left of the *x*-axis) or warmer and dryer sites (right side of the *x*-axis), relative to the optimum, they present a lower growth in basal diameter. The resulting curve regarding altitudinal transfer distance indicated clearly that the optimum growth, for all the provenances in general, was achieved at the site of intermediate elevation (3000 m a.s.l.; Figure 3c).

When comparing the average value across sites for the mean annual temperature and annual precipitation to the average value across sites of the reference period 1961–1990, it is evident that the sites have already increased in temperature (for the observed period) by an average of 0.58 °C, while precipitation has decreased by 27 mm (around 3%) (Table 5).

Table 5. Comparison of the most important climatic variables related to the survival response function, corresponding to the reference period 1961–1990 and the data obtained by averaging the test sites.

Climatic Variable	Reference Period 1961–1990	December 2019–November 2020	Difference
MAT	11.13 °C	11.71 °C	+0.58 °C
MAP	980 mm	952.7 mm	−27.3 mm

When plotting seedling mortality, monthly precipitation and mean temperature per site, and comparing among the sites (Figure 4), it is evident that precipitation differences among sites are not particularly large, but seem to have a meaningful effect, nevertheless. The low altitude site (2600 m a.s.l., 1407.4 mm total precipitation in 15 months; Figure 4a), had 16.5% lower total precipitation over the observed period than the site of intermediate altitude (3000 m a.s.l., 1685.14 mm; Figure 4b), while the site of elevated altitude (3400 m a.s.l., 1853.6 mm; Figure 4c) had 9.1% higher total precipitation than the intermediate site. The temporal patterning of the dry/rainy season seems to be similar across the three sites. Surprisingly, the apparently relative minor differences in mean temperature between the sites at 2600 m a.s.l. (14.9 °C averaging across all the 15 months observed) and 3000 m a.s.l. (13.02 °C; 1.9 °C of difference among those two sites), combined with the 16.5% difference in precipitation, seemed to be enough to trigger much higher mortality at the site at 2600 m a.s.l., particularly during the dry months, whether cold and dry (December–February) or warm and dry (March–May; Figure 4a). Moreover, an unexpected and relatively high mortality occurred just after the rainy season abruptly ended in October 2020 (Figure 4a), suggesting that several seedlings were already debilitated before the rainy season, perhaps subsequently expressing a drought stress legacy effect after the rainy season. In sharp contrast, seedling mortality at the site at 3400 m a.s.l., which was much colder and only 9.1% more rainy than the intermediate site, was nearly zero (Figure 4c).

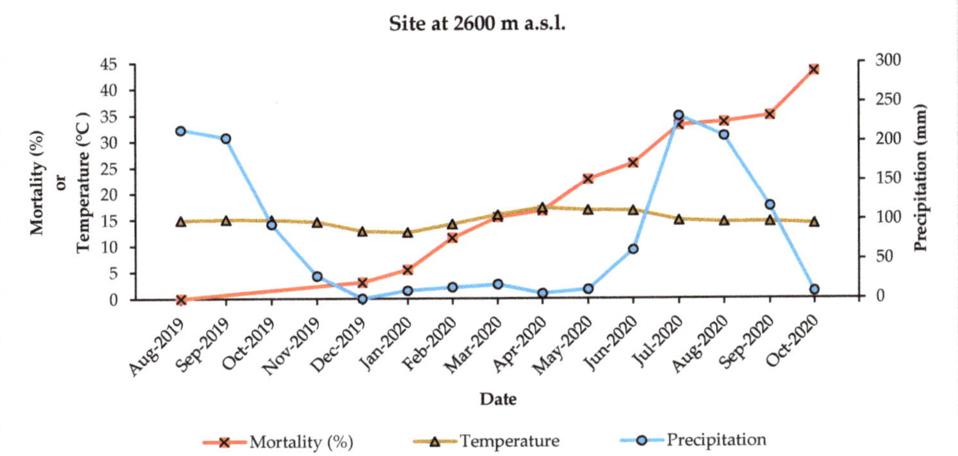

(a)

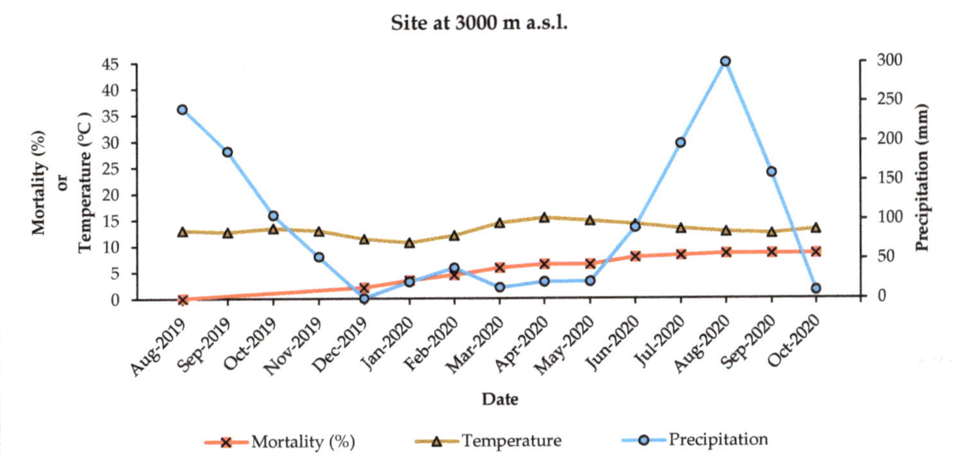

(b)

Figure 4. *Cont.*

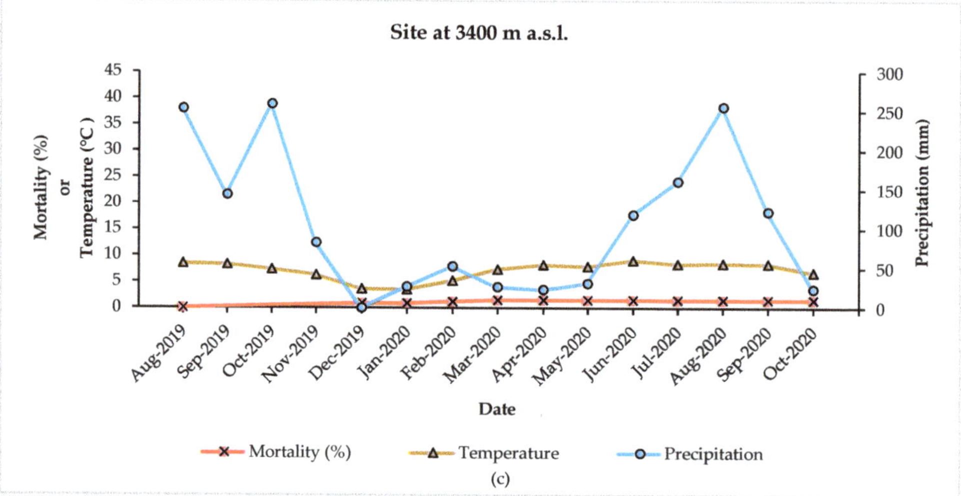

Figure 4. Mortality (%), monthly mean temperature (°C) and monthly precipitation (mm) in the period August 2019 to October 2020, for each test site, at (**a**) 2600 m a.s.l., (**b**) 3000 m a.s.l. and (**c**) 3400 m a.s.l.

4. Discussion

The most statistically significant climatic variables in terms of the response of survival and growth in *A. religiosa* seedlings to the climatic transfer distance were, in all cases, the balance between the temperature useful to growth (degree days above 5 °C) and available precipitation. That balance, expressed as dryness indices in our analysis, varied only in different seasons for the response variables: the rainy season (June–October) dryness index transfer distance was critical for survival, while that of the cold and dry season (November–February) was critical for aerial biomass, and the annual index was critical for the increase in basal diameter.

The effect of climatic transfer distance was much more negative (in terms of seedling survival) when transferring toward warmer and dryer sites, than when shifting to colder and moister sites. Apparently, there is a critical threshold that, when crossed, triggers mortality: a shift to a site 400 m lower in altitude, equivalent to +2 °C considering an overall lapse rate [34], or a difference of +1.9 °C based on the observed months, or a reduction in precipitation of 16%.

Most of the seedling mortality at the low altitude site (2600 m a.s.l.) occurred during the dry season (December–May). This is consistent with previous observations of a high mortality of emergent seedlings of natural regeneration in the core zone of the Monarch Butterfly Biosphere Reserve (MBBR) [29], and seems to support the projected decrease in climatic habitat suitable for *A. religiosa* inside the MBBR as a result of climatic change [26,30,49–51]. The apparent reason for the observed timing of this mortality is that the dry season was as dry as always, perhaps a little more so, but was warmer due to the ongoing climatic change. It was surprising to us that the relatively small differences in temperatures and precipitation between the sites at 2600 m a.s.l. and 3000 m a.s.l. made such a large difference to survival. Thus, the amount of residual moisture in the soil during the dry season, following the rainy season of the previous year, seems to be critical for survival of the young seedlings.

The high survival (around 95%) of the seed sources shifted to higher altitudes than their origin (to the site at 3400 m a.s.l.) confirms the viability of assisted migration to higher altitudes in the Trans-Mexican Volcanic Belt as a management measure for consideration, in order to reduce the projected impacts of climatic change, as shown in previous field tests [33].

Suitable climatic habitat projections for the 2060 decade, under intermediate greenhouse emission scenarios (RCP 6.0 watts/m^2), indicate that it is required to shift the *A. religiosa* seed sources upwards in altitude by between 300 and 500 m in order to realign the populations with the climate to which they are adapted [30]. Our evidence that is safe to move these sources upwards in altitude by 400 m therefore supports such a management proposal.

The mean temperature of the coldest month (MTCM) was the most important climate variable at the seed source for the increase in basal diameter. This indicates that genetic differentiation among populations for growth potential is shaped by the cold temperatures, at least in our case for those seedlings that survived the dry season. This is consistent with previous patterns of genetic variation among *A. religiosa* populations along altitudinal clines demonstrated in a common garden test under optimal conditions, where MTCM was also found to be the climatic variable that best explained the differences among populations [20].

Aerial biomass necessarily had to be measured, at the end of the observation period, in live plants that had survived the transfer toward warmer and dryer sites. It is therefore likely that those seedlings might have had some resistance or tolerance to the drought stress, and growing in a site with higher temperature (2600 m a.s.l.), they were able to grow more. In contrast, although the proportion of seedlings that survived in those that were transferred to a colder and moister site was higher (as in Figure 2a), growth was likely limited by the colder temperatures.

Implications for Management of Abies religiosa Inside the MBBR

The results indicate that increased temperatures and reduced precipitation due to climatic change would undoubtedly have a negative impact on the recruitment of *A. religiosa* seedlings at the lower altitudinal limits of the natural distribution range of the species, through increased mortality during the dry season (December–May). Active ecological restoration efforts in the MBBR must therefore consider how to increase the survival of planted *A. religiosa* seedlings in order to restore perturbed sites and ensure the maintenance of the monarch butterfly overwintering sites. In this context, perturbed sites at the low altitude of the natural range distribution of *A. religiosa* should have priority. Actions should include: (a) conducting reforestation using local shrubs (such as *Baccharis conferta* or other native understory species) as nurse plants for the *A. religiosa* seedlings [33]; and (b) conducting assisted migration to higher altitudes to realign *A. religiosa* populations with the climate to which they are adapted. Previous estimates regarding such realignment, targeting the climate projected for the 2030 decade, have suggested an upwards shift in altitude of 350 m [20]. The results of the present study suggest that it is safe to shift up to 400 m upwards in altitude, in terms of good seedling survival. Thus, an upward shift of between 350 and 400 m in altitude seems to be advisable. If such action is not undertaken, the lack of recruitment of natural regeneration seedlings is likely to continue as a result of the high mortality of young seedlings at lower altitudes (as demonstrated by ref [29]), impeding the natural renewal of adult *A. religiosa* trees within the MBBR, which will endanger the monarch butterfly overwintering sites in the long term.

5. Conclusions

The balance between the seasonal temperature available for growth (degree days > 5 °C) and the available precipitation is a determining factor for the permanence of *A. religiosa*, because the survival of young seedlings depends on the residual humidity of the soil during the dry season, following the rainy season of the previous year. A 2 °C increase in temperature and a 16% decrease in precipitation would imply a high mortality rate (~45%), negatively impacting the recruitment of naturally regenerating seedlings, mainly at the lower altitudinal limit of the natural distribution of the species. These results suggest that an upward movement of 400 m in altitude, through assisted migration, will have positive effects on the survival of *A. religiosa* seedlings, re-coupling the populations to the climate to which they are adapted and increasing the possibility of maintaining the overwintering sites of the monarch butterfly.

Author Contributions: C.S.-R., A.L.C.-V., A.B.-G. and R.L.-C. conceived the research project. A.B.-G. carried out the seed collection. A.L.C.-V., A.B.-G., M.G.-R. and C.S.-R. produced the seedlings and conducted the field experiments. A.L.C.-V. and C.S.-R. took measurements, determined the statistical analyses, and generated the response curves. R.L.-C., M.G.-R., L.L.-T. and E.d.l.B. provided important suggestions during the development of the project and for the data discussion. A.L.C.-V. and C.S.-R. led the writing of the manuscript. All authors have read and agreed to the published version of the manuscript.

Funding: Financial support was provided to C.S.-R. by The Monarch Butterfly Fund (Madison, WI, USA), the Mexican Fund for the Conservation of Nature and United States Forest Service USDA (funded by the U.S. Agency for International Development), the Fondo de Conservación del Eje Neovolcánico A.C., the American Forests (USA), the Forest Genetic Resources Working Group of the North American Forestry Commission (FAO) and the UMSNH Coordinación de la Investigación Científica; the Basic Research Fund of the Mexican Council of Science and Technology (CONACYT) (Ciencia Básica-2014-242985) was provided to A.B.-G.; the PAPIIT program (IN116218) of UNAM was provided to R.L.-C.; and a CONACyT graduate studies fellowship was provided to A.L.C.-V. (733822).

Institutional Review Board Statement: Not applicable.

Informed Consent Statement: Not applicable.

Data Availability Statement: The data presented in this study are available on request from the corresponding author.

Acknowledgments: This paper is an undertaking of the Forest Genetic Resources Working Group (FGRWG)/North American Forest Commission/Food and Agricultural Organization of the United Nations. We thank Francisco Ramírez-Cruz (deceased), Doña Petra Cruz-Cruz, Francisca Ramírez-Cruz, and A. Belén Luis-Ramírez for providing numerous facilities for seed collection, seedling nursery production, common garden test planting, maintenance and measurements at Ejido La Mesa, Municipality of San José del Rincón, Estado de México; without their help, this study would not have been possible. Thanks also go to Esmeralda Navarro-Miranda, Cecilia Zamora-Sánchez, Erika Gómez-Pineda, Verónica Osuna-Vallejo, Adanely Jiménez-Villegas, J. Sinaí Ávalos-Montes, Patricia Herrejón-Calderón and other UMSNH students for assistance in the maintenance and measurement of the nursery and field experiments. Special thanks go to Gerald E. Rehfeldt (USDA—Forest Service, Moscow, Idaho) for advice on the experimental design and data analysis. Keith MacMillan provided assistance with the English writing.

Conflicts of Interest: The authors declare no conflict of interest.

References

1. Jia, G.; Shevliakova, E.; Artaxo, P.; De Noblet-Ducoudré, N.; Houghton, R.; House, J.; Kitajima, K.; Lennard, C.; Popp, A.; Sirin, A.; et al. Land–Climate Interactions. In *Climate Change and Land: An Ipcc Special Report on Climate Change, Desertification, Land Degradation, Sustainable Land Management, Food Security, and Greenhouse Gas Fluxes in Terrestrial Ecosystems*; Shukla, P.R., Skea, J., Calvo Buendia, E., Masson-Delmotte, V., Pörtner, H.-O., Roberts, D.C., Zhai, P., Slade, R., Connors, S., van Diemen, R., et al., Eds.; Cambridge University Press: Cambridge, UK; New York, NY, USA, 2019; pp. 1–186.
2. Pearson, R.G.; Dawson, T.P. Predicting the impacts of climate change on the distribution of species: Are bioclimate envelope models useful? *Glob. Ecol. Biogeogr.* **2003**, *12*, 361–371. [CrossRef]
3. Rehfeldt, G.E.; Leites, L.P.; Joyce, D.G.; Weiskittel, A.R. Role of population genetics in guiding ecological responses to climate. *Glob. Chang. Biol.* **2017**, *24*, 858–868. [CrossRef] [PubMed]
4. Matias, L. Changes in the distribution limits of tree species as a consequence of climate variations. *Ecosistemas* **2012**, *21*, 91–96. [CrossRef]
5. Hedhly, A.; Hormaza, J.I.; Herrero, M. Global warming and sexual plant reproduction. *Trends Plant Sci.* **2009**, *14*, 30–36. [CrossRef] [PubMed]
6. Alfaro, R.I.; Fady, B.; Vendramin, G.G.; Dawson, I.K.; Fleming, R.A.; Sáenz-Romero, C.; Lindig-Cisneros, R.A.; Murdock, T.; Vinceti, B.; Navarro, C.M.; et al. The role of forest genetic resources in responding to biotic and abiotic factors in the context of anthropogenic climate change. *For. Ecol. Manag.* **2014**, *333*, 76–87. [CrossRef]
7. Rubin-Aguirre, A.; Sáenz-Romero, C.; Lindig-Cisneros, R.; del-Rio-Mora, A.; Tena-Morelos, C.; Campos-Bolaños, R.; del-Val, E. Bark beetle pests in an altitudinal gradient of a Mexican managed forest. *For. Ecol. Manag.* **2015**, *343*, 73–79. [CrossRef]
8. del-Val, E.; Sáenz-Romero, C. Bark-beetles (Coleoptera: Curculionidae) and climate change: Current situation and perspectives for temperate forests. *TIP Rev. Espec. Cienc. Químico Biológicas* **2017**, *20*, 53–60. [CrossRef]
9. Valenzuela, P.L.; López-García, J. Detection of trees damaged by pests in *Abies religiosa* forests in the Monarch Butterfly Biosphere reserve using infrared aerial photography. *Investig. Geográficas* **2017**, 1–12. [CrossRef]

10. Morales-Rangel, A.; Cambrón-Sandoval, V.H.; Soto-Correa, J.C.; Jones, R.W.; Obregón-Zúñiga, J.A. Temperature effect in populations of *Dendroctonus frontalis* Zimmerman and *Dendroctonus mexicanus* Hopkins (Coleoptera: Curculionidae: Scolytinae) under a climate change scenario in Sierra Gorda queretana. *Acta Zoológica Mex.* **2018**, *34*, 1–8. [CrossRef]
11. Flores-Nieves, P.; López-López, M.Á.; Ángeles-Pérez, G.; de la Isla-Serrano, M.d.L.; Calva-Vásquez, G. Biomass estimation and distribution models of *Abies religiosa* (Kunth) Schltdl. et Cham. In decline. *Rev. Mex. Cienc. For.* **2011**, *2*, 9–20.
12. Allen, C.D.; Breshears, D.D.; McDowell, N.G. On underestimation of global vulnerability to tree mortality and forest die-off from hotter drought in the Anthropocene. *Ecosphere* **2015**, *6*, 1–55. [CrossRef]
13. Mátyás, C.; Vendramin, G.G.; Fady, B. Forests at the limit: Evolutionary–genetic consequences of environmental changes at the receding (xeric) edge of distribution. Report from a research workshop. *Ann. For. Sci.* **2009**, *66*, 800. [CrossRef]
14. Mátyás, C. Forecasts needed for retreating forests. *Nature* **2010**, *464*, 1271. [CrossRef] [PubMed]
15. Rehfeldt, G.E.; Wykoff, W.R.; Ying, C.C. Physiologic plasticity, evolution, and impacts of a changing climate on *Pinus contorta*. *Clim. Chang.* **2001**, *50*, 355–376.
16. Leites, L.P.; Rehfeldt, G.E.; Robinson, A.P.; Crookston, N.L.; Jaquish, B. Possibilities and limitations of using historic provenance tests to infer forest species growth responses to climate change. *Nat. Resour. Model.* **2012**, *25*, 409–433. [CrossRef]
17. Rehfeldt, G.E. Ecological Genetics of *Pinus contorta* from the Rocky Mountains (USA): A Synthesis. *Silvae Genet.* **1988**, *37*, 3–4.
18. Rehfeldt, G. A model of genetic variation for *Pinus ponderosa* in the Inland Northwest (USA): Applications in gene resource management. *Can. J. For. Res.* **1991**, *21*, 1491–1500. [CrossRef]
19. Ruiz-Talonia, L.; Sánchez-Vargas, N.; Bayuelo-Jiménez, J.; Lara-Cabrera, S.; Sáenz-Romero, C. Altitudinal genetic variation among native *Pinus patula* provenances: Performance in two locations, seed zone delineation and adaptation to climate change. *Silvae Genet.* **2014**, *63*, 139–148. [CrossRef]
20. Ortiz-Bibian, M.; Blanco-García, A.; Lindig-Cisneros, R.; Gómez-Romero, M.; Castellanos-Acuña, D.; Herrerías-Diego, Y.; Sánchez-Vargas, N.; Sáenz-Romero, C. Genetic variation in *Abies religiosa* for quantitative traits and delineation of elevational and climatic zoning for maintaining monarch butterfly overwintering sites in Mexico, considering climatic change. *Silvae Genet.* **2017**, *66*, 14–23. [CrossRef]
21. Zobel, B.; Talbert, J. *Applied Forest Tree Improvement*; John Wiley & Sons: North Carolina State Univ., NC, USA, 1984.
22. Viveros-Viveros, H.; Sáenz-Romero, C.; López-Upton, J.; Vargas-Hernández, J.J. Growth and frost damage variation among *Pinus pseudostrobus*, *P. montezumae* and *P. hartwegii* tested in Michoacán, México. *For. Ecol. Manag.* **2007**, *253*, 81–88. [CrossRef]
23. Sáenz-Romero, C.; Tapia-Olivares, B. Genetic variation in frost damage and seed zone delineation within an altitudinal transect of *Pinus devoniana* (*P. michoacana*) in Mexico. *Silvae Genet.* **2008**, *57*, 165–170. [CrossRef]
24. Loya-Rebollar, E.; Sáenz-Romero, C.; Lindig-Cisneros, R.; Lobit, P.; Villegas-Moreno, J.; Sánchez-Vargas, N. Clinal variation in *Pinus hartwegii* populations and its application for adaptation to climate change. *Silvae Genet.* **2013**, *62*, 86–95. [CrossRef]
25. Oberhauser, K.; Peterson, A.T. Modeling current and future potential wintering distributions of eastern North American monarch butterflies. *Proc. Natl. Acad. Sci. USA* **2003**, *100*, 14063–14068. [CrossRef]
26. Sáenz-Romero, C.; Rehfeldt, G.E.; Duval, P.; Lindig-Cisneros, R.A. *Abies religiosa* habitat prediction in climatic change scenarios and implications for monarch butterfly conservation in Mexico. *For. Ecol. Manag.* **2012**, *275*, 98–106. [CrossRef]
27. de Pascual Pola, C.N.; Musálem, M.Á.; Ortega-Alcalá, J. Study of some characteristics of cones and seeds of *Abies religiosa* (HBK) Schl. et Cham. *Agrociencia* **2003**, *37*, 521–531. Available online: http://www.redalyc.org/articulo.oa?id=30237509 (accessed on 10 August 2020).
28. Ortiz-Bibian, M.A.; Castellanos-Acuña, D.; Gómez-Romero, M.; Lindig-Cisneros, R.; Silva-Farías, M.Á.; Sáenz-Romero, C. Variation among *Abies religiosa* (H.B.K.) Schl. et Cham populations along an altitudinal gradient. I. Seed germination capacity. *Rev. Fitotec. Mex.* **2019**, *42*, 301–308.
29. Guzmán-Aguilar, G.; Carbajal-Navarro, A.; Sáenz-Romero, C.; Herrerías-Diego, Y.; López-Toledo, L.; Blanco-García, A. *Abies religiosa* seedling limitations for passive restoration practices at the Monarch Butterfly Biosphere Reserve in Mexico. *Front. Ecol. Evol.* **2020**, *8*, 1–10. [CrossRef]
30. Gómez-Pineda, E.; Sáenz-Romero, C.; Ortega-Rodríguez, J.M.; Blanco-García, A.; Madrigal-Sánchez, X.; Lindig-Cisneros, R.; López-Toledo, L.; Pedraza-Santos, M.E.; Rehfeldt, G.E. Suitable climatic habitat changes for Mexican conifers along altitudinal gradients under climatic change scenarios. *Ecol. Appl.* **2020**, *30*, e02041. [CrossRef] [PubMed]
31. Lenoir, J.; Gégout, J.-C.; Marquet, P.; De Ruffray, P.; Brisse, H. A significant upward shift in plant species optimum elevation during the 20th century. *Science* **2008**, *320*, 1768–1771. [CrossRef] [PubMed]
32. Jump, A.S.; Mátyás, C.; Peñuelas, J. The altitude-for-latitude disparity in the range retractions of woody species. *Trends Ecol. Evol.* **2009**, *24*, 694–701. [CrossRef]
33. Carbajal-Navarro, A.; Navarro-Miranda, E.; Blanco-García, A.; Cruzado-Vargas, A.L.; Gómez-Pineda, E.; Zamora-Sánchez, C.; Pineda-García, F.; O'Neill, G.; Gómez-Romero, M.; Lindig-Cisneros, R.; et al. Ecological restoration of *Abies religiosa* forests using nurse plants and assisted migration in the Monarch Butterfly Biosphere Reserve, Mexico. *Front. Ecol. Evol.* **2019**, *7*, 421. [CrossRef]
34. Sáenz-Romero, C.; Rehfeldt, G.E.; Crookston, N.L.; Duval, P.; St-Amant, R.; Beaulieu, J.; Richardson, B.A. Spline models of contemporary, 2030, 2060 and 2090 climates for Mexico and their use in understanding climate-change impacts on the vegetation. *Clim. Chang.* **2010**, *102*, 595–623. [CrossRef]
35. Vargas, A.L.C. Variation of quantitative Characters between Procedures of *Abies religiosa* Originated in the Altitudinal Gradient in the Reserve of the Biosphere of the Monarch Butterfly: Stage of Vivero. Master's Thesis, Universidad Autónoma Chapingo, Texcoco de Mora, Estado de México, México, 2017. Available online: http://repositorio.chapingo.edu.mx:8080/bitstream/handle/20.500.12098/425/mccf-cval_17.pdf?sequence=1&isAllowed=y (accessed on 15 August 2020).

36. Sánchez, C.Z. Variación Genética en *Abies religiosa* (Kunth) Schltdl. et Cham. a lo Largo de un Gradiente Altitudinal en Respuesta al Estrés por Sequía. Master's Thesis, Universidad Michoacana de San Nicolás de Hidalgo, Morelia, Michoacán, México, 2020. Available online: http://bibliotecavirtual.dgb.umich.mx:8083/xmlui/bitstream/handle/DGB_UMICH/1912/IIAF-M-2020-0195.pdf?sequence=1&isAllowed=y (accessed on 15 August 2020).
37. Sáenz-Ceja, J.E.; Pérez-Salicrup, D.R. Modification of fire regimes inferred from the age structure of two conifer species in a tropical montane forest, Mexico. *Forests* **2020**, *11*, 1193. [CrossRef]
38. Flores-Martínez, J.J.; Martínez-Pacheco, A.; Rendón-Salinas, E.; Rickards, J.; Sarkar, S.; Sánchez-Cordero, V. Recent forest cover loss in the core zones of the Monarch Butterfly Biosphere Reserve in Mexico. *Front. Environ. Sci.* **2019**, *7*, 167. [CrossRef]
39. Prieto-Ruíz, J.A.; Bustamante-García, V.; Muñoz-Flores, H.J.; Álvarez-Zagoya, R. Análisis de conos y semillas en coníferas. In *Técnicas en el Manejo Sustentable de los Recursos Naturales*; Ocañas, F.G., González, J.A.G., Mendoza, H.V., Parra, A.C., Eds.; Facultad de Ciencias Forestales, Universidad Autónoma de Nuevo León: Monterrey, Mexico, 2014; pp. 27–44.
40. Research on Forest Climate Change: Potential Effects of Global Warming on Forests and Plant Climate Relationships in Western North America and Mexico. Available online: http://charcoal.cnre.vt.edu/climate/ (accessed on 1 November 2020).
41. Arriola, P.; Flores, G.; Gijón, H.; Pineda, O.; Jacob, C.; de Pascual, P.N. *Producción de Planta de Abies Religiosa (Kunth) Schltdl. & Cham. en Vivero*; Folleto Técnico Núm 19; CENID-COMEF, INIFAP: Ciudad de México, Mexico, 2015.
42. Martinez-Luna, J.E.; Carrillo-Anzures, F.; Acosta-Mireles, M.; Sánchez, M.E.R.; Perez-Miranda, R. Allometric equations to estimate carbon in seedlings of *Pinus hartwegii* Lindl. *Rev. Mex. Cienc. For.* **2020**, *11*, 11. [CrossRef]
43. SAS Institute. *SAS/STAT (Version 9.4) Computer Software*; SAS Institute Inc.: Cary, NC, USA, 2014.
44. SAS Institute. *Base SAS 9.4 Procedures Guide*, 5th ed.; SAS Institute Inc.: Cary, NC, USA, 2015.
45. Sáenz-Romero, C.; Rehfeldt, G.E.; Ortega-Rodríguez, J.M.; Marín-Togo, M.C.; Madrigal-Sánchez, X. *Pinus leiophylla* suitable habitat for 1961–1990 and future climate. *Bot. Sci.* **2015**, *93*, 709–718. [CrossRef]
46. Sáenz-Romero, C.; Lamy, J.B.; Ducousso, A.; Musch, B.; Ehrenmann, F.; Delzon, S.; Cavers, S.; Chałupka, W.; Dağdaş, S.; Hansen, J.K.; et al. Adaptive and plastic responses of *Quercus petraea* populations to climate across Europe. *Glob. Chang. Biol.* **2017**, *23*, 2831–2847. [CrossRef]
47. Mátyás, C. Modeling climate change effects with provenance test data. *Tree Physiol.* **1994**, *14*, 797–804. [CrossRef]
48. Leites, L.P.; Robinson, A.P.; Rehfeldt, G.E.; Marshall, J.D.; Crookston, N.L. Height-growth response to climatic changes differs among populations of Douglas-fir: A novel analysis of historic data. *Ecol. Appl.* **2012**, *22*, 154–165. [CrossRef]
49. Miranda, R.P.; Sánchez, F.M.; Hernández, A.G.; Padilla, V.J.A. Distribution of *Abies religiosa* (Kunth) Schltdl. et. Cham. and *Pinus montezumae* Lamb. in the face of climate change. *Rev. Mex. Cienc. For.* **2014**, *5*, 18–33.
50. Islas-Báez, A.; Pérez-Miranda, R.; González-Hernández, A.; Romero-Sánchez, M.E.; Velasco-Bautista, E. Risk habitat of the monarch butterfly (*Danaus plexippus*) by climate change scenarios. *Ra Ximhai* **2015**, *11*, 49–59.
51. Sánchez-Velásquez, L.R.; Pineda-López, M.R.; Ibarra-Zavaleta, S.P.; López-Serrano, Y. Fir forest demography using matrix projections: Anomaly precipitation due to climatic change decrease population viability. *For. Ecol. Manag.* **2021**, *482*, 118845. [CrossRef]

forests

MDPI

Article

Seed Sourcing Strategies Considering Climate Change Forecasts: A Practical Test in Scots Pine

Eduardo Notivol [1,2,*,†], Luis Santos-del-Blanco [3,4,†], Regina Chambel [3], Jose Climent [3,4] and Ricardo Alía [3,4,*]

1 Department of Forest Resources, Agrifood Research and Technology Centre of Aragon (CITA), Avda. Montañana 930, 50059 Zaragoza, Spain
2 ETSI de Montes, Forestal y del Medio Natural de la Universidad Politécnica de Madrid (UPM), 28040 Madrid, Spain
3 INIA-CIFOR, Department of Ecology and Forest Genetics, Ctra. Coruña km 7.5, 28040 Madrid, Spain; luisaemf@gmail.com (L.S.-d.-B.); chambel.regina@inia.es (R.C.); climent@inia.es (J.C.)
4 Sustainable Forest Management Research Institute, University of Valladolid, Av. Madrid s/n, 34004 Palencia, Spain
* Correspondence: enotivol@cita-aragon.es (E.N.); alia@inia.es (R.A.)
† Those authors contributed equally.

Received: 19 October 2020; Accepted: 17 November 2020; Published: 20 November 2020

Abstract: *Research Highlights:* We experimentally tested different seed sourcing strategies (local, predictive, climate-predictive, climate-adjusted, composite and admixture) under a climate change high emissions scenario using a Scots pine multi-site provenance test. *Background and Objectives:* There is an urgent need to conserve genetic resources and to support resilience of conifer species facing expected changes and threats. Seed sourcing strategies have been proposed to maximize the future adaptation and resilience of our forests. However, these proposals are yet to be tested, especially in long-lived organisms as forest trees, due to methodological constraints. In addition, some methods rely on the transfer of material from populations matching the future conditions of the sites. However, at the rear edge of the species, some specific problems (high fragmentation, high genetic differentiation, role of genetic drift) challenge the theoretical expectations of some of these methods. *Materials and Methods:* We used a Scots pine multi-site provenance test, consisting of seventeen provenances covering the distribution range of the species in Spain tested in five representative sites. We measured height, diameter and survival at 5, 10 and 15 years after planting. We simulated populations of 50 trees by bootstrapping material of the provenance test after removing the intra-site environmental effects, simulating different seed sourcing strategies. *Results:* We found that local and predictive methods behaved better than methods based on the selection of future climate-matching strategies (predictive-climate and climate-adjusted) and those combining several seed sources (composite and admixture seed sourcing strategies). *Conclusions:* Despite the theoretical expectations, for Scots pine, a forest tree species at its rear edge of its distribution, seed-sourcing methods based on climate matching or a combination of seed sources do not perform better than traditional local or predictive methods or they are not feasible because of the lack of future climate-matching populations.

Keywords: genetic variation; seed sourcing; forest management; genetic conservation; *Pinus sylvestris*

1. Introduction

Forests are essential to the provision of ecosystem services, products and environmental assets of a country or region, providing important incomes when globally accounting for all indicators [1]. As a result, forests, and particularly conifer forests, have been widely exploited and transformed by different human activities, namely forestry, e.g., cultivation, management [2], to increase some of these income sources. Furthermore, forests have been affected by land-use changes, e.g., agriculture, land development, or different natural hazards, e.g., forest fires, disturbances related to climate change, resulting in an enhanced vulnerability of many populations or even species. Therefore, the need for conservation and sustainable use of forest genetic resources has been increasingly claimed over the last decades [3].

There is an urgent need to conserve genetic resources and to support the resilience of conifer species facing expected changes and threats. Past changes could have already affected the adaptability of the populations to the ongoing rapid climatic changes [4], but also could have resulted in adaptive lags of some European forest tree species [5].

Reforestation or restoration are major forestry activities, where seeds or plants are used to establish a future forest. For the last decades, there has been an increasing concern to improve resilience of future forests. A key tool to achieve this aim is the use of genetically-variable material, as this is related to future adaptation and provision of goods and services [6]. As a consequence, great efforts are being done to define and take into account genetic features of the material deployed in forestry.

Selecting the most productive seed source for a given area has been a traditional approach by forest managers. At present, information on genetic variability and the predictions provided by climate change science enable different seed sourcing strategies to maximize the future adaptation and resilience of our forests [7–11]. Among the possible strategies, local, predicted, composite, admixture, and climate-adjusted seed sourcing have been widely promoted (see Materials and Methods for a detailed description).

These different strategies are difficult to test and compare against each other in forest conifers. Their long lifespan and wide distribution ranges impose some limitations in field tests in regards to obtaining seedlots based on the different strategies, establishing the material in common garden tests, and recording the results after a significant testing period. However, multi-site provenance tests have demonstrated their value to address many evolutionary and ecological questions related to adaptation [12,13], as different populations can be studied under common environmental conditions, and therefore taking advantage to simulate different seed sourcing strategies based on the results at a given time after planting.

Here, we used Scots pine (*Pinus sylvestris* L.) provenance tests located in Spain as a study case. We departed from the hypothesis that provenance tests constitute a representative sample to analyze the performance of different populations of a given species. Then, we used a set of Scots pine provenance tests covering the natural range of the species in Spain to evaluate different seed sourcing strategies under climate change scenarios.

Scots pine in Spain is at the southern edge of its distribution [14], and this in turn has implications for the fragmentation of the populations and genetic differentiation among populations [15–17], but also for the genetic variation in adaptive traits [17,18]. The species has a low degree of domestication and remarkable phenotypic and genetic differences in important traits among populations, despite high levels of gene flow [19]. Scots pine is a very important species in afforestation and restoration programs in Spain. It is the third species in Spain in terms of planted area, with 20,000 kg/seed/year and 1218 ha/year and with 17 regions of provenance for marketing of reproductive material [20]. Previous studies have reported genetic variation for traits related to growth, survival and branching among the Spanish provenances and a significant genotype–environment interaction for height and diameter [18]. Therefore, the study area provides an excellent case study to address how the species genetic variation in growth traits affects the outcome of seed sourcing strategies. In addition, it can potentially inform us of the implications for genetic conservation at the distribution limit of a widely distributed species.

The aims of this work are: (i) to analyse the levels of variation of the species for height, diameter and survival, (ii) to analyse different seed sourcing strategies (local, predicted composite, admixture and climate-adjusted) under different environments and climate change scenarios, and (iii) to propose measures for the conservation and sustainable use of the populations in Spain. Firstly, we analyzed the gain of the different strategies in traits related to adaptation (height growth and their annual components, and survival) in five different provenance tests. Secondly, we integrated this information and the existing knowledge regarding Scots pine, to analyse the strategies for the conservation of its genetic resources. We believe this case study would be useful for other widely distributed species in their distribution edge.

The main aim of this work is to experimentally test different seed sourcing strategies under the climate change high emission scenario RCP 8.5 (business as usual) in order to increase adaptation and resilience of future forests. Our conclusion is that, for Scots pine, a forest tree species in the rear-edge of distribution, the proposed methods do not provide better results than the ones traditionally used in forestry (local and predictive methods). We show that depending of the structure of variation of the species and the future climatic predictions, the theoretical expectations of the climate-adjusted, composite and admixture methods would not be met.

2. Materials and Methods

2.1. Plant Material

We used information from a multi-site provenance test (Table 1), with populations covering the distribution range in Spain as well as populations from Germany (see [18] for details on the setting up and early results). In this study we use the populations from Spain only (Figure 1, and Online Supplementary Material Table S1). Each provenance test included a population that was considered as "local".

Table 1. Location and two climatic indices of the sites for different periods (see Appendix A for details).

				Period							
				1974–1989		1990–2005		2025–2050		2075–2098	
Site	ALT [1]	LAT [2]	LONG [3]	MAT [4]	PDQ [5]	MAT	PDQ	MAT	PDQ	MAT	PDQ
ARA	1370	42°44′ N	00°37′ W	7.5	135.1	8.1	140.5	9.6	115.5	13.6	57.9
BAZ	1850	37°21′ N	02°56′ W	9.3	50.3	9.7	36.9	11.3	14.5	15.1	5.9
CUR	1150	42°46′ N	06°21′ W	9.7	87.6	10.1	88.4	11.4	57.1	15.2	28.3
GUD	1700	40°27′ N	00°35′ W	7.4	77.3	7.9	77.9	9.7	38.5	13.7	17.2
NAV	1600	41°02′ N	03°49′ W	8.1	66.5	8.6	63.6	10.2	32.0	14.3	16.5

[1] ALT: Altitude (m), [2] LAT: Latitude (degrees), [3] LONG: Longitude (degrees), [4] MAT: Mean Annual Temperature (°C), [5] PDQ: Precipitation Driest Quarter (mm).

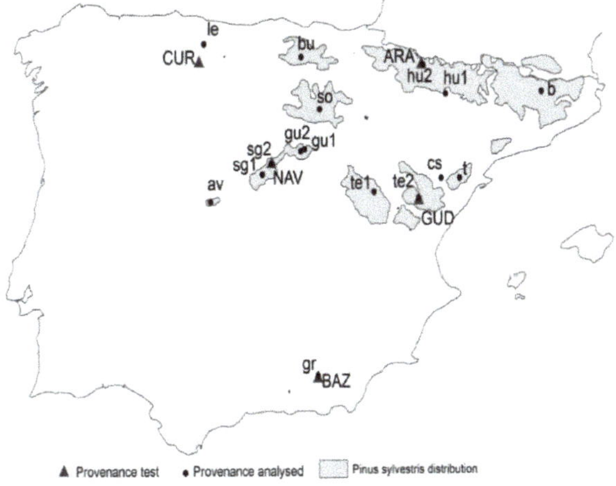

Figure 1. Location of population and sites used in this study.

Some seed sourcing strategies rely on genetic similarity indices as a criterion for the selection of provenances. A seedlot (25 seeds) of each population (except for 151-Castell de Cabrés) was genotyped using 7 CpSSR (see [16] for details). CpSSR, chloroplast simple sequence repeat markers, have a paternal inheritance in conifers, and are suitable for the estimation of pollen gene flow among populations [21]. We computed the pairwise differentiations (PhiST), obtained by AMOVA using Genalex 6.5 software [22].

2.2. Climatic Scenarios

For each provenance and site, we obtained the average value of five climatic indices commonly used and defined in the WorldClim database [23] for the periods 1974–1989 (representing the conditions when the seed was collected), 1990–2005 (corresponding to the available trial measurements) and for the period 2025–2050 and 2075–2098 using the high emissions scenario representative concentration pathway (RCP8.5). These five indices were: MAT (bio1, Mean annual Temperature), MDR (bio2: Mean Diurnal Range, i.e., Mean of monthly (max temp-min temp)), TS (bio4: Temperature Seasonality (standard deviation ×100)), MTWQ (bio10: Mean Temperature of Warmest Quarter) and PDQ (bio17: Precipitation of Driest Quarter) (See Appendix A for details).

Climatic data of sites and provenances were extracted from the B4EST-DT geo-web service [https://ibbr.cnr.it/b4est/] according to their corresponding coordinates. This tool provides data by downscaling algorithms of the climatic indices on a 5 km grid using the UKCP18 (United Kingdom Climatic Projections) datasets (1900–2098) [24]. The dataset was firstly generated by combining the CRU-TS (Climatic Research Unit gridded Time Series) [25].

2.3. Traits Analyzed and Common Garden Analysis

We used information from total height (*HT*), diameter (*DBH*, at 1.3 m) and survival (*SUR*) measurements of each tree assessed at age 5, 10 and 15 years after planting (two years more after seeding).

For each trait and site we used the remlf90 procedure of BreedR [26] in order to remove fine-scale spatial variation and generate phenotypic data devoid of that source of variation. We used the spline correction option, and whenever this was not significant (ARA, BAZ and CUR), a block correction was performed.

Then, we performed a combined analysis of sites using spatially-corrected individual measurements [18] according to the model:

$$Y_{ijkl} = \mu + S_i \times P_j + B(S)_k + U(S/B)_l + e_{ijkl}$$

where Y is the vector containing the variable values, μ is the general mean, S is the study site, P is the provenance, B is the block within the site, U is the experimental unit within the block and site, which consists of four contiguous trees, and e is the random error term. All variables were coded as random. Significance of random factors was tested by means of Likelihood Ratio Tests (LRTs). We also obtained age–age correlations and trait correlations at the individual tree level after corrections.

2.4. Seed Sourcing Strategies

We considered six different seed-sourcing strategies for each site (see Figure 2):

S1. *Local seed sourcing*: We identified the local provenance corresponding to each trial site: ARA (hu2), BAZ (gr), CUR (le), GUD (cu) and NAV (sg2).

S2. *Predictive seed sourcing*. Proposes the use of naturally occurring genotypes that are experimentally determined to be adapted to projected conditions. We selected the best performing provenance at each site, provided that this was not the local one. Whenever this was the case, scenarios S1 and S2 concurred.

S3. *Climate-predictive seed sourcing*: We selected seed sources from areas where current climate matches that of the testing sites in future scenarios (2050 predictions) [29].

S4. *Climate-adjusted seed sourcing*. We combined trees from provenances sourced from across the predicted gradient of climate change, including genetic material that is likely to contribute to future climate resilience. The contribution of different provenances was not even but biased toward the direction of predicted climatic change [27], i.e., provenances sourced from places with predicted climates closer to future environmental conditions than the local provenance itself contributed to the mix, but the contribution was proportional to future climatic suitability. Smoothed fits were conducted for the period 1974–2098 for each of the five climatic indices in all the testing sites and provenances to define the expected climatic gradients and the suitability of the populations to the future climatic conditions. Then, each provenance was classified as suitable or not for each climatic index in the future (Table A3). Provenances with three or more matches were retained and their final contribution was proportional to the times that each provenance was classified as suitable.

S5. *Composite seed sourcing*: We combined trees from different provenances in order to increase genetic diversity and to reflect gene flow among populations. This method aims to mimic natural patterns of gene flow by mixing seeds from multiple provenances, but with a contribution inversely proportional to the genetic pair-wise distance between each provenance and the local provenance, i.e., progressively fewer trees as the distance of the collection site from the planting site increases, but ecogeographically matching the sources [30]. Pairwise population differentiation (Online Supplementary Material Table S2) was used as a proxy of gene flow.

S6. *Admixture seed sourcing*. We combined trees from all available provenances to increase genetic diversity regardless of gene flow. We constructed a simulated population with equal number of individuals from each provenance. Similar to composite seed sourcing, this strategy combines provenances, but simply aims to sample a wide variety of genetic sources, without considering the location of the source population relative to the planting site [11].

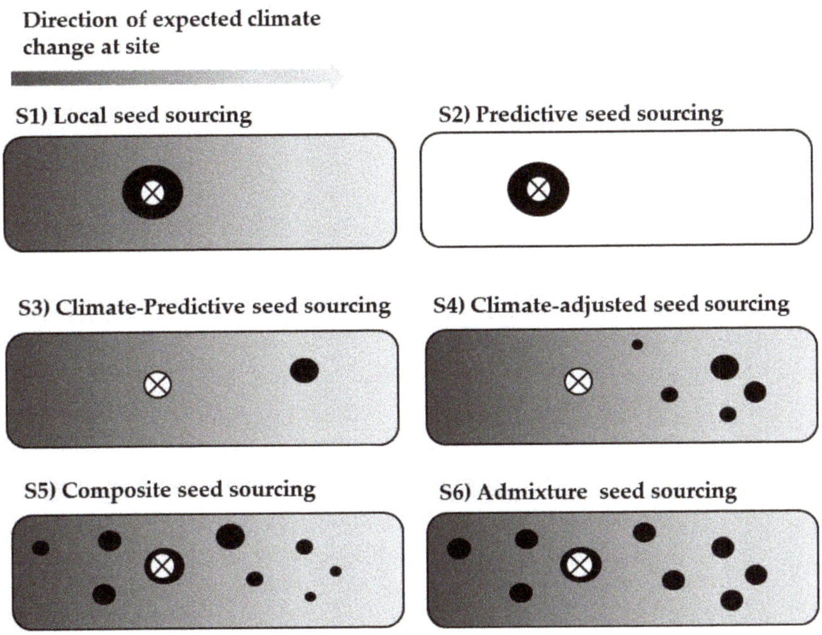

Figure 2. Seed sourcing strategies used in this study (modified from [27,28]). The cross indicates the site (provenance test) and the circles indicate the populations used as seed sources. The size of the circle indicates the contribution of each population to the seed sourcing strategy.

For each strategy (S1–S6), we created 1000 synthetic populations of $n = 50$ trees, according to the restrictions of the strategy. Individual tree values corrected by environmental effects, were bootstrapped (with replacement) among all the suitable individuals of the site we were evaluating. The procedure was carried out with an ad hoc script in R. For each synthetic population, we computed the mean and variance for each of the traits (HT, DBH and SUR) at age 5 (height only), 10 and 15. We also estimated the 5% and 95% percentiles and the percentile of values of each scenario that were greater than the mean of scenario S1. This was done for all possible combinations of trial site (5), phenotypic variable (6) and scenario (5 or 6), with a total of 172 combinations. The comparison among scenarios was studied according to the distribution of the values defined by the percentiles.

2.5. Data Accessibility

Raw data belongs to the GENFORED database (www.genfored.es) and they are accessible at zenodo repository (https://doi.org/10.5281/zenodo.4280004).

3. Results

3.1. Combined Analysis and Genetic Variation of Scots Pine Populations

We found strong poristive correlations (significant at $p < 0.001$) among variables, and among years (Table 2), both at the individual level and plot levels.

Table 2. Correlation among traits at the plot (above diagonal) and the individual (below diagonal) level.

	HT5	HT10	HT15	DBH10	DBH15	SUR15
HT5	-	0.739	0.640	0.692	0.651	0.453
HT10	0.743	-	0.912	0.849	0.866	0.659
HT15	0.577	0.927	-	0.772	0.905	0.698
DBH10	0.664	0.946	0.938	-	0.832	0.350
DBH15	0.540	0.882	0.923	0.845	-	0.549

The combined analysis of the five sites and for all the variables considered indicates a significant variation among populations and sites for height and diameter (Table 3), but not for survival where the population term was not significant ($p = 0.50$). For survival there were significant differences among populations in the BAZ site, but only for the first evaluation at 5 years. Site was the most important factor influencing the performance of different provenances, with bold site-to-site differences. ARA provided a better performance than the other ones, and GUD had a higher mortality and a reduced height and diameter for all provenances. Site by provenance interaction was significant for height and diameter at all ages. This can be interpreted as differences in phenotypic plasticity of the provenances. As the values at the different ages were consistently similar (see Annex 2), here we only included the results for the last measurement at age 15 after planting.

The sites differed greatly in survival after planting (Table 4) indicating a different proportion of selected trees among sites (from 0.445 in GUD at age 15 to 0.906 in BAZ). However, this selected proportion is similar for the different populations tested.

3.2. Seed Sourcing Strategies

According to the predictions by 2050, the expected future conditions will increase the mean annual temperature, the mean diurnal range, the temperature seasonality, the mean temperature of the warmest quarter, and will decrease the precipitation of the driest quarter (See Annex 1 for details). In general, by 2050, some of the populations will still be suitable for the future conditions of the different sites (Tables 5 and A3) considering at least three of the climatic indices, but not by 2100. It is interesting to notice that at BAZ, the southernmost population, there will not be any suitable provenance. On the contrary, at ARA and CUR, there will still be different populations suitable under these conditions.

Provenance "gr" from the Southern limit of distribution would be suitable for all the sites except its own (the local one). This included the GUD testing site, where provenance "gr" was not included in the trial.

Table 3. Variance explained per variable and corresponding percentage relative to total variance, HT, height measured in cm at 15 years; DBH15, diameter at breast height measured in cm at 15 years, SUR15, survival at 15 years. Significant factors ($p < 0.05$) are indicated in in bold. Residual variance for survival fitted with a binomial model with logistic link function is fixed to $\pi^2/3$.

	HT15		DBH15		SUR15	
Factor	**Variance**	**%**	**Variance**	**%**	**Variance**	**%**
Site	**22,356.5**	60.5%	**1409.13**	63.7%	**0.811**	16.2%
Provenance	**539.4**	1.5%	**19.66**	0.9%	0.009	0.2%
Site x provenance	**834.9**	2.3%	**44.63**	2.0%	0.004	0.1%
Block	**714.3**	1.9%	**31.24**	1.4%	**0.128**	2.5%
Experimental unit	**1854.5**	5.0%	116.41	5.3%	**0.779**	15.5%
Residual	10,631.8	28.8%	591.63	26.7%	3.290	65.5%

Table 4. Survival rates at five Scots pine trial sites in Spain along 15 years after planting.

Site	*N*	Age 5	Age 10	Age15
ARA	960	0.92	0.89	0.88
BAZ	896	0.98	0.91	0.91
CUR	1024	0.97	0.80	0.68
GUD	640	0.60	0.48	0.45
NAV	960	0.91	0.89	0.88
Total	4480	0.90	0.81	0.78

Table 5. Provenances within the range of future conditions up to 2050 of each experimental site for at least three of the climatic indices (in brackets, number of indices matching future conditions. See Appendix A for details and Table A3).

Site	Provenances Where Current Climate Matches Future Scenarios
ARA	hu1 (4), b (4), te1 (3), te2 (3), gu1 (3), gu2 (3), cs (5), t (3), gr (4)
BAZ	-
CUR	hu1 (3), b (3), gu2 (3), av (3), cs (4), t (3), gr (4)
GUD	-
NAV	cs (3), gr (3)

Threrefore, no populations can be considered in the scenarios climate predicted and climate-adjusted in the BAZ site nor in the GUD site. In this last case because the provenance from gr was not tested.

The overall analysis of the means and variances of bootstrapped synthetic populations indicated that seed sourcing methods differed for the different traits analysed ($p < 0.001$). Considering the five trial sites, predictive and local seed sourcing was superior to the other methods for height and diameter mean (Table 6). Composite and admixture seed sourcing did not behave in a superior fashion for the mean of any variable, but did have a greater variance for height. Survival was not considered for the analysis as there were no significant differences among populations, and therefore we assumed that all the populations behaved similarly, even under constant site conditions.

Table 6. Comparison of different methods for height and diameter at age 15.

	HT15				DBH15			
	Mean	**Group**	**Variance**	**Group**	**Mean**	**Group**	**Variance**	**Group**
Local	339.5	b	12073.84	c	64.33	b	635.55	d
Predictive	358.0	a	11948.67	c	76.07	a	708.32	a
Climate-predictive	327.1	c	12891.22	a	60.37	c	586.42	e
Climate-adjusted	322.2	d	12286.24	b	59.26	d	698.58	a
Composite	320.1	e	12335.81	b	58.66	e	678.08	b
Admixture	320.7	e	12723.13	a	58.88	e	658.29	c

However this general pattern varied somewhat depending on the trial site, so that the optimal sampling strategy could be better defined by taking into account site effects (Online Supplementary Material Tables S3 and S4, and Figure 3 for HT15). There were two sites (ARA and NAV) where local seed sourcing performed better in terms of height and diameter. At the CUR testing site, local seed sourcing had a very similar performance to predictive seed sourcing. Then at two other testing sites (BAZ and GUD), local seed sourcing was outperformed by the predictive strategy, and behaved similarly to the other tested strategies (Figure 3). Climate adjusted seed sourcing behaved better than the other two methods based on the combiantion of seed sources (composite and admixture), having a greater mean height and similar variance in the estimations.

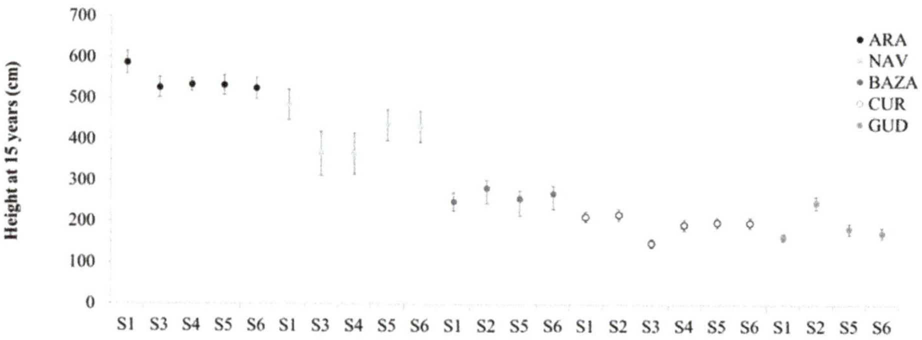

Figure 3. Values for the different scenarios of the Height at 15 years in the different sites. [Code for the seed-sourcing scenarios, S1: Local, S2: Predictive, S3: Predictive-Climate, S4: Climate-Adjusted, S5: Composite, S6: Admixture].

4. Discussion

This paper analyses for the first time the outcome of different seed sourcing strategies under climate change scenarios in Scots pine by using information on provenance tests established in multi-site common environments. The age of evaluation (15 years after planting) is an adequate period for the material to acclimatise to the local conditions of each experimental site. The results also indicate the stability of the estimation across years, and the high correlation among the traits considered. These traits are important to estimate the production and future adaptation of the species, and therefore are essential in the selection of seed sources in afforestation and restoration programs.

The interpretation of the results should consider different aspects of the study, related to the type of adaptation/acclimation evaluated, namely phenotypic plasticity. In addition, the evaluation period and its relationship with the expected climatic changes in the future, the area of study and the influence of the sample sizes should be considered in the estimation of seed sourcing strategies.

Phenotypic plasticity is a process mainly related to the future response of the species to different climatic conditions when the plants are already established in common garden experiments. The results indicate the existence of a significant site effect for all the traits considered, and a significant provenance and site x provenance interaction for height and diameter, but not for survival. This is in accordance with previous results reported with the same material at age 5 [18]. These results confirm the importance of phenotypic plasticity in growth traits. That is, the acclimation to site conditions at the species level as a whole, but also the differences in phenotypic plasticity at the provenance level [31].

A second process of adaptation related to the survival of better adapted trees is of minor importance in this study. We need to take into consideration that mortality, like in most provenance tests, was artificially reduced by planting nursery-cultivated healthy seedlings, and therefore blurring any selection effect at very early stages of development (1–2 years). We have reduced the mortality artificially in this stage, because as in most of the provenance tests, the focus was placed on the evaluation of established seedlings during the duration of the experiment. Firstly, the selected proportion of trees, indicated by the survival in the different sites, differed among sites, with values close to 0.44 in one of the sites (GUD), and close to 0.90 in the other sites. Therefore the selection intensity was too low—compared to that observed at the seedling stage in some conifers [32] and also in artificial selection programs of the species [33]—to expect significant future adaptation. Secondly, natural selection will operate within populations, where standing genetic variation (intra-population genetic variation for the strategies based on one population, or a combination of among and within population genetic variation for the strategies based on the mix of different populations) should take into consideration such levels of variation [4,34–36] that are quite important in Scots pine [37]. The effects of this selection would be observed in a second generation from the material established in the sites. In Scots pine, this period is around 120 years in the study area [38], similar to the furthest climatic scenario predicted (2100). Thirdly, because there are no differences among provenances in survival (there was no bias in provenance selection due to mortality), seed sourcing strategies will not depend on these differences for the future adaptation of this species at its southern range.

While the time scale of this work comparing strategies of seed sourcing was set at 2050, it is noteworthy to warn that no provenance will be suitable for the next century if the climate change follows the high emission scenario. This situation has severe implications and negative consequences for the conservation of genetic diversity, not only for Scots pine but also for a wider set of biodiversity when losing a key component of the forest habitat.

Scots pine has a different pattern of variation at its northern edge of distribution, where a clinal pattern has been detected for neutral and adaptive traits, compared to the rear edge of distribution in Spain, where the populations are highly fragmented and genetic drift may have played an important role in the distribution of neutral and adaptive variation [16,17,39]. These contrasting patterns influence seed transfer recommendations. In northern countries like Sweden and Finland, a movement of seeds from southern to northern areas has been suggested based on the future performance [40,41]. In Spain, however, this strategy is not recommended as local provenances are only showing the best performance in the areas where the conditions for local adaptation [42] are met. These are ARA and NAV sites where large population sizes and extensive gene flow is detected [21] in contrast to other marginal populations where genetic drift could have played a major role [43,44]. Therefore, local seed sourcing methods are superior to the other methods in only two of the sites.

Predictive seed sourcing is based on the selection of the best performing provenance (whenever it is not local), and therefore it is based on the results already available. This is the method most used based on provenance tests. By definition, this method will always provide the best results, sometimes paired with other single-source strategies like local or climate-predictive. However, there are two major drawbacks that limit its practical application. First, as it is not known *a priori* which provenance will be the best at each particular afforestation or restoration site, provenance tests with a comprehensive set of provenances need to be extensively replicated. Second, this is a *post hoc* method, for which phenotypic data collected several years after planting are needed.

Climate-based seed sourcing methods behave intermediately compared to the local/predictive seed sourcing methods. The predicted effects of climate change on the species in the Iberian peninsula include an increase in both tree growth and mortality [45]. When considering intra-population variation, data-driven models indicate that Scots pine populations from southern Spain would have higher relative survival in northern territories (that will be warmer) under future climate scenarios than under current conditions, and Scots pine populations from the northern Iberian Range would have the largest decrease in suitable area and the lowest survival over different climatic conditions [46]. The provenance tests reported in this study have been measured in the period 1995–2005, where the climates in the different sites have followed the trend expected in the climate change scenarios (Table 1). However, these changes are still small compared to those predicted at the 2100 period, and therefore, we can expect an underestimation of the results with these two methods (Climate-predictive and Climate-adjusted).

Composite, Admixture and Climate-adjusted seed sourcing have been suggested in the context of rapid climate change [27] but this has not been tested against other methods [28]. This is, to our knowledge, the first attempt to test different seed sourcing strategies using experimental data in a forest tree species. We would expect a higher variance in the traits, as a proxy of the standing genetic variation. Our results suggest that the combination of different seed sources does not provide a clear advantage to the methods based on one sole population. Another aspect not considered in our study is the possibility of a combination of alleles from different populations to produce new combinations for natural selection to occur. However, it is possible that some outbreeding depression might occur as a result of these new combinations [47,48]. Therefore we need to take these results with caution, as the context (early estimation under the climate change scenario, non-significant variation among provenances in survival) can hinder some of the advantages of these methods as proposed by different authors [7–11]. These methods can also have some risks in the conservation of the genetic resources of the species in the long term [49], and therefore we will need to show clear advantages over traditional methods to implement them at an operational scale in conditions like the ones tested in our study.

5. Conclusions

We tested different seed sourcing strategies in Scots pine by using a multi-site provenance test established in Spain. Our results suggest that in forest tree species such as Scots pine, with high levels of intra-population variability, and low inter-population variation, the methods based on the selection of one population (local and predictive seed sourcing methods) provide better performance (mean and variance in height and diameter), compared to the admixture and composite methods based on the combination of different seed sources. In addition, we propose that at the rear edge of distribution of the species, climate-adjusted methods are not suitable due to the absence of populations that follow the predicted climate gradient. This information should be taken into account in forest policies at national and/or EU level by adapting the transfer guidelines of FRM as in the ones provided by EUFORGEN [50]. In Spain, the National Strategy of FGR [51] should take into consideration those results in order to improve the recommendations for future climatic scenarios.

Supplementary Materials: The following are available online at http://www.mdpi.com/1999-4907/11/11/1222/s1, Table S1: Information on the origin of seed samples, the representation at the trial sites, and characteristics of the climate in the four periods considered in the study, Table S2: Pairwise population differentiation. The estimation was made using the Phist estimate implemented in Genalex (Peakall and Smouse, 2012) based in 7 CpSSR data in the populations of Scots pine, Table S3: Bootstrap analysis of the scenarios for the variable HT15. (50-trees samples), Table S4: Bootstrap analysis of the scenarios for the variable DBH15 (50-trees samples).

Author Contributions: R.A. and E.N.: conceived and design the research; R.C. and J.C.: contributed materials; L.S.-d.-B., R.A. and E.N. analyzed the data. All the authors wrote the paper. All authors have read and agreed to the published version of the manuscript.

Funding: This research was supported by the Project AEG 17-048 established in the frame of the measure 15.2 "support to the conservation and use of forest genetic resources" and under Regulation (EU) No 1305/2013 of the European Parliament and of the Council of 17 December 2013 on support for rural development by the European Agricultural Fund for Rural Development (EAFRD) with 75% co-financing. This study was funded by the European Union Horizon 2020 research and innovation programme under grant agreement No 773383 (B4EST project), and the Ministry of Science (RTI2018-094691-B-C32).

Acknowledgments: We are thankful to Maurizio Marchi (CNR-IBBR, Italy) and the B4EST climate downscaling tool for providing climate data. Most data used in this study are integrated in the Spanish Network of Forest Genetic Trials (GENFORED). We thank all persons and institutions involved in establishing and measuring field trials and maintaining the network.

Conflicts of Interest: The authors declare no conflict of interest. The funders had no role in the design of the study; in the collection, analyses, or interpretation of data; in the writing of the manuscript, or in the decision to publish the results.

Appendix A

Climatic Scenarios and Suitability of the Provenances

The B4EST Downscaling Tool (B4EST-DT) [https://ibbr.cnr.it/b4est/] outputs more than 80 climatic variables and indices over the whole time period (1900–2098). The climatic variables comprise raw climatic variables Tmin, Tmax, Prec with monthly resolution and the 19 bioclimatic parameters from the WorldClim database. Data served is produced by downscaling procedures on a 5 km grid using the UKCP18 datasets. For the five selected indices in this work, average values in four periods were obtained for the corresponding coordinates of the provenances analysed and plantation sites of the provenance tests. The four periods are: 1974–1989 representing the conditions when the seed was collected, 1990–2005 corresponding to the available trial measurements and for the period 2025–2050 and 2075–2098 using the high emissions scenario representative concentration pathway (RCP8.5).

Figure A1 and Table A1 show the trend of climate in the periods considered for the locations of the five trials. Provenance data are available in Table A2.

Table A1. Mean of the climatic indices for the different periods considered in the study for the location of the provenance test sites.

	Period			
	1974–1988	**1990–005**	**2025–2050**	**2075–2098**
Site	**bio1 (MAT) [1]**			
ARA	7.55	8.08	9.61	13.61
BAZ	9.35	9.69	11.28	15.13
CUR	9.67	10.16	11.42	15.17
GUD	7.42	7.91	9.66	13.68
NAV	8.15	8.61	10.18	14.33
	bio2 (MDR) [2]			
ARA	9.85	9.81	11.42	13.19
BAZ	12.32	12.56	14.51	16.10
CUR	10.11	10.25	12.35	14.41
GUD	11.92	12.059	14.08	15.81
NAV	10.57	10.85	12.77	14.61
	bio4 (TS) [3]			
ARA	630.39	649.8625	680.38	755.14
BAZ	712.13	732.525	799.72	848.90
CUR	572.41	589.04	579.98	641.05
GUD	654.38	676.74	719.36	776.99
NAV	641.02	658.67	677.21	755.74
	bio10 (MTWQ) [4]			
ARA	15.82	16.54	18.25	23.22
BAZ	19.00	19.59	21.90	26.31
CUR	17.23	17.87	18.80	23.20
GUD	16.20	16.97	19.04	23.82
NAV	16.91	17.54	19.25	24.19

Table A1. *Cont.*

Site	Period 1974–1988	1990–005	2025–2050	2075–2098
	bio17 (PDQ) [5]			
ARA	135.16	140.53	115.53	57.94
BAZ	50.35	36.87	14.48	5.90
CUR	87.62	88.45	57.08	28.32
GUD	77.35	77.94	38.53	17.17
NAV	66.52	63.58	32.062	16.47

[1] biol, MAT: Mean Annual Temperature, [2] bio2, MDR: Mean Diurnal Range, i.e., mean of monthly (max temp − min temp)), [3] bio4, TS: Temperature Seasonality (standard deviation ×100)), [4] bio10, MTWD: Mean Temperature of Warmest Quarter), [5] bio17, PDQ: Precipitation of Driest Quarter.

Table A2. Mean of the Provenance's climatic indices for the different period considered in the study.

Code	bio1 (MAT) [1]		bio2 (MDR) [2]		bio4 (TS) [3]	
	1974–1988	1990–2005	1974–1988	1990–2005	1974–1988	1990–2005
Le	6.95	7.43	9.59	9.73	555.66	571.01
Bu	8.9	9.4	10.35	10.56	549.19	566.78
Hu1	11.91	12.45	10.54	10.48	662.38	683.51
Hu2	6.75	7.28	9.69	9.64	631.6	650.66
B	9.68	10.16	11.91	12.12	651.59	672.24
So	7.76	8.26	11.41	11.67	624.99	644
Gu1	9.19	9.7	11.27	11.56	648.29	668.82
Gu2	9.17	9.67	11.13	11.41	649.4	669.87
Sg1	8.53	8.95	10.34	10.6	640.99	657.41
Sg2	8.16	8.61	10.57	10.85	642.04	659.67
Av	8.51	8.89	10.88	11.05	651.16	665.38
Te1	7.7	8.16	12.89	13.14	654.57	675.45
Te2	7.43	7.92	11.92	12.06	654.4	676.7
Cs	10.08	10.6	11.18	11.27	648.45	669.28
T	11.51	12.04	10.12	10.14	628.75	650.1
Gr	8.32	8.66	12.44	12.68	718.01	738.23
	bio10 (MTWQ) [4]		bio17 (PDQ) [5]			
Le	14.37	14.97	118.57	122.16		
Bu	16.15	16.78	115.28	113.16		
Hu1	20.52	21.3	81.41	87.51		
Hu2	15.05	15.77	141.39	147.94		
B	18.31	19.02	70.47	71.36		
So	16.15	16.83	92.22	92.4		
Gu1	17.91	18.64	67.8	67.38		
Gu2	17.9	18.64	67.34	66.46		
Sg1	17.3	17.87	61.6	57.67		
Sg2	16.93	17.55	67.08	63.96		
Av	17.29	17.78	52.99	46.81		
Te1	16.49	17.21	73.74	71.41		
Te2	16.21	16.97	77.49	78.14		
Cs	18.69	19.47	59.53	58.18		
T	19.79	20.56	63.48	64.68		
Gr	18.08	18.67	55.07	40.47		

[1] biol, MAT: Mean Annual Temperature, [2] bio2, MDR: Mean Diurnal Range, i.e., mean of monthly (max temp − min temp)), [3] bio4, TS: Temperature Seasonality (standard deviation ×100)), [4] bio10, MTWD: Mean Temperature of Warmest Quarter), [5] bio17, PDQ: Precipitation of Driest Quarter.

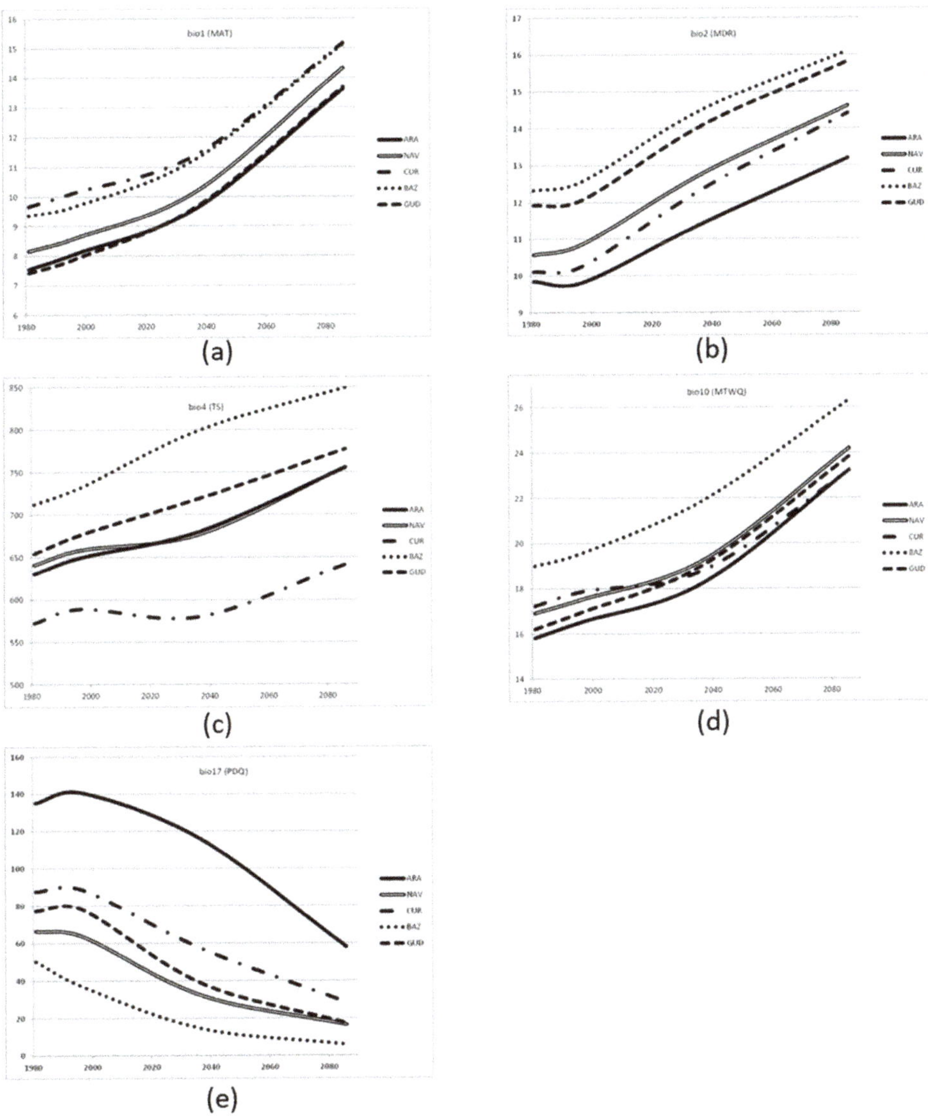

Figure A1. Projection of the climatic indices for the period 1980–2100 using the B4EST-DT geo-web service [https://ibbr.cnr.it/b4est/]. (**a**) bio1, MAT: Mean Annual Temperature, (**b**) bio2, MDR: Mean Diurnal Range, i.e., mean of monthly (max temp-min temp)), (**c**) bio4, TS: Temperature Seasonality (standard deviation ×100)), (**d**) bio10, MTWD: Mean Temperature of Warmest Quarter), (**e**) bio17, PDQ: Precipitation of Driest Quarter.

For *Climate-adjusted seed sourcing* the contribution of the different provenances are based on the suitability of the populations in the light of current conditions for the future climate conditions of the plantation sites. The mix of provenances is proportional to the number of matches (suitability) of the five climatic indices. The suitability is accepted for every index if the current value is better or the same as the value predicted for the year 2050, (i.e., if the bio1 (MAT) of a provenance in 2020 is 10 °C and in a testing site in 2050 the predicted value of this index is 8 °C, in this case the provenance computes one positive match because the Tª adaptation is assured, but if the predicted value were 11 °C, it would not).

The values for these sets of comparisons have been obtained from smoothed fits for the period 1974–2098 (and zoomed for the 2020–2025 time lapse) for each of the five climatic indices in all the testing sites and provenances from the raw data obtained from the B4EST Downscaling Tool (B4EST-DT) [https://ibbr.cnr.it/b4est/]. Figure A2 shows examples for the index bio1 (MAT).

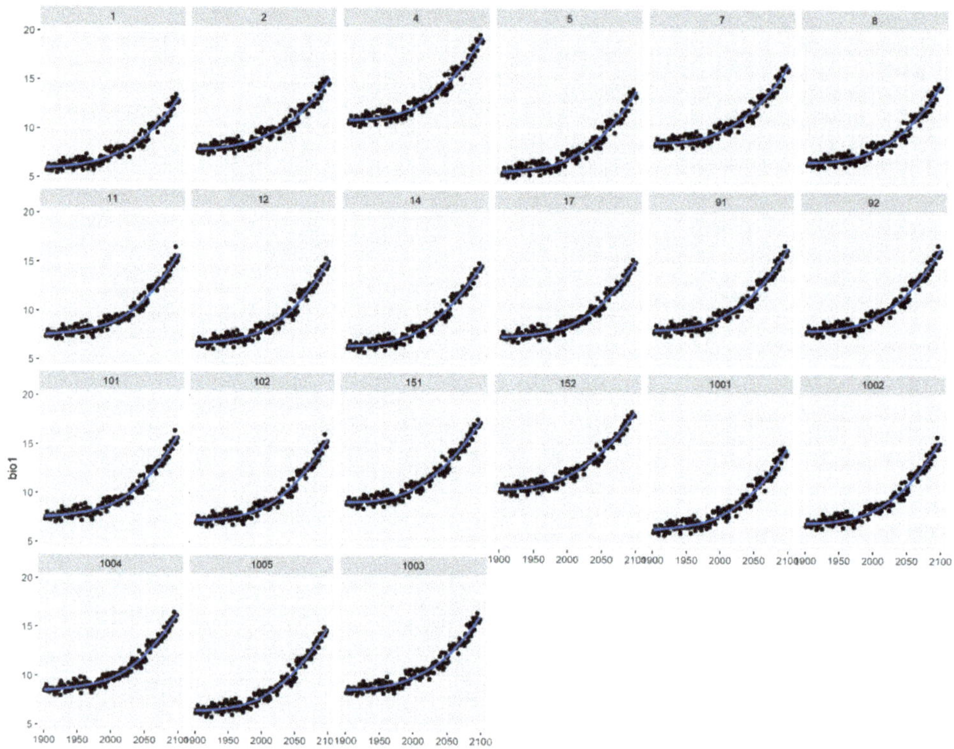

Figure A2. Smooth fits of bio1 (MAT) annual raw values from B4EST-DT for the period 1900–2098 for all provenances and test sites.

Table A3. Provenances matching the future scenario (2050) for each site. Asterisks indicate provenances not included in the trial site, but are included here for easier comparisons among sites.

		Provenance															
Site	**Index**	**Le**	**Bu**	**Hu1**	**Hu2**	**B**	**So**	**Gu1**	**Gu2**	**Sg1**	**Sg2**	**Av**	**Te1**	**Te2**	**Cs**	**T**	**Gr**
ARA	bio1 [1]			1		1									1	1	
	bio2 [2]					1	1	1	1			1	1	1	1		1
	bio4 [3]			1									1	1	1		1
	bio10 [4]			1		1		1	1						1	1	1
	bio17 [5]			1		1	1	1	1	1	1	1	1	1	1	1	1
	sum	0	0	4	0	4	2	3	3	1	1	2	3	3	5	3	4
BAZ	bio1																
	bio2																
	bio4																
	bio10			1													
	bio17																
	sum	0	0	1	0	0 *	0	0	0	0	0	0	0 *	0	0	0	0

Table A3. *Cont.*

Site	Index	Provenance															
		Le	Bu	Hu1	Hu2	B	So	Gu1	Gu2	Sg1	Sg2	Av	Te1	Te2	Cs	T	Gr
CUR	bio1			1											1	1	
	bio2					1	1	1	1			1	1	1	1		1
	bio4			1	1	1	1		1	1	1	1	1	1	1	1	1
	bio10			1		1		1	1						1	1	1
	bio17									1		1					1
	sum	0	0	3	1	3	2	2	3	2	1	3	2	2	4	3	4
GUD	bio1			1		1									1	1	
	bio2																
	bio4																1
	bio10			1											1	1	1
	bio17											1					1
	sum	0	0	2	0 *	1 *	0	0	0	0 *	0 *	1	0 *	0 *	2	2	3 *
NAV	bio1			1		1									1	1	
	bio2												1	1			1
	bio4			1									1	1	1		1
	bio10			1											1	1	1
	bio17																
	sum	0	0	3 *	0	1	0	0	0	0	0	0	2	2	3	2	3

[1] bio1, MAT: Mean annual Temperature, [2] bio2, MDR: Mean Diurnal Range, i.e., Mean of monthly (max temp-min temp)), [3] bio4, TS: Temperature Seasonality (standard deviation ×100)), [4] bio10, MTWD: Mean Temperature of Warmest Quarter), [5] bio17, PDQ: Precipitation of Driest Quarter.

References

1. Campos, P.; Caparrós, A.; Oviedo, J.L.; Ovando, P.; Álvarez-farizo, B.; Díaz-balteiro, L.; Carranza, J.; Beguería, S.; Díaz, M.; Herruzo, A.C.; et al. Bridging the Gap Between National and Ecosystem Accounting Application in Andalusian Forests, Spain. *Ecol. Econ.* **2019**, *157*, 218–236. [CrossRef]
2. Ratnam, W.; Rajora, O.P.; Finkeldey, R.; Aravanopoulos, F.; Bouvet, J.M.; Vaillancourt, R.E.; Kanashiro, M.; Fady, B.; Tomita, M.; Vinson, C. Genetic effects of forest management practices: Global synthesis and perspectives. *For. Ecol. Manag.* **2014**, *333*, 52–65. [CrossRef]
3. Loo, J.; Souvannavong, O.; Dawson, I.K. Seeing the trees as well as the forest: The importance of managing forest genetic resources. *For. Ecol. Manag.* **2014**, *333*, 1–8. [CrossRef]
4. Alberto, F.J.; Aitken, S.N.; Alía, R.; González-Martínez, S.C.; Hänninen, H.; Kremer, A.; Lefèvre, F.; Lenormand, T.; Yeaman, S.; Whetten, R.; et al. Potential for evolutionary responses to climate change—Evidence from tree populations. *Glob. Chang. Biol.* **2013**, *19*, 1645–1661. [CrossRef]
5. Fréjaville, T.; Vizcaíno-Palomar, N.; Fady, B.; Kremer, A.; Benito Garzón, M. Range margin populations show high climate adaptation lags in European trees. *Glob. Chang. Biol.* **2020**, *26*, 484–495. [CrossRef]
6. Alía, R.; Alba, N.; Chambel, M.R.; Barba, D.; Iglesias, S. Genetic quality of forest reproductive materials in Land restoration programmes. In *Innovative Approaches in Forest Restoration*; Vallejo, V.R., Bautista, S., Aaronson, J., Eds.; CEAM: Valencia, Spain, 2009; pp. 89–103.
7. Thomas, E.; Jalonen, R.; Loo, J.; Boshier, D.; Gallo, L.; Cavers, S.; Bordács, S.; Smith, P.; Bozzano, M. Genetic considerations in ecosystem restoration using native tree species. *For. Ecol. Manag.* **2014**, *333*, 66–75. [CrossRef]
8. Havens, K.; Vitt, P.; Still, S.; Kramer, A.T.; Fant, J.B.; Schatz, K. Seed Sourcing for Restoration in an Era of Climate Change. *Nat. Areas J.* **2015**, *35*, 122–133. [CrossRef]
9. Bucharova, A.; Bossdorf, O.; Hölzel, N.; Kollmann, J.; Prasse, R.; Durka, W. Mix and match: Regional admixture provenancing strikes a balance among different seed-sourcing strategies for ecological restoration. *Conserv. Genet.* **2019**, *20*, 7–17. [CrossRef]
10. Kramer, A.T.; Crane, B.; Downing, J.; Hamrick, J.L.; Havens, K.; Highland, A.; Jacobi, S.K.; Kaye, T.N.; Lonsdorf, E.V.; Ramp Neale, J.; et al. Sourcing native plants to support ecosystem function in different planting contexts. *Restor. Ecol.* **2019**, *27*, 470–476. [CrossRef]
11. Breed, M.F.; Stead, M.G.; Ottewell, K.M.; Gardner, M.G.; Lowe, A.J. Which provenance and where? Seed sourcing strategies for revegetation in a changing environment. *Conserv. Genet.* **2013**, *14*, 1–10. [CrossRef]

12. Matyas, C. Climatic adaptation of trees: Rediscovering provenance tests. *Euphytica* **1996**, *92*, 45–54. [CrossRef]
13. Matyas, C. Modelling climate change effects with provenance test data. *Tree Physiol.* **1994**, *14*, 797–804. [CrossRef] [PubMed]
14. Hampe, A.; Petit, R.J. Conserving biodiversity under climate change: The rear edge matters. *Ecol. Lett.* **2005**, *8*, 461–467. [CrossRef] [PubMed]
15. Prus-Glowacki, W.; Stephan, B.R.; Bujas, E.; Alía, R.; Marciniak, A. Genetic differentiation of autochthonous populations of *Pinus sylvestris* (Pinaceae) from the Iberian peninsula. *Plant Syst. Evol.* **2003**, *239*, 55–66. [CrossRef]
16. Soto, A.; Robledo-Arnuncio, J.J.; González-Martínez, S.C.; Smouse, P.E.; Alía, R. Climatic niche and neutral genetic diversity of the six Iberian pine species: A retrospective and prospective view. *Mol. Ecol.* **2010**, *19*, 1396–1409. [CrossRef]
17. Notivol, E.; Garcia-Gil, M.R.; Alía, R.; Savolainen, O. Genetic variation of growth rhythm traits in the limits of a latitudinal cline in Scots pine. *Can. J. For. Res.* **2007**, *37*, 540–551. [CrossRef]
18. Alía, R.; Moro, J.; Notivol, E.; Moro-serrano, J. Genetic variability of Scots pine (*Pinus sylvestris* L.) provenances in Spain: Growth traits and survival. *Silva Fenn.* **2001**, *35*, 27–38. [CrossRef]
19. Savolainen, O.; Pyhäjärvi, T.; Knürr, T. Gene Flow and Local Adaptation in Trees. *Annu. Rev. Ecol. Evol. Syst.* **2007**, *38*, 595–619. [CrossRef]
20. Alía, R.; Garcia del Barrio, J.M.; Iglesias, S.; Mancha, J.A.; de Miguel, J.; Nicolás, J.L.; Pérez, F.; Sánchez de Ron, D. *Regiones de Procedencia de Especies Forestales de España;* O.A. Parques Nacionales: Madrid, Spain, 2009; ISBN 9788480147590.
21. Robledo-Arnuncio, J.J.; Smouse, P.E.; Gil, L.A.; Alía, R. Pollen movement under alternative silvicultural practices in native populations of Scots pine (*Pinus sylvestris* L.) in central Spain. *For. Ecol. Manag.* **2004**, *197*, 245–255. [CrossRef]
22. Peakall, R.; Smouse, P.E. GenALEx 6.5: Genetic analysis in Excel. Population genetic software for teaching and research-an update. *Bioinformatics* **2012**, *28*, 2537–2539. [CrossRef]
23. Fick, S.E.; Hijmans, R.J. WorldClim 2: New 1-km spatial resolution climate surfaces for global land areas. *Int. J. Climatol.* **2017**, *37*, 4302–4315. [CrossRef]
24. Met Office. *UK Climate Projections: Headline Findings;* Met Office: Exeter, UK, 2019; pp. 1–12.
25. Harris, I.; Osborn, T.J.; Jones, P.; Lister, D. Version 4 of the CRU TS monthly high-resolution gridded multivariate climate dataset. *Sci. Data* **2020**, *7*, 1–18. [CrossRef] [PubMed]
26. Muñoz, F.; Sanchez, L. BreedR: Statistical Methods for Forest Genetic Resources Analysts; R-Package software. Available online: https://github.com/famuvie/breedR (accessed on 19 November 2020).
27. Prober, S.M.; Byrne, M.; McLean, E.H.; Steane, D.A.; Potts, B.M.; Vaillancourt, R.E.; Stock, W.D. Climate-adjusted provenancing: A strategy for climate-resilient ecological restoration. *Front. Ecol. Evol.* **2015**, *3*, 1–5. [CrossRef]
28. Byrne, M.; Prober, S.; McLean, L.; Steane, D.; Stock, W.; Potts, B.; Vaillancourt, R. *Adaptation to Climate in Widespread e Ucalypt Species;* Gold Coast: National Climate Change Adaptation Research Facility: Gold Coast, Australia, 2013.
29. Sgrò, C.M.; Lowe, A.J.; Hoffmann, A.A. Building evolutionary resilience for conserving biodiversity under climate change. *Evol. Appl.* **2011**, *4*, 326–337. [CrossRef] [PubMed]
30. Broadhurst, L.M.; Lowe, A.; Coates, D.J.; Cunningham, S.A.; McDonald, M.; Vesk, P.A.; Yates, C. Seed supply for broadscale restoration: Maximizing evolutionary potential. *Evol. Appl.* **2008**, *1*, 587–597. [CrossRef]
31. Chambel, M.R.; Climent, J.M.; Alía, R.; Valladares, F. Phenotypic plasticity: A useful framework for understanding adaptation in forest species. *For. Syst.* **2005**, *14*, 334–344. [CrossRef]
32. Vizcaíno-Palomar, N.; Revuelta-Eugercios, B.; Zavala, M.A.; Alía, R.; Gonzalez-Martínez, S.C.; González-Martínez, S.C. The Role of Population Origin and Microenvironment in Seedling Emergence and Early Survival in Mediterranean Maritime Pine (*Pinus pinaster* Aiton). *PLoS ONE* **2014**, *9*, e109132. [CrossRef]
33. Mullin, T.J.; Lee, S. *Best Practice for Tree Breeding in Europe;* Mullin, T.J., Lee, S., Eds.; Skogforsk: Uppsala, Sweden, 2013.
34. Aitken, S.N.; Yeaman, S.; Holliday, J.A.; Wang, T.; Curtis-McLane, S. Adaptation, migration or extirpation: Climate change outcomes for tree populations. *Evol. Appl.* **2008**, *1*, 95–111. [CrossRef]
35. Chevin, L.-M.; Lande, R.; Mace, G.M. Adaptation, plasticity, and extinction in a changing environment: Towards a predictive theory. *PLoS Biol.* **2010**, *8*, e1000357. [CrossRef]

36. Chevin, L.-M.; Collins, S.; Lefèvre, F. Phenotypic plasticity and evolutionary demographic responses to climate change: Taking theory out to the field. *Funct. Ecol.* **2013**, *27*, 967–979. [CrossRef]
37. Bastien, C.; Alía, R. What might be useful measures of genetic variability for adaptive tratis within populations of Scots pine? *For. Syst.* **2000**, *9*, 97–110.
38. Rojo-Alboreca, A.; García-Villabrille, J.D.; Corral-Rivas, J.J.; Alía, R.; Montero, G. A new approach to defining rotation ages on the basis of productive and technological aspects. Application to natural Pinus sylvestris L. stands in central Spain. *For. Syst.* **2017**, *26*, 13. [CrossRef]
39. Robledo-Arnuncio, J.J.; Collada, C.; Alía, R.; Gil, L.A. Genetic structure of montane isolates of Pinus sylvestris L. in a Mediterranean refugial area. *J. Biogeogr.* **2005**, *32*, 595–605. [CrossRef]
40. Berlin, M.; Persson, T.; Jansson, G.; Haapanen, M.; Ruotsalainen, S.; Bärring, L.; Gull, B.A. Scots pine transfer effect models for growth and survival in Sweden and Finland. *Silva Fenn.* **2016**, *50*, 1–21. [CrossRef]
41. Berlin, M.; Almqvist, C.; Haapanen, M.; Högberg, K. Common Scots Pine Deployment Recommendations for Sweden and Finland. Skogforsk Report. Uppsala, 2019. Available online: https://www.skogforsk.se/english/news/2019/common-scots-pine-deployment-recommendations-for-sweden-and-finland (accessed on 19 November 2020).
42. Leimu, R.; Fischer, M. A Meta-Analysis of Local Adaptation in Plants. *PLoS ONE* **2008**, *3*, e4010. [CrossRef] [PubMed]
43. Robledo-Arnuncio, J.J.; Alía, R.; Gil, L.A. Increased selfing and correlated paternity in a small population of a predominantly outcrossing conifer, Pinus sylvestris. *Mol. Ecol.* **2004**, *13*, 2567–2577. [CrossRef]
44. Robledo-Arnuncio, J.J.; Gil, L.A. Patterns of pollen dispersal in a small population of Pinus sylvestris L. revealed by total-exclusion paternity analysis. *Heredity* **2005**, *94*, 13–22. [CrossRef]
45. Benito-Garzón, M.; Ruiz-Benito, P.; Zavala, M.A. Interspecific differences in tree growth and mortality responses to environmental drivers determine potential species distributional limits in Iberian forests. *Glob. Ecol. Biogeogr.* **2013**, *22*, 1141–1151. [CrossRef]
46. Benito-Garzón, M.; Alía, R.; Robson, T.M.; Zavala, M.A. Intra-specific variability and plasticity influence potential tree species. *Glob. Ecol. Biogeogr.* **2011**, *20*, 766–788. [CrossRef]
47. Unger, G.M.; Heuertz, M.; Vendramin, G.G.; Robledo-Arnuncio, J.J. Assessing early fitness consequences of exotic gene flow in the wild: A field study with Iberian pine relicts. *Evol. Appl.* **2016**, *9*, 367–380. [CrossRef]
48. Robledo-Arnuncio, J.J.; Unger, G.M. Measuring Viability Selection From Prospective Cohort Mortality Studies: A Case Study in Maritime Pine. *Evol. Appl.* **2019**, *12*, 863–877. [CrossRef] [PubMed]
49. Koskela, J.; Lefèvre, F.; Schueler, S.; Kraigher, H.; Olrik, D.C.; Hubert, J.; Longauer, R.; Bozzano, M.; Yrjana, L.; Alizoti, P.; et al. Translating conservation genetics into management: Pan-European minimum requirements for dynamic conservation units of forest tree genetic diversity. *Biol. Conserv.* **2013**, *157*, 39–49. [CrossRef]
50. Konnert, M.; Fady, B.; Gömöry, D.; A'Hara, S.; Wolter, F.; Ducci, F.; Koskela, J.; Bozzano, M.; Maaten, T.; Kowalczyk, J. *Use and Transfer of Forest Reproductive Material: In Europe in the Context of Climate Change*; European Forest Genetic Resources Programme (EUFORGEN), Bioversity International: Rome, Italy, 2015; ISBN 9789292550318.
51. MIMAM. *Estrategia de Conservación y uso Sostenible de los Recursos Genéticos Forestales*; Dirección General para la Biodiversidad, Ministerio de Medio Ambiente: Madrid, Spain, 2006; ISBN D.L.: BI-2.749-07.

Publisher's Note: MDPI stays neutral with regard to jurisdictional claims in published maps and institutional affiliations.

forests

MDPI

Article

Climate Warming Impacts on Distributions of Scots Pine (*Pinus sylvestris* L.) Seed Zones and Seed Mass across Russia in the 21st Century

Elena I. Parfenova [1], Nina A. Kuzmina [2], Sergey R. Kuzmin [2] and Nadezhda M. Tchebakova [1,*]

[1] Laboratory of Forest Monitoring, Sukachev Institute of Forest FRC KSC SB RAS, 660036 Krasnoyarsk, Russia; lyeti@ksc.krasn.ru

[2] Laboratory of Forest Genetics and Breeding, Sukachev Institute of Forest FRC KSC SB RAS, 660036 Krasnoyarsk, Russia; kuz@ksc.krasn.ru (N.A.K.); skr_7@mail.ru (S.R.K.)

* Correspondence: ncheby@ksc.krasn.ru

Abstract: *Research highlights:* We investigated bioclimatic relationships between Scots pine seed mass and seed zones/climatypes across its range in Russia using extensive published data to predict seed zones and seed mass distributions in a changing climate and to reveal ecological and genetic components in the seed mass variation using our 40-year common garden trial data. *Introduction:* seed productivity issues of the major Siberian conifers in Asian Russia become especially relevant nowadays in order to compensate for significant forest losses due to various disturbances during the 20th and current centuries. Our goals were to construct bioclimatic models that predict the seed mass of major Siberian conifers (Scots pine, one of the major Siberian conifers) in a warming climate during the current century. *Methods:* Multi-year seed mass data were derived from the literature and were collected during field work. Climate data (January and July data and annual precipitation) were derived from published reference books on climate and climatic websites. Our multiple regression bioclimatic models were constructed based on the climatic indices of growing degree days > 5 °C, negative degree days < 0 °C, and annual moisture index, which were calculated from January and July temperatures and annual precipitation for both contemporary and future climates. The future 2080 (2070–2100) January and July temperatures and annual precipitation anomalies were derived from the ensemble of twenty CMIP5 (the Coupled Model Intercomparison Project, Phase 5) global circulation models (GCMs) and two scenarios using a mild RCP (Representative Concentration Pathway) 2.6 scenario and an extreme RCP 8.5 scenario. *Results:* Site climate explained about 70% of the seed mass variation across the Scots pine range. Genetic components explained 30% of the seed mass variation, as per the results from our common garden experiment in south central Siberia. Seed mass varied within 3.5 g (min) and 10.5 g (max) with the mean 6.1 g (n = 1150) across Russia. Our bioclimatic seed mass model predicted that a July temperature elevated by 1 °C increased seed mass by 0.56 g, and a January temperature elevated by 5 °C increased seed mass by 0.43 g. The seed mass would increase from 1 g to 4 g in the moderate RCP 2.6 and the extreme RCP 8.5 climates, respectively. Predicted seed zones with heavier seed would shift northwards in a warming climate. However, the permafrost border would halt this shifting due to slower permafrost thawing; thus, our predicted potential for Scots pine seed zones and seed mass would not be realized in the permafrost zone in a warmed climate. Our common garden experiment in central Siberia showed that trees of northerly origins produced lighter seeds than local trees but heavier ones than the trees at the original site. Trees of southerly origins produced heavier seeds than local trees but lighter seeds than the trees at the original site. *Conclusions:* The findings from this study could serve as blueprints for predicting new landscapes with climatic optima for *Pinus sylvestris* to produce better quality seeds to adjust to a warming climate.

Keywords: scots pine seed mass and seed zones; a provenance trial; bioclimatic models; an ensemble of general circulation models; RCP 2.6 and RCP 8.5 scenarios; Russia

Citation: Parfenova, E.I.; Kuzmina, N.A.; Kuzmin, S.R.; Tchebakova, N.M. Climate Warming Impacts on Distributions of Scots Pine (*Pinus sylvestris* L.) Seed Zones and Seed Mass across Russia in the 21st Century. *Forests* **2021**, *12*, 1097. https://doi.org/10.3390/f12081097

Academic Editor: Adele Muscolo

Received: 13 July 2021
Accepted: 12 August 2021
Published: 16 August 2021

1. Introduction

Forest restoration is a key issue in world forestry. Climate change, numerous wildfires, massive damage by fungi, pathogens, insects, and clear-cutting have resulted in a significant reduction of forest lands, the loss of valuable tree species populations, and a decline in tree species biodiversity and productivity. Forest regeneration and restoration in Russia has become especially urgent since the end of the 19th century and during the 20th century, which has been caused by intensive clear-cutting for enlarged wood requirements. To compensate for forest land losses that are both natural and man-made, restoration became a principal issue in both theoretical and applied forestry research. In this regard, research directed to improve measures for restoring forest land losses, and, in particular, the seed productivity of conifers, which are the main boreal trees in Russia, has become especially relevant nowadays.

Pinus sylvestris L. (Scots pine) is one of the main forest-forming tree species that covers 16.2% of the forest lands of Russia (14% within the former Soviet Union), covering about two thirds of northern Eurasia. The reproductive activity of Scots pine across its geographic range, specifically, at the north, is of the foremost interest of forest scientists. The first studies on pine fruiting were accomplished from the end of the 19th century to the mid-20th century (1890–1955) in Scandinavia (Norway, Sweden, Finland), northern Karelia, and the Cola Peninsula (within Russia). In the second half of the 20th century, studies extended their research in Russia to deepen our knowledge of Scots pine cone and seed dynamics, factors controlling seed mass, and quality. Scots pine history research, distribution geography, genetics, and polymorphism were published in detail in monographs [1–5].

Seed mass is one of the indicators of seed quality. Seed mass is a variable trait that depends on a complex combination of ecological and geographical factors and is a stable tree population feature that exposes an inherited evolutionary-adaptive nature of a species.

Quantitative and qualitative biogeography relationships between seed mass and site environmental variables have been studied around the world. As a rule, the environmental variables that were studied were geographical latitude, longitude, and elevation (as an indirect heat measure); monthly temperature (mean, max, min), growing-degree days, or negative degree days (as a direct heat measure); or many other variables. Moles et al. (2003) [6] discovered that latitude explained 21% of the seed mass of 2706 plant species variation globally, with a significant seed mass decrease along the latitudinal direction, increasing to the Pole Circle. Scots pine experiment results in the Sierra Nevada, Spain, suggested that the relationship between the seed mass of the maternal trees and the relative growth rate of their seedlings was controversial; thus, it was not a direct causal [7]. Liu et al. (2013) [8] studied the seed mass of three major forest-forming tree species within their ranges across Canada: black spruce (*Picea mariana* (Mill.) B.S.P.), white spruce (*Picea glauca* (Moench) Voss), and jack pine (*Pinus banksiana* Lamb), depending on their geography (latitude, longitude, and elevation) and 96 climatic variables. Climatic variables explained 34, 14, and 29% of the seed mass variation in black spruce, white spruce, and jack pine, respectively. Himanen et al. (2016) [9] noted the possible impact of different ripening years on spruce seed mass through a different spruce seed moth activities. Freire et al. (2019) [10] discussed the complex connections between climate and the morphological characteristics of *Pinus pinea* L. and its seed production. The important influence of humidity characteristics on the seed yield of *P. pinea* in its Turkish native habitats was stated [11]. The abiotic influence of nutrient and water stress on Mediterranean pine (*Pinus pinaster* Ait) was studied in pine populations of various origins [12]. The authors explored the effect of seed mass on seedling traits such as germination, survival, growth, and biomass partitioning [13]. Populations from more stressful sites were found to produce smaller seeds.

Cherepnin (1980) [3] conducted intensive research regarding the relationship between seed mass and climate in Russia. He collected data on the seed mass and germination of some 1100 provenances from forest units and specialized forest seed stations over the former USSR. He then related part of the data to regional heat variables (growing degree days above 10 °C) and found that the determination coefficients (e.g., Kazakhstan, the European part of Russia) were as high as 0.8. In the 1970s, common garden trials were started all over the Soviet Union [4,14,15]. Trees tested in some of these provenance trials reached the fruiting age and became seed sources for studying the relationship between seed mass and climatic conditions at these test sites.

This study's goals are 1. to construct a bioclimatic regression model that relates *P. sylvestris* seed mass to habitat climatic variables across Russia to predict this integral indicator of the seed quality in a changing climate by the end of the century; 2. to relate Scots pine seed mass with its seed zones/climatypes across its range at present and future; 3. to reveal how ecological and genetic factors are partitioned to impact seed mass using our common garden trial data.

2. Data and Methods

2.1. Seed Mass Data

There are three databases of seed mass and environmental variables (geography, climate, soil) that were created for the bioclimatic modeling of *P. sylvestris* seed mass across the former Soviet Union. The first database included multi-year seed mass data collated by Cherepnin [3] from regularly collected and measured by many forestry units of the All-Union Forestry Agency. This study chose 166 signature sites with seed mass (mass of 1000 seeds) from [3] and assigned geographic coordinates (Figure 1) and multi-year means of climatic variables for each site from a close weather station within a 30 km radius on plains and calculated climate variables based on lapse rates in mountains. In Russian forestry practices, tree and shrub seed mass is measured based on the instructions of the State Standardization System (GOST in Russian, #13056.4-67). In particular, Scots pine seeds are harvested under the clearcutting of mature trees of 100–120 years old. All of the seeds are mixed, and two portions of 500 seeds are selected, weighted, averaged, and calculated for 1000 seed. This method is statistically significant and is used in various countries: all former Soviet Union countries, Eastern European countries, Turkey, Scandinavia, Mongolia, etc. [16–18].

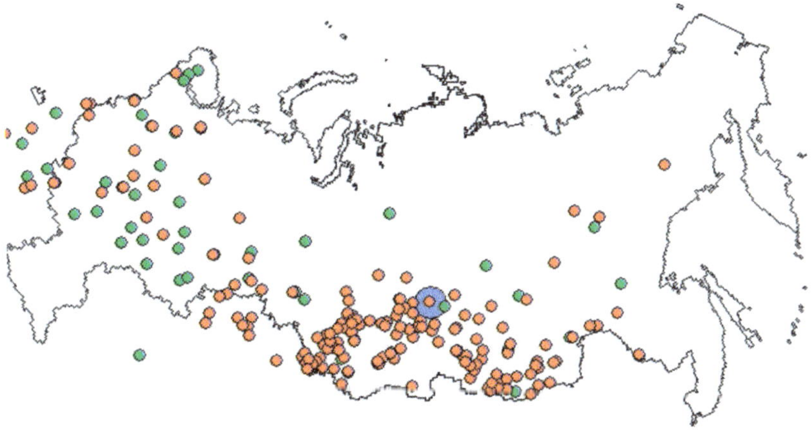

Figure 1. The distribution of 42 *P. sylvestris* provenances across Russia (green points) that were used in our common garden trial in the test site of the Boguchany settlement (blue circle), central Siberia. All 166 provenances (green and red points) were used to construct our bioclimatic seed mass model.

To analyze and identify the partial impacts of site climates and genetic traits on seed mass, the second database was composed based on the seed mass of trees from 82 Scots pine provenances of the former Soviet Union tested in two common garden trials on loamy and sandy soils near the Boguchany settlement, which is located in the southern taiga zone of the near-Angara Central Siberian Plateau, Siberia, in 1976–1977. The seeds were collected in 1974 so that the climatic variables (monthly temperatures of the warmest and coldest months, annual precipitation) of this year and the provenance geography coordinates and multi-year climate means were included in this database. Regular measurements of tree growth in height and diameter, stem volume, needle morphology, cone and seed productivity, and resistance to fungi and pathogens were regularly conducted for about 40 years. Study results were published by Kuzmina and Kuzmin from 1978 to 2020 (see the References section) [19–24].

The third seed mass database was designed from the seed mass measurements of 25–35-year-old trees (the second class in the Russian age classification) of 42 Scots pine provenances from the common garden trial near the Boguchany settlement, central Siberia. The seeds of 5–10 randomly selected trees from various provenances were collected and weighed in 1999, 2000, 2001, 2002, 2003, and 2010 to determine the influence of a Scots pine original provenance (a genetic trait) on seed mass in the introduction site and to determine the influence of the climatic conditions on the seed mass of various provenances in the introduction site (an ecological trait).

2.2. Climate Data

The climatic variables included mean January and July temperatures, and annual precipitation levels were derived from climatic websites (Supplementary 2) and were calculated. The climatic indices included: growing degree days >5 °C, GDD_5; negative degree days < 0 °C, NDD; and annual moisture index, AMI, calculated as the ratio of GDD_5/annual precipitation. The future 2080 (for 2070–2100) climate variables such as January and July temperatures and annual precipitation were calculated as sums of climatic means for the basic period (1960–90) and their mean anomalies for 2070–2100. Anomalies were calculated as anomaly means derived from the ensemble of twenty general circulation models, GCMs, of the CMIP5 from the IPCC Data Distribution Centre (Supplementary 2), and two scenarios using a mild climate (RCP 2.6 scenario) and an extreme climate (RCP 8.5 scenario); anomalies were interpolated on a 0.25° resolution grid. The future climate variables for the January and July temperatures and the annual precipitation were employed to calculate the resultant climatic indices at the end of the century.

Additionally, the current permafrost border position was explicitly included in the analyses as a factor controlling the Scots pine distribution across Russia. The northern and northeastern border of Scots pine distribution across Russia is limited by the southern permafrost border. However, Scots pine may break through further north up, to 70° N, along the broad valleys of the big Siberian rivers, the Yenisei and the Lena, in eastern Siberia, where a deeper active layer thaws in the summer, allowing for Scots pine to succeed [25]. The permafrost border distribution across Russia was calculated using growing GDD_5, NDD, and annual precipitation (R^2 = 0.78), as was also the case with our bioclimatic Siberian vegetation model [26,27]. Seed mass was predicted within the potential *P. sylvestris* range and was overlapped with the permafrost border to separate actual and non-realized (north-eastwards of the border) seed zones in both contemporary and future climates.

2.3. *P. sylvestris* Range and Seed Zones across Russia

Climatic envelopes of the *P. sylvestris* range were subdivided into seed zones/ climatypes using the results of Rehfeldt et al. [28–30]. These results were based on common garden studies that had been established across the former Soviet Union. The differential performance of populations reflects the adaptive differences that have accrued from natural selection in the climate of the provenance where the seeds originated. The results of such studies can be used in defining a climatype as the climatic space occupied by a group of populations whose individuals are adapted to the same or a similar climate. The analyses of Rehfeldt et al. [28–30] used published data across the former Soviet Union within 46–86°N and 24–150°E on the height and survival of 313 populations of *P. sylvestris* that had been planted on 36 sites. These data were used to develop transfer functions that predicted

12-year height and survival from the difference in climate between the provenance of a population and the planting site. The functions were based on a Weibull model. There were three transfer functions driven by GDD_5, NDD, and AMI that were developed for *P. sylvestris* growth and survival. The transfer distances were ±240 degrees for GDD_5, ±575 degrees for NDD, and ±0.6 units for AMI, which were applied for Siberia [31]. It follows, therefore, that populations separated by the breadth of these transfer intervals tend to be genetically different for the traits controlling growth and survival. All possible combinations of these classes produced the maximal number of pine seed zones/climatypes 180 (Table 1). However, not all seed zones were realized in current and future climates. Of the 180 potential seed zones, only 39 realized seed zones, which make up more than 1% of the total Scots pine range, were found in both the current and future climates (Table 1), and these zones were used for this study's analyses. The arrow In Table 1 indicates the realized seed zones and seed mass change along the climatic gradient: from low seed mass in moist climates with extreme cold/long winters and cool/short summers (upper left) to intermediate seed mass in sufficiently moist climates with cool winters and warm summers (middle) to high seed mass in dry climates with warm winters and hot/long summers (lower right).

Table 1. Climatic limits (GDD5, NDD, and AMI) of 180 potential seed zones (#), 39 realized seed zones (>1% of the total range, bold), and their modeled seed mass (g/1000 seed, italic) in current and CMIP5 RCP2.6 and RCP 8.5 climates in Russia.

GDD_5, °C	**AMI**	**NDD, °C**					
		−6000–−4850	**−4850–−3700**	**−3700–−2550**	**−2550–−1400**	**−1400–−250**	**>−250**
600–1080	0.6–1.8	1 *3.4–5.9*	2 *3.9–6.4*	3 *4.3–6.8*	4 *4.8–7.3*	5 *5.3–7.8*	6
	1.8–3.0	**7** ***4.1–6.6***	**8** ***4.6–7.1***	**9** ***5.0–7.5***	10	11	12
	3.0–4.2	13	14	15	16	17	18
	4.2–5.4	19	20	21	22	23	24
	5.4–6.6	25	26	27	28	29	30
	6.6–7.0	31	32	33	34	35	36
1080–1560	0.6–1.8	37	38	**39** ***5.7–8.2***	**40** ***6.1–8.6***	**41** ***6.6–9.1***	42
	1.8–3.0	43	**44** ***5.9–8.4***	**45** ***6.4–8.9***	**46** ***6.8–9.3***	**47** ***7.3–9.8***	48
	3.0–4.2	**49** ***6.1–8.6***	**50** ***6.6–9.1***	**51** ***7.1–9.6***	**52** ***7.5–10.0***	53	54
	4.2–5.4	55	56	57	58	59	60
	5.4–6.6	61	62	63	64	65	66
	6.6–7.0	67	68	69	70	71	72
1560–2040	0.6–1.8	73	74	75	76	77	78
	1.8–3.0	79	80	**81** ***7.7–10.2***	**82** ***8.2–10.7***	**83** ***8.6–11.1***	**84** ***9.1–11.2***
	3.0–4.2	85	**86** ***7.9–10.4***	**87** ***8.4–10.9***	**88** ***8.9–11.4***	**89** ***9.3–11.8***	90
	4.2–5.4	91	92	**93** ***9.1–11.6***	94	95	96
	5.4–6.6	97	98	99	100	101	102
	6.6–7.0	103	104	105	106	107	108

Table 1. *Cont.*

GDD$_5$, °C	AMI	NDD, °C					
		−6000– −4850	−4850– −3700	−3700– −2550	−2550– −1400	−1400– −250	>−250
2040–2520	0.6–1.8	109	110	111	112	113	114
	1.8–3.0	115	116	117	**118** ***9.5–12.0***	**119** ***9.9–12.4***	**120** ***10.4–12.5***
	3.0–4.2	121	122	123	**124** ***10.2–12.7***	**125** ***10.7–13.2***	126
	4.2–5.4	127	128	129	130	**131** ***11.4–13.9***	132
	5.4–6.6	133	134	135	136	**137** ***12.1–14.6***	138
	6.6–7.0	139	140	141	142	143	144
2520–3000	0.6–1.8	145	146	147	148	149	150
	1.8–3.0	151	152	153	154	155	156
	3.0–4.2	157	158	159	160	161	162
	4.2–5.4	163	164	165	166	**167** ***12.7–15.2***	168
	5.4–6.6	169	170	171	172	**173** ***13.4–15.9***	**174** ***13.9–16.0***
	6.6–7.0	175	176	177	178	179	180

2.4. Experiments with Seed Mass during the Common Garden Trial at the Test Site

We defined a genetic component in the seed mass as the difference between the local and a *j*-provenance seed mass in a certain year of measurements (1999–2003 and 2010) at the Boguchany test site. This difference in seed mass indicates the genetic component of the provenance traits because trees grown from the provenances and local seeds presently grow in the same environmental conditions.

We defined an ecological component in the seed mass as the difference between the seed mass of a *j*-provenance collected in 1974 in the original provenance and a *j*-provenance seed mass in a certain year of measurements (1999–2003 and 2010) at the Boguchany test site, and we then related these differences to the climatic conditions of the corresponding years for 1999–2003 and 2010.

2.5. Mapping Seed Zones and Seed Mass across Russia

Finally, the seed zones of Scots pine were mapped for Russia by coupling the climate maps of GDD$_5$, NDD, and AMI for the contemporary climate and the climates predicted by an ensemble of twenty GCMs for 2070–2100 with our seed zone and seed mass bioclimatic models. Multiple regression bioclimatic models of seed mass were constructed to simulate seed mass based on climatic indices using STATISTICA v. 8.0, and the seed mass distributions in the present and future climates were mapped using raster software TerrSet v. 18.21.

3. Results

3.1. Bioclimatic Models of P. sylvestris Seed Mass

There are two bioclimatic models (Equations (1) and (2)) that are linear multiple regressions that relate to Scots pine seed mass (M) for the January (T_1) and July (T_7) temperatures and for the annual precipitation (R_{mm}) (Equation (1)) and for the annual moisture index (AMI), growing degree days (GDD$_5$), and Negative degree days (NDD) (Equation (2)) that were constructed:

$$M = -1.35 + 0.085\ T_1 + 0.557\ T_7 - 0.0021\ Rmm$$

$$N = 166, R^2_{adj} = 0.61, \text{Std. Err.} = 0.72, p < 0.00000; \quad (1)$$

$$M = 3.9 + 0.59\ AMI + 0.00042\ NDD + 0.00274\ GDD_5$$
$$N = 166;\ R^2_{adj} = 0.68;\ Std.\ Err. = 0.65,\ p < 0.00000; \quad (2)$$

Both bioclimatic seed mass models showed reasonable results over the Russian territory. A July temperature elevated by 1 °C increased the seed mass by 0.56 g, and a January temperature elevated by 5 °C increased the seed mass by 0.43 g (Figure 2). Coupled with climatic layers of AMI, GDD_5, and NDD, the model resulted in the Scots pine seed mass distribution maps across Russia in contemporary and future climates (Figure 3). Our modeled seed mass map is good visual agreement with a real seed mass distribution map by Cherepnin [3].

The two maps of the seed mass distribution in future climates (Figure 3) showed that the seed mass would increase by 1 g under the moderate RCP 2.6 and to 4 g under the extreme RCP 8.5, in which the July temperatures were predicted to increase by 1.5–2.0 and 4.0–6.0 °C, respectively, and in which the January temperatures were predicted to elevate by 3.0–4.0 and 8.0–12.0 °C, respectively; the annual precipitation was expected to increase by 30–70 и60–100 mm accordingly. As expected, seed mass decreased along the southwest to the northeast, with a declining temperature gradient and increasing climate severity and continentality along with the decreasing of the active layer depth, which is a crucial factor for Scots pine survival and distribution in interior northern Asia.

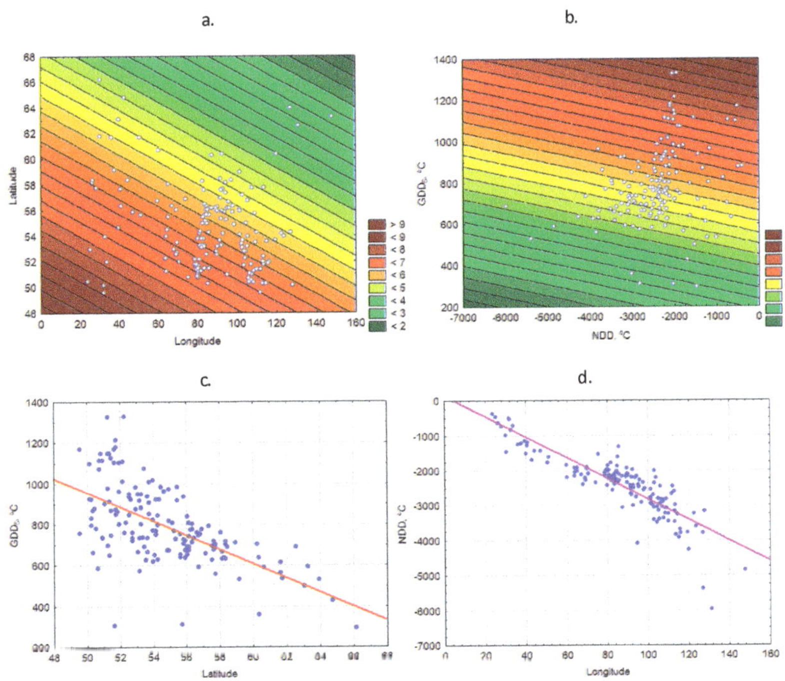

Figure 2. The dependence of Scots pine seed mass on latitude and longitude (**a**); GDD_5 and NDD (**b**); the dependence of GDD_5 on latitude (**c**); the dependence of NDD on longitude (**d**).

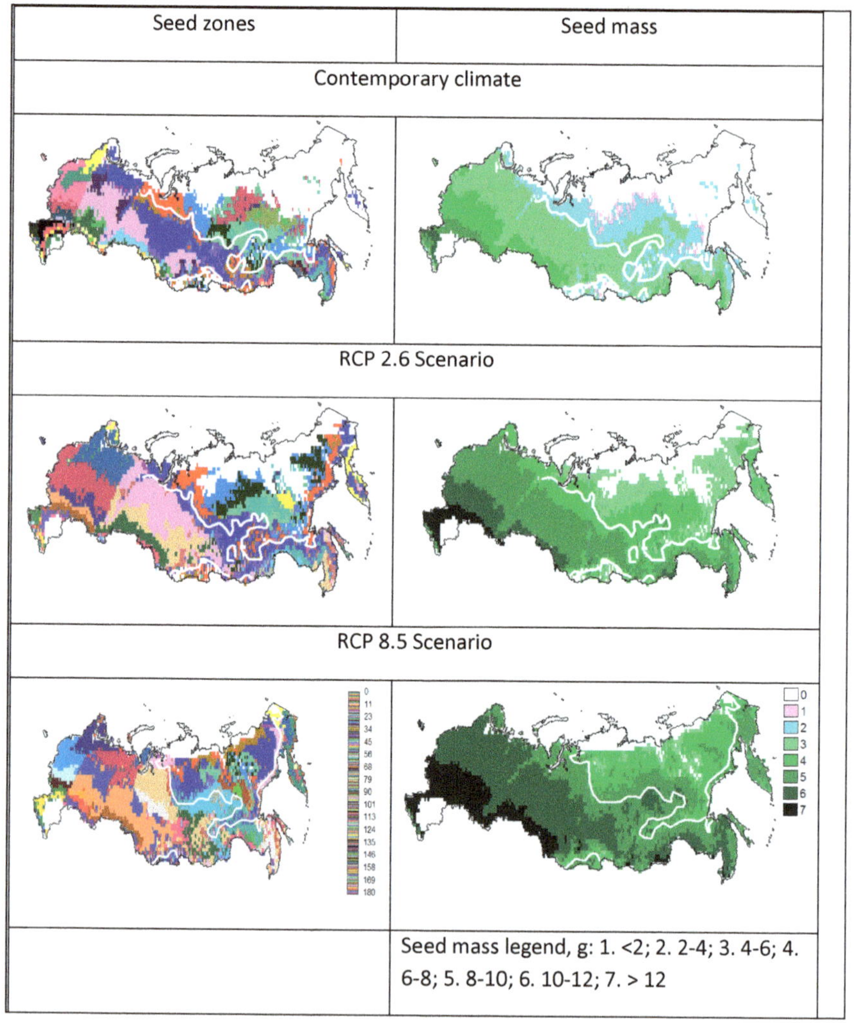

Figure 3. Distributions of potential seed zones (**Left**) and seed mass (**Right**) across Russia in contemporary climate (Upper) and future RCP 2.6 (Middle) and RCP 8.5 (Lower) climates by 2080. The white line is the permafrost border that divides the realized (below the border) and non-realized (above the border) seed zones with the corresponding seed mass. Seed zone legend see Table 1.

3.2. Distribution of *P. sylvestris* Seed Zones in a Warming Climate by 2080

Of the 180 potential seed zones in the climatic envelope of *P. sylvestris*, most were small: 39 seed zones, with areas >1% accounting for 90–92% of the species' envelope over Russia in both the present and future climates; 141 potential seed zones accounted for the remaining <10% of the envelope. Of these 39 large seed zones, 24 were found at present and 19 and 21 were found in the RCP 2.6 and RCP 8.5 climates, respectively. Some zones occurred in the present and two future climates; some zones did not (Tables 1 and 2).

Table 2. Scots pine seed zone coverage (%, >5% of the total range in bold) of various seed mass grades (the prevailing grade, % of a seed zone, is in bold) in contemporary and future climates in Russia.

Seed Zone	Current Climate Seed Zone, % of Each Mass Grade						Seed Zone, % of Total Area	Scenario RCP 2.6 Seed Zone, % of Each Mass Grade					Seed zone, % of Total Area	Scenario RCP 8.5 Seed Zone, % of Each Mass Grade				Seed Zone, % of Total Area
	Seed Mass Grades							Seed Mass Grades						Seed Mass Grades				
	<2	2–4	4–6	6–8	8–10	10–12		2–4	4–6	6–8	8–10	10–12		6–8	8–10	10–12	>12	
1	52.9	47.1					3.36											
2	16.9	82.3	0.8				**6.75**	96.6	3.4				**5.43**					
3	2.1	87.9	10.0				**7.78**	51.8	48.2				4.59					
4		39.6	60.4				**6.53**	3.4	96.6				1.24					
5		9.8	90.2				1.99											
7	12.8	87.2					**5.22**											
8	1.8	87.6	10.6				2.36	62.4	37.6				4.61					
9		36.0	64.0				1.89											
39		28.1	71.9				1.88		98.7	1.3			3.29	98.9	1.2			1.22
40		0.6	99.4				2.29		88.2	11.8			2.60					
41			100				1.74											
44		28.8	71.2				**5.97**		100				3.01	97.4	2.6			2.17
45		0.4	98.8	0.7			**13.06**		89.2	10.8			**11.10**	68.7	31.3			**5.43**
46			76.3	23.7			**11.29**		35.3	64.7			**15.66**					
47			24.5	75.5			3.74		16.7	83.3			**8.14**					
49		50.9	49.1				3.19											
50		3.9	96.1				2.47		82.5	17.5			3.36					
51			80.2	19.8			2.25		19.6	80.4			2.07					
52			4.8	95.2			1.74											
81			30.8	69.2			1.08			100			1.17		100			**4.97**
82				100			1.15			97.4	2.6		**8.21**		78.9	21.1		**12.08**

Table 2. *Cont.*

Seed Zone	Current Climate Seed Zone, % of Each Mass Grade						Seed Zone, % of Total Area	Scenario RCP 2.6 Seed Zone, % of Each Mass Grade					Seed zone, % of Total Area	Scenario RCP 8.5 Seed Zone, % of Each Mass Grade				Seed Zone, % of Total Area
	Seed Mass Grades							Seed Mass Grades						Seed Mass Grades				
	<2	2–4	4–6	6–8	8–10	10–12		2–4	4–6	6–8	8–10	10–12		6–8	8–10	10–12	>12	
83				98.8	1.2		1.18			76.5	23.5		**9.22**		26.5	73.5		**7.79**
84															10.3	89.7		**4.62**
86															100			1.23
87															80.1	19.9		**7.27**
88				92.6	7.4		2.25			55.6	44.4		3.97		35.9	64.1		3.52
89										0.6	99.4		2.35					
93															19.8	80.2		1.21
118															4.5	95.5		4.38
119																100		**7.34**
120																93.8	6.2	4.06
124																100		3.13
125											92.4	7.6	1.29			70.7	29.3	**12.88**
126																9.0	91.0	1.71
131											18.3	81.7	1.30				100	2.15
137					55.1	44.9	1.08											
167																	100	1.16
173																	100	1.05
174																	100	0.98

Not all potential seed zones could be realized in both contemporary and future climates because of impossible climate combinations and the occurrence of permafrost, e.g., extremely cold winters (lowest NDD) never intermix with extremely hot (high GDD_5) and dry (high AMI) summers (Table 1). The climate gradient along which the seed zones were distributed started from the permafrost border in interior north Asia, which varied from being extremely cold in the winter and moderately warm and sufficiently moist in the summer in the extreme continental climate in the interior Asian continent, to being moderately warm in the summer and cold in winter and sufficiently moist in the moderate climate of Eastern Europe and west central Siberia, to the hot and dry climates at the southern border of the Scots pine range in Eurasia. Northeast of the permafrost border, Scots pine can only survive in specific geomorphology that allows for active layer depth (ALD) thawing at 1.5–2 m during the summer [26].

As the climate becomes warmer in the RCP 2.6 scenario, 2/3 (16) of the seed zones would remain, and 1/3 (8) would be lost compared to the present. In the much warmer RCP 8.5 climate, the picture is the opposite: 17 out of 24 contemporary seed zones would be lost, as the climate for which they are best suited disappears, but 14 novel seed zones should appear in association with the appearance of climates currently not found in Russia that will cover about 1/2 of the Scots pine range. Only 1/3 of the lands within the climatic envelope of today would remain, but they would change their geographic position, shifting northeastwards along the leading climate change gradient direction (Figure 3). The seven largest (each >5% of the total range) seed zones that currently dominate > 60% of the Scots pine forests in Russia are expected to be reduced to 6% of their contemporary distributions in the extreme RCP 8.5 climate (Table 2 and Figure 3). On the other hand, of the seven largest seed zones expected in the future, three would be absent, three would be minor (each ~1%), and one would be twice as large as the present. The moderate RCP 2.6 scenario performs a transition climate between the contemporary and the extreme RCP 8.5 climate. Of the seven largest seed zones that occur nowadays, six will be present in the RCP 2.6 climate. However, their size will be smaller in cold climates and larger in warm climates. Compared to the extreme RCP 8.5 climate, sizes of the largest seed zones in the moderate RCP 2.6 climate will be larger in warm climates and smaller (or be absent) in hot climates (Table 2)

3.3. *Coupled Seed Zones and Seed Mass Distributions in Contemporary and Future Climates*

Seed mass was calculated from Equation (2) for the realized seed zones (Table 1). In one seed zone, the seed mass varied 2–2.5 g/1000 seed. From the 1st seed zone to seed zone #174, the seed mass was predicted to vary from <2 g to 16 g. Below, we analyze the shifting of the largest seed zones (>5% of the total Scots pine range) jointly with their seed mass (Table 2) in contemporary (Figure 4 Left) and future climates (Figure 4 Central and Right).

The future location of climates inhabited by contemporary seed zones is expected to shift geographically. In cool summers (under GDD_5 < 1080 °C) in forest-tundra, northern, and middle taiga, seed zones ## 2, 3, 4, and 7 with a prevailing seed mass of 2–4 g, were predicted in both the current and future climates. However, they would not be realized in the future because they were predicted to be in the permafrost zone (north of the permafrost border marked by the white line) where Scots pine cannot survive except for in warmer wide floodplains [32]. The warmer seed zones, ## 44, 45, 46, and 47, with a prevailing mass 4–6 g occurring in southern taiga, subtaiga, and forest-steppe (under GDD_5 = 1080–1600 °C) would be realized in the current and the moderate RCP 2.6 climates, wherein the seed mass may reach 8 g. These seed zones were predicted to be south of the permafrost border. However, in the RCP 8.5 climate, these seed zones were predicted to cross the permafrost border so that they would not be suitable for the survival of Scots pines on cold soils with an ALD less than the 1.5 m (Figure 3). The warmer seed zones, ## 81, 82, 83, 84, and 87, would be realized in the southern taiga, subtaiga, and forest-steppe (under GDD_5 = 1600–2000 °C) in the current and future climates, including the extreme RCP 8.5 climate scenario. The prevailing seed mass was predicted to be as high as 6–8 g under the moderate scenario RCP 2.6 and 8–12 g under the extreme scenario RCP 8.5. These seed zones were predicted south of the permafrost border where warm soils would be suitable for Scots pine. However, these predictions could only be realized under sufficient moisture conditions at the southern border of the Scots pine range. The warmest seed zones, ##118, 119, 120, 124, and 125, with a high seed mass of 10–12 g and higher would mostly be realized in the dry conditions of the forest-steppe (under GDD_5 = 2000–2500 °C) in the extreme RCP 8.5 climate. Small coverage of the warmest seed zones, ##167, 173, 174, with a seed mass greater than 12 g were predicted in hot (under GDD_5 > 2500 °C) and dry (AMI > 4.0) steppe habitats that could be suitable for Scots pine only in the permafrost zone, where the melting of the permafrost provides additional water.

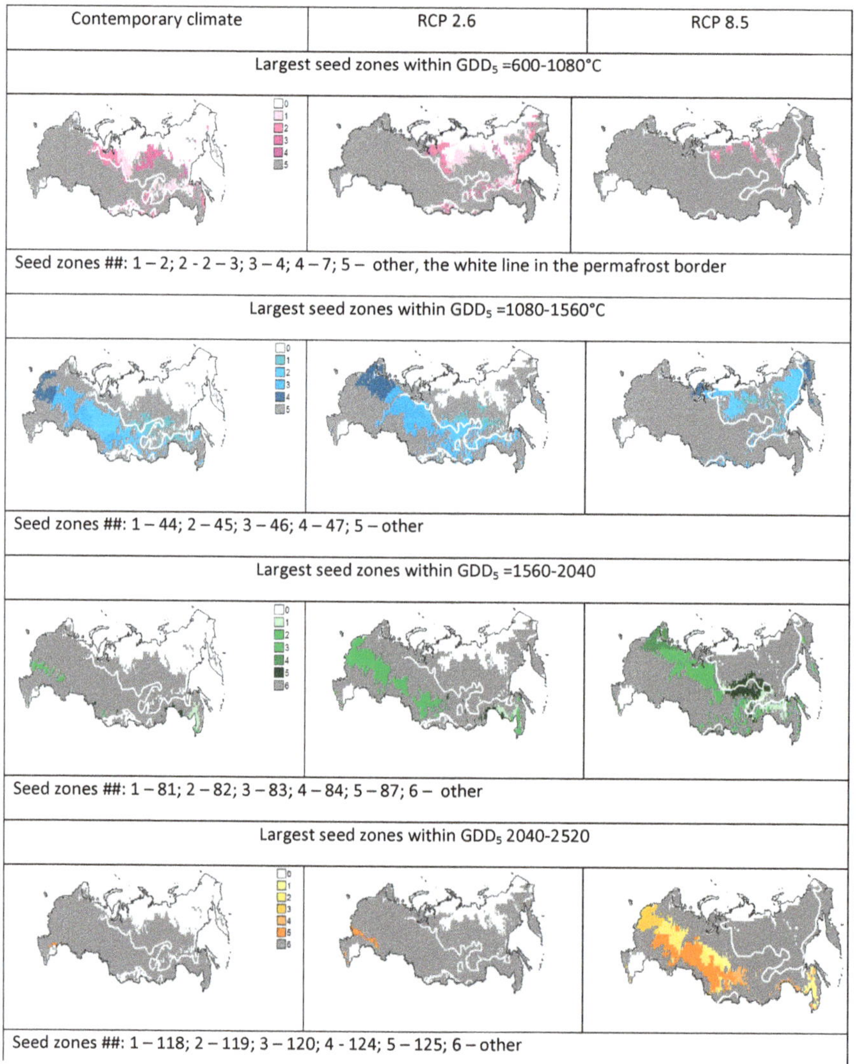

Figure 4. Distributions of large seed zones (>5% of the total seed zone area) across Russia in contemporary climate (**Left**) and future RCP 2.6 (**Central**) and RCP 8.5 (**Right**) climates in 2080. The white line is the permafrost border that divides the realized (below the border) and non-realized (above the border) seed zones.

The seed zone shift may amount within 500 to 1000 km, following shifting vegetation zones in a warming climate [31] that would bring new plant species from Europe to Asia over the Ural Mountain barrier and change the future flora in southern Siberia, as was reconstructed for the mid-Holocene at about 5500 years BP [33,34].

3.4. The 40-Year Common Garden Trial in the Boguchany

Time series data of seed mass from 42 provenances collected and measured for the 1999–2010 period at the Boguchany test site allowed for the bioclimatic modeling of seed mass depending on the weather conditions of each year (departures from the multi-year climatic norm).

We started from modeling the relationship between local provenance (Boguchany) seed mass and provenance seed mass that arrived in 1974, which was originally used for the establishment a common garden trial at the Boguchany settlement (Figure 5). From Figure 5, it followed that trees grown from the lighter seeds of 1974 in the Boguchany set site produced heavier seeds than in their home provenances for 1999–2003 and 2010. On the contrary, trees grown from the heavier seeds from 1974 produced lighter seeds than in their home provenances for 1999–2003 and 2010.

$$\text{Mi, g} = 4.1 + 0.412\ \text{M}_{1974};$$

$$\text{N} = 70;\ r = 0.44;\ R^2 = 0.196;\ p < 0.0001;\ \text{Std. Er. of est.: } 0.88 \quad (3)$$

The seed mass of 25–37-year old trees from 42 provenances in the site test in central Siberia was found to vary from seed mass in home provenances due to the difference in environmental conditions. Otherwise, in the absence of the environmental impact, the seed masses at the test site were the same as in the home provenances.

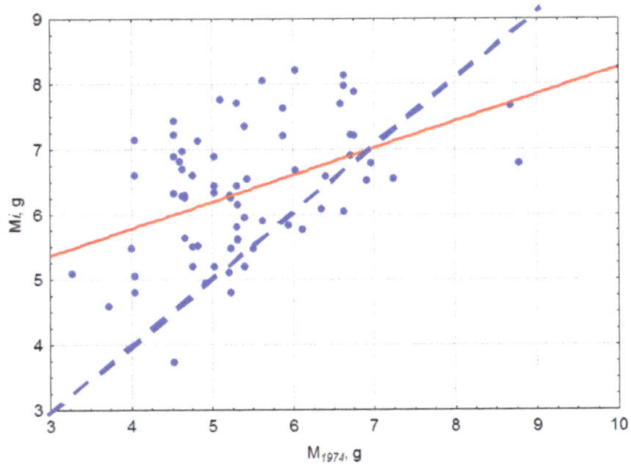

Figure 5. The relationship between provenance seed mass (M_i) for 1999–2003 and 2010 in the Boguchany common garden trial and provenance seed mass in 1974.

Seed mass at the test site was analyzed depending on a) the latitude of provenance (Figure 6a) and on b) the differences between the July temperature of the i-th year of seed measurements in the Boguchany test site and the multi-year mean (1960–1990) of provenance (Figure 6b) and on the c) differences between the January temperature of the *i*-th year of seed measurements in Boguchany and the multi-year mean (1960–1990) of provenance. The determination coefficients were weak (R^2 = 0.14 and R^2 = 0.21) for both the July and January temperature simple regressions; however, the multiple linear regression including both temperatures was good (R^2 = 0.30, Equation (4), Figure 7).

$$\text{Mi} = 7.09 - 0.153\ \Delta\text{T}_7 + 0.044\ \Delta\text{T}_1$$

$$r = 0.56;\ R^2_{adj} = 0.30;\ n = 72; \tag{4}$$

The evident outcome of this analysis was that seeds from the southerly (warmer) provenances are heavier than the local provenance seeds from Boguchany, and seeds from northerly (cooler) provenances are lighter than local provenance seeds from Boguchany. Kuzmina and Kuzmin came to the same conclusion by analyzing the previously obtained seed mass data of 25-year-old Scots pines from the same common garden experiment [20].

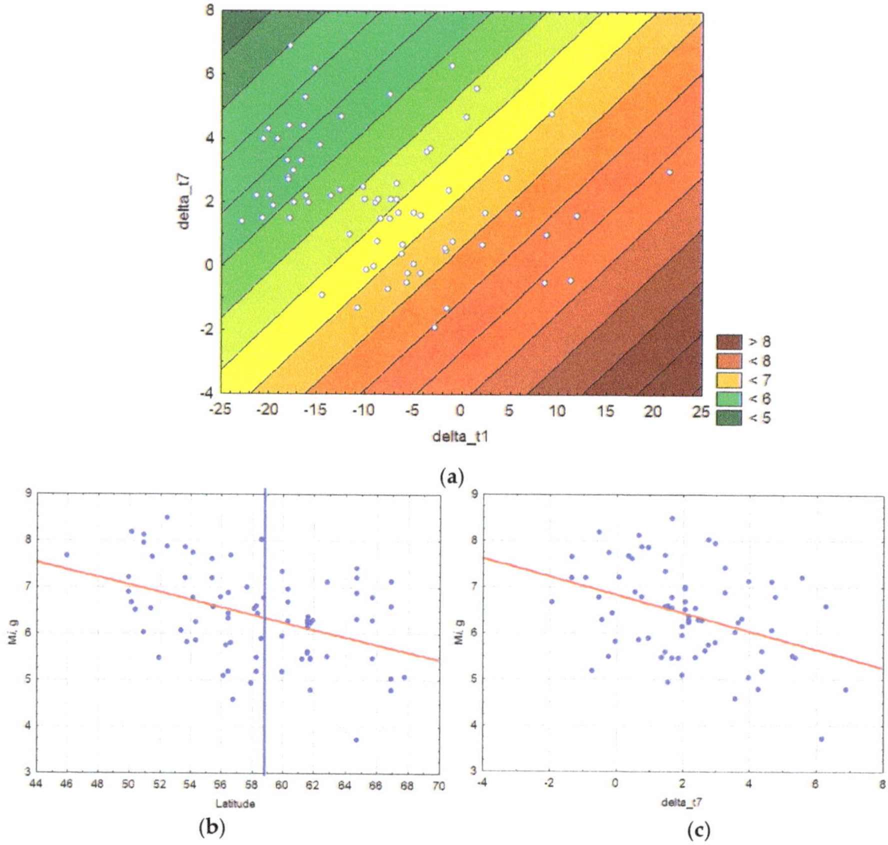

Figure 6. The dependence of M_i on a provenance latitude (**a**) and the difference between July temperatures for the *i*-year and a multi-year mean of provenance (**b**) for 1999–2003 and 2010 (a year of seed measurements in the test site). The blue line is the latitude of the test site.

The difference between provenance seed mass and that of the Boguchany test site (Δ M) is the indicator of the genetic component of the variation in the seed mass in the same ecological conditions. This difference indicates how favorable the local climate is for the tested seed provenance: the less seed mass difference, the more favorable climate is (Equation (5)):

$$\Delta Mi = -1.23 + 0.19\ T_7,$$

$$N = 55;\ r = 0.47;\ R^2 = 0.22;\ \text{St. err.} = 0.71;\ p < 0.00026 \tag{5}$$

The difference between provenance and local seed mass (ΔM_i) in a given year at the test site characterized the genetic component of provenance (Equation (6)):

$$\Delta M_i = -0.79 - 0.09\ \Delta LAT$$

$$N = 59, r = 0.54; R^2 = 0.30; \text{St. Err.} = 0.68; p < 0.00008 \quad (6)$$

The difference of the July temperatures (ΔT_7) in a given year in the test site and the multi-year mean of provenance explain 22% of the seed mass variation by the genetic component (Figure 8a). The difference between the latitudes (ΔLAT) of provenance and the test sites explained 30% of the seed mass variation (Figure 8b), likely due to the impact of photoperiodism in addition to heat resources.

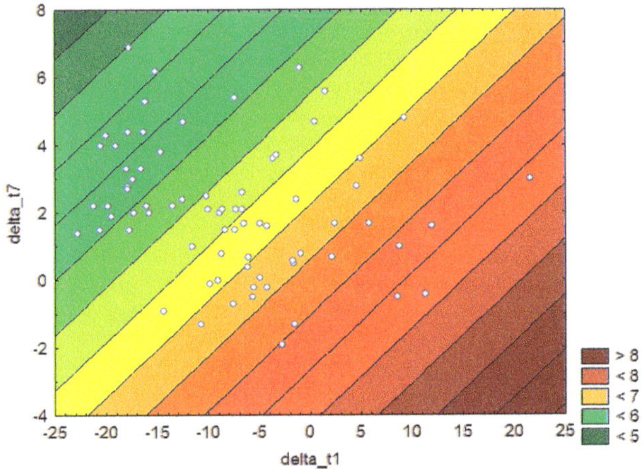

Figure 7. The dependence of a *j*-provenance seed mass on the differences of the July (ΔT_7) and January temperatures (ΔT_1) in the *i*-th year for 1999–2003 and 2010 (a year of seed measurements in the test site) and the multi-year T_1 and T_7 means of *j*-provenance, respectively.

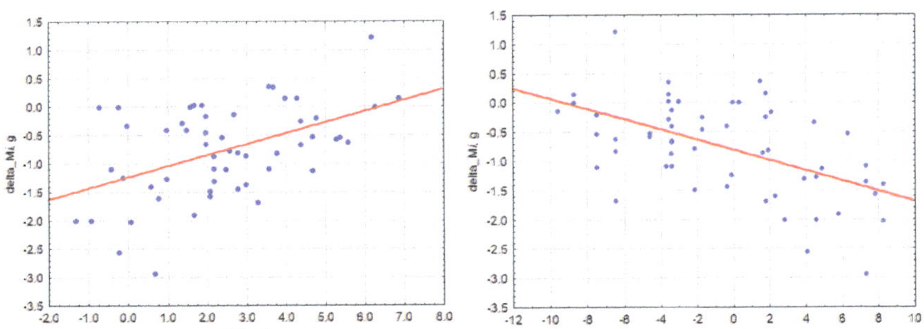

(a) (b)

Figure 8. The dependence of the seed mass provenance differences and local seed mass (ΔM_i) in measurement years 1999–2003 and 2010 in the Boguchany test site (the genetic component) on the July temperature difference in the measurement year at Boguchany and the multi-year mean of provenance (**a**) and the difference of Boguchany latitudes and provenance (**b**).

4. Discussion

A bioclimatic seed mass model of *P. sylvestris* was built depending on heat, cold and moisture conditions within its range across Russia. As the most common plastic tree species, P. sylvestris stretches through all vegetation zones except for treeless tundra at the very north and the semidesert at the very south and the forests laid by permafrost in the heart of inner Eurasia. *P. sylvestris* genetic and morphology variability has been studied in detail by Russian and European foresters from the 19th century [1–4]. Due to its plasticity, Scots pine is the most used tree species across Russia for restoring and regenerating lands of failed forests after natural and man-made disturbances, primarily caused by forest fire and clearcutting.

Seed mass is the most important grade indicator of the State Standardization System (GOST in Russian) in Russian forestry. Scots pine seed mass varies 2–3 times along its range: from 3 g (1000 seeds) in the north to 10–11 g in the pure pine forests in Kazakhstan at the very southern edge of the range.

The Scots pine range spreads extensively over Eurasia, more extensively than all other species of the genus *Pinus*, and even of the whole *Pinaceae* family. The Scots pine range covers 14 thousand km from Spain in the west to interior Yakutia in the east and 2.7 thousand km from northern Scandinavia to the Sierra Nevada Mountains in Spain [35]. In the former Soviet Union, its range covered the distance between 40 to 70° N and from 20 to 138° E [1]. Thus, as the *P. sylvestris* range is mostly located in the former Soviet Union, it was better studied in the Russian literature as a major forest-forming conifer of the Russian boreal forest. Studies of the seed mass variation in other regions beyond the former Soviet Union showed that the seed mass only exceeded the seed mass range across the former Soviet Union in the southernmost sites. In Turkey, at the elevation 1160 m, seed mass varied in 8.6–13.2 g intervals among the clones and in 7.1–15.9 g intervals within the clones [16]. The seed mass economic seed stands and seed orchards in Poland ranged from 5.2 to 9.1 g [36]. In 1970, Antosiewicz [36] found the mean seed mass of 1000 seeds to be 6.2 g within the range from 4.5 to 8.5 g. Przybylski [37] revealed that the mean seed mass in six Scots pine populations in Poland ranged from 5.49 to 6.60 g, with the averaged value being 6.01 g. The seed mass of southern populations of Scots pine in Spanish mountains was much higher than that of northern populations in Finland [17]. The seed mass of Scots pine provenances from a wide geographic area (19 from Finland, 4 from former Soviet Union (Arkhangelsk, Latvia, Novosibirsk, Yakutsk), 6 from Central or Western Europe (Poland, Scotland, France, Belgium, Hungary, Bulgaria) and 1 from Turkey), varied from 3.1 g to 9.2 g, with average 5.2 g [38]. The seed mass for 16 years (1973–1988) in Swedish Scots pine seed orchards varied from 5.31 g to 6.93 g in the north, from 5.43 g to 7.15 g in the middle, and from 5.86 g to 7.31 g in the south [39]. Thus, the seed mass was heavier in the south than in the north. In northern Mongolia [18], seed mass varied between 4.58 до 6.97 g, which was within the range of our seed mass data in Russia.

Seed mass depends on tree age; however, the dependence is ambiguous. There is no consensus on this issue in the literature. Pravdin [1] reviewed studies that related seed mass to tree age. In [1], Kurdiani (1932) and Kapper (1954) made conclusions on a stable seed mass decline from 8.0 g to 5.5 g with ages from 50 to 160 years old. Others found the seed mass decrease from 20 to 60 years old followed by an increase. Mamaev et al. [2] mentioned that many researchers noted the seed mass decrease with aging at mature to overmature age. Additionally, Mamaev et al. [2] concluded that coniferous tree species maintained their ability to produce high-quality seeds for an indefinitely long time. Debain et al. [40] found no effect of age on seed dimensions and mass while comparing two Scots pine stands in southern France that were of the ages of 22–27- and 30–40-years-old, although both seed dimension and mass varied between and within trees. Significant correlations (R^2 = 0.96) were found between the age of the parents stands (124 to 180 years old) and the average seed weight, which decreased by approximately 15% as the Scots pine trees grew older [41].

In [3], Azniev (1956), Pikhelgas (1971), and Shulga (1972) came to the conclusion that there was no seed mass dependence on tree age and found no difference between seed quality of trees at various agea.

In our study, we have not aimed to find a relationship between seed mass and tree age because our data was limited by the age range of 25–36 years, which was not sufficient to draw robust conclusions. However, assuming that seeds planted in 1974 in our common garden experiment were collected from trees of 100 years of age and older that had been cut during clearcutting, our data demonstrated some seed mass decreases at mature age (Figure 5) compared to seed mass at 25–36 yr^{-1} of age in our experiment.

A qualitative relationship between seed mass and latitude across Russia was established in the 1960s [1]. Cherepnin [3] established a quantitative relationship between seed mass and heat resources (temperature sums for the period with temperatures >10 °C) for large ecoregions. This study established robust regressions that related *P. sylvestris* seed mass with average climatic conditions: heat resources (mean July and January temperatures, growing degree days, 5 °C), and climate severity (negative degree days, 0 °C). Noting that moisture conditions (annual precipitation, annual moisture index) did not contribute to Scots pine seed mass much, it was still significant (Equation (1)). This fact confirms that Scots pine is tolerant to low moisture and poor soils. To adapt to those conditions, it developed a long taproot that was able to reach the water table [42].

Thus, heat resources explained about 70% of the seed mass variation over Russia. In this study, Scots pine seed mass over Russia was mapped in contemporary and future climates using our bioclimatic seed mass model. At the end of the century, seed mass may increase from 1 g (a moderate scenario RCP 2.6) to 4 g (an extreme scenario RCP 8.5). Seed mass isolines would shift 500 km north and northeast and would follow the southern permafrost border that controls the northern edge of the Scots pine range. The northerly shift would be slow because the summer thawing of the active layer at the depth of 1.5–2 m is necessary for Scots pine to thrive [32] and would be inertial in substrates (soils) due to a larger substrate heat capacity than that of the ambient air. Thus, the permafrost border would follow near-surface air warming with some lag [43].

Another regression that related to the provenance seed mass and the climates of 1974 (the year of seed collecting) was constructed to determine how much the climate of 1974 rather than the average multi-year climate was important for seed mass in each provenance. The correlation between seed mass and the provenance climates of 1974 was low but significant ($r = 0.30$; $p < 0.000$). Adding provenance latitudes to this regression significantly improved the correlation ($r = 0.78$; $p < 0.000$) because the latitude is not only a proxy of heat, but photoperiodism (daylight and dark periods), a phenomenon of equal importance that Russian forestry classics [1,44] emphasized in seed transfer practices improved the correlation as well.

The third section of this study was to infer the ecological and genetic components of the seed variability of 25–37-year-old trees that produce seeds at the test site. In doing so, this study considered the seed mass of the Boguchany provenance as a normal reaction for the environment; the positive or negative seed mass differences of 42 provenances from the norm were considered as a genetic response to an environmental change. These differences were ordinated around 0 for the Boguchany norm, which was in the center. The main message delivered was that four quadrants of heat/cold differences between provenances and the test site demonstrated that the seed mass of colder provenances were lighter than that of the local provenance Boguchany but were heavier than that at the original provenance; the seed mass of warmer provenances was heavier than that of the local provenance Boguchany but were lighter than that at the original provenance.

Using seed mass data dated by the year 1974 of seed collection, a reverse ordination of the seed mass difference in original provenances and those in the Boguchany test site was examined. A greater seed mass in the test site compared to a local provenance was demonstrated by northerly provenances, and vice versa—a smaller seed mass in the trial site compared to a local provenance was demonstrated by southerly provenances. Hence, the outcome of these findings is that we recommend seeds from southerly provenances to improve seed mass and quality over reforestation/afforestation lands. However, this study has not considered the impacts of fungi and pathogens on seed mass and quality (survival) from various provenances in our common garden trial in the Boguchany site.

In our Russian seed mass models (Equations (1) and (2)), seed mass varied between 3.5–10 g in the current climate. We used our model to predict seed mass in the extremely warm RCP 8.5 climate (Table 2). We extended the predictive power of our seed mass model based on the assumption that models with high explanatory power are inherently of high predictive power [45]. Research in southerly landscapes beyond our study area confirmed our model predictions that the seed mass should be heavier in warmer conditions, e.g., in Turkey, which is south of Russia, the overall mean of 1000 seed masses was 10.9 g, with a range of 8.6 to 13.2 g in a 13 year-old *P. sylvestris* L. clonal seed orchard that included 30 clones [16].

Additionally, this study needs to take into account the understanding of seed transfer and seed zones/climate types in forestry practices under climate warming. Rehfeldt et al. [28–30] established the number, size, and distribution of *P. sylvestris* seed zones/climate types across Russia using transfer functions for growing degree days, negative degree days, and annual moisture index developed from common garden data across Russia. The GDD_5 seed zone breadth was found to be 480 °C or 3.5 °C of the July temperatures, which corresponds to ~ 5 degrees of latitude [3]. Forestry practitioners recommended transferring seed from the origin to planting sites within this distance of 5 degrees of latitude [23,44].

At present, the least productive seed zones with seed mass being less than 2 g and 2–4 g dominate the forest-tundra and northern taiga in cold climates. More productive seed zones with seed mass of 4–6 up to 6 g dominate the middle, southern, and subtaiga in moderately warm climates. In a warming climate under the RCP 2.6 scenario, more productive seed zones with a seed mass from 2–4 to 6–8 g would dominate the southern landscapes. The most productive seed zones, with seed mass 8–10 g, could be found in small areas of the temperate forest-steppe in East Europe. In the extremely warm climate under the RCP 8.5 scenario, seed zones with a seed mass 8–12 g were predicted to dominate larger temperate forest-steppe lands in Eastern Europe; however, the water availability would be a limiting factor at the southern border of the Scots pine range.

5. Conclusions

For the whole range of *P. sylvestris* L. that is distributed over the former Soviet Union, its bioclimatic model of seed mass explains 70% of the variation by site climatic variables. A one degree increase of the July temperature caused a 0.5 g increase of 1000 seeds, and a one degree increase of the January temperature caused a 0.1 g increase of 1000 seeds. The bioclimatic model was applied to the future moderate (RCP 2.6) and extreme (RCP 8.5) climate change scenarios. Climate warming would result in seed mass increase in each site by 1 to 4 g on average in both scenarios accordingly by the end of the century. Scots pine seed mass measurements in a common garden experiment in south central Siberia showed that trees of northerly origins produced lighter seeds than local trees but heavier ones than trees at the original site. Trees of southerly origins produce heavier seeds than local trees but lighter seeds than those at the original site. This study's findings would serve as blueprints for predicting new landscapes with climatic optima for *P. sylvestris* to produce better quality seeds to adjust to a warming climate.

Supplementary Materials: The following are available online at https://www.mdpi.com/article/10.3390/f12081097/s1, Table S1: Comparison of the temperature sums TSt < 0 and TS t > 10 calculated from a sine function and linear regressions between temperature sums and January and July mean temperature.

Author Contributions: Conceptualization, E.I.P. and N.M.T.; methodology, N.M.T. and N.A.K.; validation, N.M.T., N.A.K. and S.R.K.; formal analysis, E.I.P.; investigation, N.M.T. and E.I.P.; resources, N.A.K. and S.R.K.; data curation, N.A.K. and S.R.K.; writing—original draft preparation, E.I.P.; writing—review and editing, N.M.T., N.A.K. and S.R.K.; visualization, E.I.P.; funding acquisition, N.M.T. All authors have read and agreed to the published version of the manuscript.

Funding: The authors acknowledge the support from the RFBR Project #20-05-00540 and partially from Project #19-45-240004, funded by the Russian Foundation for Basic Research, Government of Krasnoyarsk Territory, Krasnoyarsk Regional Fund of Science.

Institutional Review Board Statement: Not applicable.

Informed Consent Statement: Not applicable.

Data Availability Statement: Not applicable.

Acknowledgments: We thank Jerry Rehfeldt for his useful comments; we are very grateful to Jane Bradford for her help with English editing.

Conflicts of Interest: The authors declare no conflict of interest.

References

1. Pravdin, L.F. *Pinus sylvestris*; Nauka: Moscow, Russia, 1964; p. 192.
2. Mamaev, S.A. *Intraspecies Variability of Woody Plants (A Case Study of the Family Pinaceae in the Ural mountains)*; Nauka: Moscow, Russia, 1973; p. 284.
3. Cherepnin, V.L. *Pinus sylvestris Seeds Variability*; Nauka: Novosibirsk, Russia, 1980; p. 182.
4. Giertych, M.; Matyas, C. *Genetics of Scots Pine. Developments of Plant Genetics and Breeding*; Elsevier: Amsterdam, The Netherlands, 1991; Volume 3, p. 260.
5. Mátyás, C.; Ackzell, L.; Samuel, C.J.A. *EUFORGEN Technical Guidelines for Genetic Conservation and Use for Scots Pine (Pinus sylvestris)*; International Plant Genetic Resources Institute: Rome, Italy, 2004; p. 6.
6. Moles, A.T.; Westoby, M. Latitude, seed predation and seed mass. *J. Biogeogr.* **2003**, *30*, 105–128. [CrossRef]
7. Castro, J.; Reich, P.B.; Sanchez-Miranda, A.; Guerrero, J.D. Evidence that the negative relationship between seed mass and relative growth rate is not physiological but linked to species identity: A within-family analysis of Scots pine. *Tree Physiol.* **2008**, *28*, 1077–1082. [CrossRef]
8. Liu, J.; Bai, Y.; Lamb, E.G.; Simpson, D.; Liu, G.; Wei, Y.; Wang, D.; McKenney, D.W.; Papadopol, P. Patterns of Cross-Continental Variation in Tree Seed Mass in the Canadian Boreal Forest. *PLoS ONE* **2013**, *8*, e61060. [CrossRef] [PubMed]
9. Himanen, K.; Helenius, P.; Ylioja, T.; Nygren, M. Intracone variation explains most of the variance in *Picea abies* seed weight: Implications for seed sorting. *Can. J. For. Res.* **2016**, *46*, 470–477. [CrossRef]
10. Freire, J.A.; Rodrigues, G.C.; Tome, M. Climate change impacts on *Pinus pinea* L. silvicultural system for cone production and ways to contour those impacts: A review complemented with data from permanent plots. *Forests* **2019**, *10*, 169. [CrossRef]
11. Balekoglu, S.; Caliskan, S.; Dirik, H. Effects of geoclimatic factors on the variability in *Pinus pinea* cone, seed, and seedling traits in Turkey native habitats. *Ecol. Process.* **2020**, *9*, 55. [CrossRef]
12. Suarez-Vidal, E.; Sampedro, L.; Zas, R. Is the benefit of larger seed provisioning on seedling performance greater stress? *Environ. Exp. Bot.* **2017**, *134*, 45–53. [CrossRef]
13. Loewe, M.V.; Balzarini, M.; Alvarez, C.A.; Delard, R.C.; Navarro, C.R. Fruit productivity of Stone pine (*Pinus pinea* L.) along a climatic gradient in Chile. *Agric. For. Meteorol.* **2016**, *223*, 203–216. [CrossRef]
14. Iroshnikov, A.I. *Provenance Trials of Conifers in South Siberia. Provenance Trials and Plantations of Conifers in Siberia*; Nauka: Novosibirsk, Russia, 1977; pp. 4–110.
15. Kuzmina, N.A. *Pinus sylvestris* variability in Angara region. *Lesovedenie* **1985**, *2*, 40–46.
16. Sivacioğlu, A.; Ayan, S. Evaluation of seed production of scots pine (*Pinus sylvestris* L.) clonal seed orchard with cone analysis method. *African J. Biotechnol.* **2008**, *7*, 4393–4399. [CrossRef]
17. Matías, L.; Jump, A.S. Impacts of predicted climate change on recruitment at the geographical limits of Scots pine. *J. Exp. Bot.* **2014**, *65*, 299. [CrossRef] [PubMed]
18. Batkhuu, N.; Udval, B.; Jigjid, B.; Jamyansuren, S.; Fischer, M. Seed and cone morphological variation and seed germination characteristics of scots pine populations (*Pinus sylvestris* L.) in Mongolia. *Mong. J. Biol. Sci.* **2020**, *18*, 41–54. [CrossRef]
19. Kuzmina, N.A. Specific features of Scots pine provenance trials in Angara river basin. *Lesovedenie* **1999**, *4*, 23–29. (In Russian)
20. Kuzmina, N.A.; Kuzmin, S.R. Features of generative organs of Scots pine of different origin in the provenance trial. *Conifers Boreal Area* **2007**, *24*, 225–234.
21. Kuzmina, N.A.; Senashova, V.A.; Kuzmin, S.R. Distribution of *Lophodermium* needle cast agents in Scots pine stands in Central Siberia. *Contemp. Probl. Ecol.* **2015**, *8*, 909. [CrossRef]
22. Kuzmin, S.R.; Kuzmina, N.A. Choosing of Promising Climatypes of Scots Pine in Geographic Cultures in Different Forest Growth Conditions. *Lesovedenie* **2020**, 451–465. [CrossRef]
23. Kuzmin, S.R.; Kuzmina, N.A. Forest Seed Regions of Scots Pine Based on Growth Assessement in the Provenance Trial in Siberia. *Sibirskij Lesnoj Zurnal* **2020**, *6*, 3–15. [CrossRef]
24. Kuzmin, S.R. Response of Annual Ring Width and Latewood Content of Scots pine to Weather Conditions in Provenance Trials. *Lesn. Zhurnal* **2020**, *5*, 64–80. [CrossRef]
25. Pobedinsky, A.V. *Pinus sylvestris*; Forestry Publishers: Moscow, Russia, 1979; p. 128.
26. Tchebakova, N.M.; Parfenova, E.; Soja, A.J. The effects of climate, permafrost and fire on vegetation change in Siberia in a changing climate. *Environ. Res. Lett.* **2009**, *4*, 045013. [CrossRef]
27. Parfenova, E.I.; Tchebakova, N.M.; Soja, A.J. Assessing landscape potential for human sustainability and 'attractiveness' across Asian Russia in a warmer 21st century. *Environ. Res. Lett.* **2019**, *14*, 065004. [CrossRef]
28. Rehfeldt, G.E.; Tchebakova, N.M.; Barnhardt, L.K. Efficacy of climate transfer functions: Introduction of Eurasian populations of Larix into Alberta. *Can. J. For. Res.* **1999**, *29*, 1660–1668. [CrossRef]
29. Rehfeldt, G.E.; Tchebakova, N.M.; Parfenova, E.I.; Wykoff, W.R.; Kuzmina, N.A.; Milyutin, L.I. Intraspecific responses to climate in *Pinus sylvestris*. *Glob. Chang. Biol.* **2002**, *8*, 912–929. [CrossRef]

30. Rehfeldt, G.E.; Tchebakova, N.M.; Milyutin, L.I.; Parfenova, E.I.; Wykoff, W.R.; Kouzmina, N.A. Assessing population responses to climate in *Pinus sylvestris* and Larix spp. of Eurasia with Climate-Transfer Models. *Eurasian J. Res.* **2003**, *6*, 83–98.
31. Tchebakova, N.M.; Rehfeldt, G.E.; Parfenova, E.I. From vegetation zones to climatypes: Effects of climate warming on Siberian ecosystems. In *Permafrost Ecosystems. Siberian Larch Forests*; Osawa, A., Zyryanova, O.A., Matsuura, Y., Kajimoto, T., Wein, R.W., Eds.; Springer: Berlin/Heidelberg, Germany, 2010; pp. 427–447.
32. Pozdnyakov, L.K. *Permafrost Forestry*; Nauka: Novosibirsk, Russia, 1986; p. 192.
33. Khotinsky, N.A. *The Holocene of the Northern Eurasia*; Nauka: Moscow, Russia, 1977; p. 200.
34. Tchebakova, N.M.; Blyakharchuk, T.A.; Parfenova, E.I. Reconstruction and prediction of climate and vegetation change in the Holocene in the Altai-Sayan mountains, Central Asia. *Environ. Res. Lett.* **2009**, *4*, 045025. [CrossRef]
35. Boratynski, A. Range of natural distribution. In *Developments in Plant Genetics and Breeding*; Giertych, M., Mâtyâs, C., Eds.; Elsevier: Amsterdam, The Netherlands, 1991; pp. 19–30.
36. Hauke-Kowalska, M.; Borowiak, E.; Barzdajn, W.; Kowalkwski, W.; Korzeniewicz, R.; Wawro, T. Cone and seeds variability in seed orchards and seed stands of *Pinus sylvestris* L. *Balt. For.* **2019**, *25*, 187–192. [CrossRef]
37. Przybylski, P.; Masternak, K.; Jastrzębowski, S. Isozyme polymorphism and seed and cone variability of Scots pine (*Pinus sylvestris* L.) in relation to local environments in Poland. *Folia For. Pol. Ser. A* **2020**, *62*, 88–99. [CrossRef]
38. Mikola, J. The effect of seed size and durarion of growth on the height of Scots pine (*Pinus sylvestris* L.) and Norway spruce (*Picea abies* (L.) Karst.) provenances and progenies at the nursery stage. *Silva Fenn.* **1980**, *14*, 84–94. [CrossRef]
39. Prescher, F.; Lindgren, D.; Wennstrom, U.; Almqvist, C.; Ruotsalainen, S.; Kroon, J. Seed production in Scots pine seed orchards. In *Status, Monitoring and Targets for Breeding Programs, Proceedings of the Meeting of Nordic Forest Tree Breeders and Forest Genetics, Syktyvkar, Russia, 13–15 September 2015*; Fedorkov, A., Ed.; Russian Academy of Sciences: Moscow, Russia, 2005; pp. 65–71.
40. Debain, S.; Curt, T.; Lepart, J.; Prevosto, B. Reproductive variability in *Pinus sylvestris* in southern France: Implications for invasion. *J. Veg. Sci.* **2003**, *14*, 509–516. [CrossRef]
41. Kaliniewicz, Z.; Rawa, T.; Tylek, P.; Markowski, P.; Anders, A.; Fura, S. The effect of the age of Scots pine (*Pinus sylvestris* L.) stands on the physical properties of seeds and the operating parameters of cleaning machines. *Tech. Sci. Univ. Warm. Mazury Olszt.* **2013**, *16*, 63–72.
42. Orlov, A.Y.; Koshelkov, S.P. *Soil ecology of Scots Pine*; 'Nauka' Publ.: Moscow, Russia, 1971; p. 324. (In Russian)
43. Dostovalov, B.N.; Kudryavtsev, V.A. *Basics of Geocryology*; Obshcheye Merzlotovedeniya (General Permafrost Science): Moscow, Russia, 1967; p. 404. (In Russian)
44. Sannikov, S.N.; Petrova, I.V.; Sannikova, N.S.; Afonin, A.N.; Chernodubov, A.I.; Egorov, E.V. Genetic-climatologic-geographical principles of seed zoning of pine forests in Russia. *Sib. Lesn. Zhurnal (Sib. J. For. Sci.)* **2017**, *2*, 19–30, (In Russian with English abstracts).
45. Shmueli, G. To Explain or to Predict? *Stat. Sci.* **2010**, *25*, 289–310. [CrossRef]

forests

MDPI

Review

Assisted Migration Field Tests in Canada and Mexico: Lessons, Limitations, and Challenges

Cuauhtémoc Sáenz-Romero [1,*], Greg O'Neill [2], Sally N. Aitken [3] and Roberto Lindig-Cisneros [4]

1 Instituto de Investigaciones sobre los Recursos Naturales (INIRENA), Universidad Michoacana de San Nicolás de Hidalgo (UMSNH), Av. San Juanito Itzícuaro s/n, Col. Nueva Esperanza, Morelia 58337, Michoacán, Mexico

2 British Columbia Ministry of Forests, Lands and Natural Resource Operations and Rural Development, Vernon, BC V1T2X5, Canada; greg.oneill@gov.bc.ca

3 Centre for Forest Conservation Genetics and Department of Forest and Conservation Sciences, Faculty of Forestry, University of British Columbia, 3041-2424 Main Mall, Vancouver, BC V6T1Z4, Canada; sally.aitken@ubc.ca

4 Instituto de Investigaciones en Ecosistemas y Sustentabilidad, Universidad Nacional Autónoma de México (UNAM), Morelia 58190, Michoacán, Mexico; rlindig@iies.unam.mx

* Correspondence: csaenzromero@gmail.com

Abstract: Assisted migration of forest tree populations through reforestation and restoration is a climate change adaptation strategy under consideration in many jurisdictions. Matching climates in which seed sources evolved with near future climates projected for plantation sites should help reduce maladaptation and increase plantation health and productivity. For threatened tree species, assisted migration outside of the species range could help avert extinction. Here, we examine lessons, limitations, and challenges of assisted migration through the lens of three assisted migration field trials of conifers in Canada and Mexico: *Pinus albicaulis* Engelm., an endangered subalpine tree species in the mountains of western North America; the *Picea glauca* (Moench) Voss × *P. engelmannii* Parry ex Engelm hybrid complex, of great economic and ecological importance in western Canada, and *Abies religiosa* (Kunth) Schltdl. & Cham., a tree species that provides overwintering sites for the monarch butterfly. We conclude that: (a) negative impacts of climate change on productivity of *Picea glauca* × *P. engelmannii* may be mitigated by planting seed sources from locations that are 3 °C mean coldest month temperature warmer than the plantation; (b) it is possible to establish *Pinus albicaulis* outside of its current natural distribution at sites that have climates that are within the species' modelled historic climatic niche, although developing disease-resistant trees through selective breeding is a higher priority in the short term; (c) *Abies religiosa* performs well when moved 400 m upward in elevation and local shrubs (such as *Baccharis conferta* Kunth) are used as nurse plants; (d) new assisted migration field trials that contain populations from a wide range of climates tested in multiple disparate climates are needed, despite the costs; and (e) where naturalization of a migrated tree species in recipient ecosystem is viewed as undesirable, the invasive potential of the tree species should be assessed prior to large scale establishment, and stands should be monitored regularly following establishment.

Keywords: provenance test; genecology; reforestation; restoration; conservation; assisted migration; climatic change

Citation: Sáenz-Romero, C.; O'Neill, G.; Aitken, S.N.; Lindig-Cisneros, R. Assisted Migration Field Tests in Canada and Mexico: Lessons, Limitations, and Challenges. *Forests* **2021**, *12*, 9. https://dx.doi.org/10.3390/f12010009

Received: 28 November 2020
Accepted: 21 December 2020
Published: 23 December 2020

1. Introduction

Climate change is expected to decouple forest tree populations from the climates to which they have become locally adapted over many generations [1–4]. Local adaptation can involve critical trade-offs; frequently, populations from colder sites are more frost resistant, although they have less growth potential than populations from warmer locations that have genetically extended growing seasons [5–7], and populations from drier locations may be more drought resistant or tolerant [8,9]. Suitable climatic habitat for given biomes, tree species, and their populations is predicted to occur at more polar latitudes and higher elevations as climate change progresses [10–14]. Indeed, natural migration of some tree populations via seed dispersal to more polar latitudes and higher elevations has already been observed [15–17], although this is not happening at the speed needed to keep populations coupled with the climatic habitat to which they have evolved. Such ongoing decoupling is creating increasing stress for tree populations. Perhaps most evident is drought stress, which makes trees weaker and less able to defend against insect and pest attacks [18]. This cascade of abiotic and biotic stresses is resulting in massive mortality of tree populations around the world [19–21]. The largest negative impacts will likely happen at the rear-edge (lower latitudinal, elevational, or xeric limit) of the natural distribution of each tree species [22–25].

The decoupling of forest tree populations from their suitable climatic habitat presents an evolutionary dilemma: populations must adapt or migrate; otherwise, the likely outcome is extirpation [3]. In the very short term, individual trees may tolerate climate change through phenotypic plasticity [26], for example, by reducing leaf area and growth rate to tolerate increasing temperatures and decreasing precipitation. In the long run, adaptation could involve natural selection of individuals more tolerant of new stresses and evolutionary rescue, when genetic adaptation allows a population to recover from demographic effects initiated by environmental change that would otherwise cause extirpation [27]. Natural populations of forest trees usually have high levels of within-population genetic variation, due to their generally large effective population sizes and outcrossing mating systems [28]. Consequently, in most tree species, there is a wide range of different genotypes where natural selection can act to confer more resistance or tolerance to drought stress, pest outbreaks, or high temperatures in subsequent generations. However, if adaptation does not occur quickly enough and decoupling continues for several generations, then the average fitness of individuals within the population or the size of the population may drop too low for it to recover through adaptation and evolutionary rescue [27].

Due to the high likelihood that climate change will negatively impact forest health and productivity [13,29,30], assisted migration—the human assisted relocation of genotypes through reforestation and restoration intended to mitigate future impacts of climate change on forest health and productivity—is a climate change adaptation strategy that warrants critical examination.

Even the most critical views against assisted migration concede that more field experimental evidence of assisted migration is needed [31]. We agree that the risks and benefits of assisted migration need to be informed with more experimental evidence, particularly with well-designed provenance trials. Here, we present three case studies of field experiments established to evaluate assisted migration in forest trees. Two of these studies pertain to tree species of conservation concern (*Pinus albicaulis* (whitebark pine) in Canada and *Abies religiosa* (Sacred fir) in Mexico) and the other deal with economically important tree species (*Picea glauca* × *engelmannii* hybrid complex known as interior spruce, in Canada). These examples were selected as examples of assisted migration both within and outside the species' native geographic range. These field tests were designed expressly as tests of assisted migration, with the experimental design intended to assess the growth and survival of seed sources from climates (in general) warmer than the planting site. The detailed results of each of these case studies have already been published or are in preparation and are summarized here to provide examples of a range of experimental approaches and results, with the hopes of encouraging similar experiments with additional tree species globally.

The aim of this paper is to summarize and discuss what we have learned, including limitations and challenges to the establishment and maintenance of the field tests, based on cases that we consider illustrative from Canada and México, in order to document our current understanding and identify past pitfalls and future research priorities for new assisted migration experiments.

2. Key Assisted Migration Definitions

Assisted migration (AM), as an "umbrella" term, has several different names, some of which have different meanings, but all of which refer to the translocation of individuals in anticipation of or in response to environmental change. General terms include managed relocation, managed translocation, and conservation translocation [32]. Terminology varies somewhat between fields, whether for commercial forestry management, ecological restoration, or endangered species conservation [33–45]. For the purpose of our discussion, we consider that in a forestry or tree species conservation context, assisted migration is usually considered to involve planting tree seeds from sources with climates that are slightly warmer than that of the planting site. By nudging tree populations in the direction of climate change, assisted migration is expected to help maintain forest health and productivity by restoring populations to climates where they are best adapted [37,46–48].

Assisted migration in general can be categorized into two forms (after Peterson St-Laurent et al. [49]):

(a) *Assisted migration within current geographic range*, for simplicity referred to herein as assisted migration within range, when it involves the movement of populations within a species' current natural distribution. It has also been referred to as assisted gene flow [40] or assisted population migration [37].
(b) *Assisted migration outside of current geographic range*, referred herein as assisted migration outside of range, when it involves the movement of populations to locations just beyond a species' current distribution. This form of assisted migration has also been called assisted colonization [50] or assisted range expansion [37].

A third form of assisted migration, *exotic translocation*, involves long distance movement (sometimes trans-continental) of a population or species to locations where it is not likely to migrate naturally within the foreseeable future [37]. While intensive plantation forestry in many parts of the world is focused on non-native species (e.g., pines in the southern hemisphere), there is little demand for exotic translocations in North America because of the large number of tree species in most locations, particularly in biodiverse regions of Mexico. Consequently, we have few data upon which to evaluate long distance transfers and will not address this topic further in this report.

While assisted migration outside of range may be the form of assisted migration most commonly considered by the general public, and indeed by many forest scientists and foresters, the wide natural distribution of most North American tree species combined with the relatively short geographic and climate distances actually proposed for assisted migration [51] suggest that most AM in forestry will likely take the form of assisted migration within range.

3. Field Tests Examined

3.1. Canada. *Pinus albicaulis*

Materials and Climate Data

Pinus albicaulis (whitebark pine) is a high-elevation species foundational to treeline ecosystems in mountainous areas of western Canada and the United States. It has been listed as endangered under Canada's Species at Risk Act as populations are in steep decline over much of its range, primarily due to the introduced pathogen causing white pine blister rust (*Cronartium ribicola*), but also from mountain pine beetle (*Dendroctonus ponderosae*), fire suppression, and climate change. A species distribution model (SDM) projected a loss of climatically favorable habitat over most of its range in the USA under future climate scenarios [52]. A separate SDM also identified potential climate refugia in northwestern British Columbia (BC) outside of the current species range that already have contemporary climates suitable for this species and are projected to remain favorable to the end of this century (Figure 1) [53]. The apparent underfilling of the potential contemporary range of this species is typical of many North American tree species [54]. A field experiment was initiated in 2007 to (1) test the climate suitability projections of the latter model, and (2) evaluate the potential for assisted migration to establish populations of this species outside of the current range in areas predicted to be habitable [53].

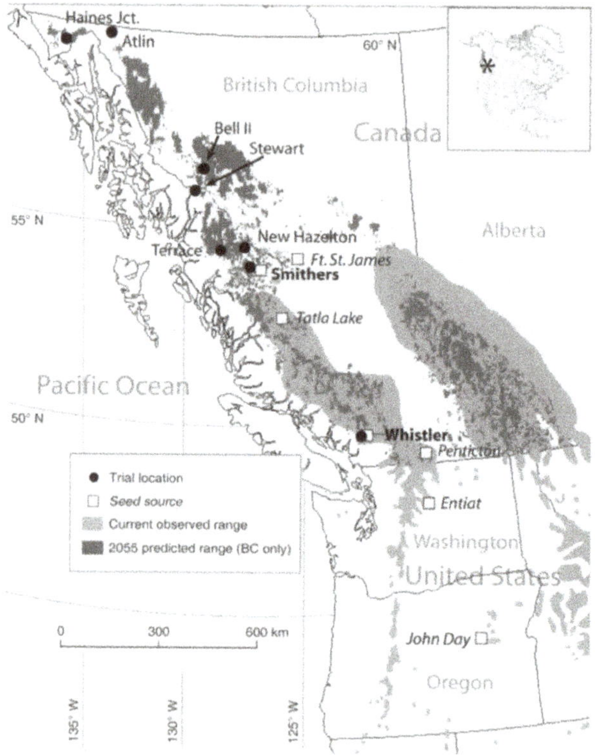

Figure 1. Location of *Pinus albicaulis* provenances and common garden locations for the assisted migration experiment, with current and 2055 predicted *Pinus albicaulis* ranges shown (from [53]).

Briefly, seed from six provenances of *Pinus albicaulis* was directly sown into common gardens in eight subalpine locations in BC, including sites within the current species range, sites north of the range predicted to be climatically suitable in 2055, and sites north of the species range predicted to be unsuitable in 2055 (Figure 1). Seeds rather than seedlings were sown to test whether sites were within the species ecological niche without circumventing life history stages. Half the seeds planted were extracted from cones and sown in fall 2007, untreated, while half were subject to a maturation and dormancy-breaking treatment, then sown in the spring of 2008. Data on seedling germination, survival, and growth were collected in 2008, 2009, 2010 (see [53]), and in 2017 (Reid, McLane, Chourmouzis and Aitken, in prep.). iButton temperature recorders were placed at the soil surface at each common garden location and provided records of air temperatures in the absence of snow, as well as duration of snowpack. In alpine treeline environments, microclimate varies greatly with microsite, so data regarding slope, soil depth and type, and surrounding vegetation were collected at each common garden location. Multinomial logistic models were used to analyze germination and survival data, while general linear models were used to analyze growth data. Predictor variables tested in models included provenance climate estimates for temperature and precipitation (obtained from ClimateBC), temperatures and snowmelt dates at common garden sites (from iButtons), and planting microsite characteristics.

3.2. Canada. *Picea glauca × engelmannii* Hybrid Complex (Interior Spruce)

Materials and Climate Data

As *Picea glauca* (Moench) Voss, *Picea engelmannii* Parry ex Engelm., and their hybrids (known locally as interior spruce, but herein referred to as *Picea glauca × engelmannii*) are among the most ecologically and economically important species in BC and throughout Canada [55], a large, long-term field provenance trial of *Picea glauca × engelmannii* was established by the BC government in 2005. The trial consisted of 128 natural stand populations and orchard seedlots from Canada and USA planted at each of 17 sites in BC, Yukon, and Alberta. Details regarding the populations, test sites and experimental design can be found in O'Neill et al. [56] (Figure 2).

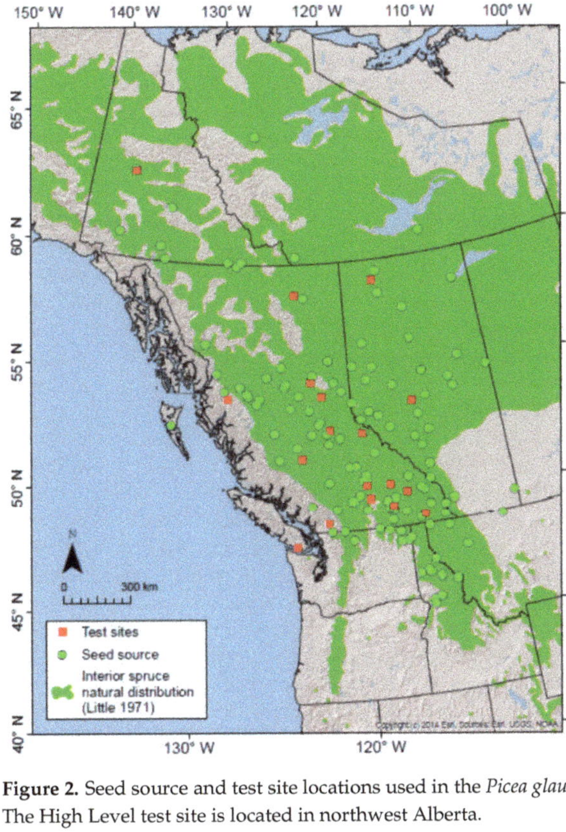

Figure 2. Seed source and test site locations used in the *Picea glauca × engelmannii* provenance trial. The High Level test site is located in northwest Alberta.

Height and survival were assessed at the end of the 10th field season. At each site, a productivity index (PI10) was obtained for each of the 97 natural stand populations by multiplying population mean height by population survival proportion. Values of mean coldest month temperature (MCMT) for the 1961–1990 period (hereafter referred to by its center year, 1975) were obtained for each provenance and test site from ClimateBC or ClimateNA (http://climatebc.ca/) version 6.22 [57]. Mean coldest month temperature was selected because it is frequently identified as being among the most important of climate variables in accounting for population differentiation in *Picea glauca × engelmannii* [51] and other North American temperate conifer species [58,59].

A universal transfer function (UTF), which predicts the productivity of any population at any location, was developed following methods of O'Neill et al. (2008; Appendix A). Individual transfer functions were first fitted to each site by relating population PI10 to population MCMT_tx (MCMT of the test site minus MCMT of the seed source) using a Cauchy function. Linear and quadratic relationships were then developed between the fitted coefficients and test site MCMT. Fitting of peaked transfer functions at the warmest and coldest sites was facilitated by the imputation of a single warm and cold anchor point population using an approach similar to that of Wang et al. [46]. Likewise, fitting of the scale parameter was enabled by the imputation of a single warm ghost test site using an approach similar to that of Rehfeldt, Leites, Joyce and Weiskittel [7]. For details of the imputation procedures see the Appendix A.

The impact of assisted migration was assessed in two ways. First, the climate of the vertex of each individual transfer function (i.e., the climate of the population expected to have the greatest productivity) was determined at each site. Second, fitted equations for the three Cauchy function parameters were substituted back into the transfer function, and response functions were estimated for a "local" population from MCMT = −18 °C, and a "migrated" population from MCMT = −15 °C, representing climates of northern and central BC, respectively. As response functions illustrate population productivity across a range of climates, responses of alternative seed sources over time can be compared to assess their relative merit over time when planted in a common climate.

3.3. *México. Abies religiosa (Sacred Fir, Oyamel)*

Abies religiosa (Sacred fir, oyamel) forms a high altitude (mostly 2800 to 3500 m), dense, shade-tolerant, moist, temperate forest in the Trans-Mexican Volcanic Belt, an east-west mountain chain containing the highest elevations in Mexico, between 19 and 20° N latitude [60–62], on which eastern migrating populations of monarch butterfly exclusively overwinter [63]. Some stands of this species have been severely fragmented due to their proximity to large urban areas [64]. Two high-elevation field provenance tests of *Abies religiosa* were established inside the core zone of the Monarch Butterfly Biosphere Reserve in México State to assess assisted migration. Specifically, the adaptation of populations from an elevation transect from the lowest to the highest elevational limit of the natural distribution of the species in the region was tested at the two high-elevation test sites. Ten populations represented by 2 year-old seedlings originating from seed collected between 3000 and 3450 m were planted in 2015 at Las Palomas (3440 m). Similarly, in 2017, six populations of 2 year-old seedlings originating from seed collected between 2960 and 3450 m were planted at Los Ailes (3360 m). Due to the extreme solar radiation and the large differences between day and night temperatures at these elevations, the effect of nurse plant cover provided by pre-existing natural shrubs (mostly *Baccharis conferta*) on survival and growth of the planted *Abies religiosa* seedlings was also assessed. Survival and growth were assessed for 3.5 and 1.5 years at Las Palomas and Los Ailes, respectively, providing results for 5.5 and 3.5 year-old seedlings (from germination), respectively [65].

4. Lessons

4.1. *Pinus albicaulis Can Be Migrated North of Its Current Species Distribution in Canada within the Predicted Climatic Niche If Adequate Snowpack Is Present*

Germination and 9 or 10 year survival of seedlings varied greatly among sites. Establishment from seed was expected to be low relative to seedlings as seed germination is often poor for *Pinus albicaulis*, because seeds of this species are eaten by many rodents and birds. In 2017, 7.4% of all seeds produced seedlings that survived. Survival was highest at the northernmost site within the species range (Smithers, BC, Canada), at the southernmost site outside of the species range (Hazelton, BC, Canada), and surprisingly, at the northernmost site (Haines, BC, Canada), approximately 600 km north of the species' current distribution limit near the Yukon border. Germination rates varied with population but were more a function of environmental effects on seed maturation and condition at the time of harvest than genetic differences among populations. Early results in 2009 and 2010 showed that pre-treated seeds germinated faster and at higher rates than untreated seeds, but this advantage had disappeared by 2017. In the longer term, while most untreated seeds had germination delayed by a year while they completed maturation in the soil, a characteristic of *Pinus albicaulis*, they were better synchronized with site seasonality than treated seeds. Treated seeds germinated immediately following planting regardless of site conditions and suffered greater mortality post-germination. Snow cover and snow-melt timing played a key role in average seedling survival. Sites with little or no protective snow cover in winter and those with deep snowpacks that melted late in the year both had poorer survival than those on sites with continuous snowpacks in winter that melted in April or May.

Pinus albicaulis is an extremely slow-growing species. Average height of living seedlings after eight (untreated seeds) or nine years (treated seeds) was less than 20 cm, although individual seedlings varied greatly in height. Seedlings at the northernmost site (Haines) were approximately twice as tall as seedlings at all other sites, on average, indicating that latitude per se is not critical to the species' ecological niche, but rather, temperature and snowpack play critical roles. While this site is far north, it is not far from the Pacific Coast, and has intermediate snowfall (equivalent to 1020 mm of rain), relatively mild temperatures (0.4 °C mean annual temperature), and the warmest average summer temperature of all common garden sites [53]. These results collectively suggest that *Pinus albicaulis* could be migrated to climatically favourable sites well north of its current species distribution. These are likely areas of colonization debt, i.e., post-glacial migration has been too slow to fill the available niche. If seedlings rather than seeds were used to nucleate conservation populations, the survival rate would likely be higher than the current study.

4.2. *Picea glauca* × *engelmannii* Hybrid Complex

Negative impacts of climate change on plantation productivity may be mitigated by planting seed sources from locations that are 3 °C MCMT warmer than the plantation.

The potential benefit of assisted migration is illustrated in the interior spruce genecology function for the test site at High Level, Alberta, where the most productive populations are those that originate from locations having a MCMT approximately 3 °C warmer than the plantation (i.e., populations that have been transferred coldward 3 °C MCMT) (Figure 3). The strong relationship between productivity and seed source climate at this site and others provides convincing evidence of the degree to which evolution has molded populations to be locally adapted.

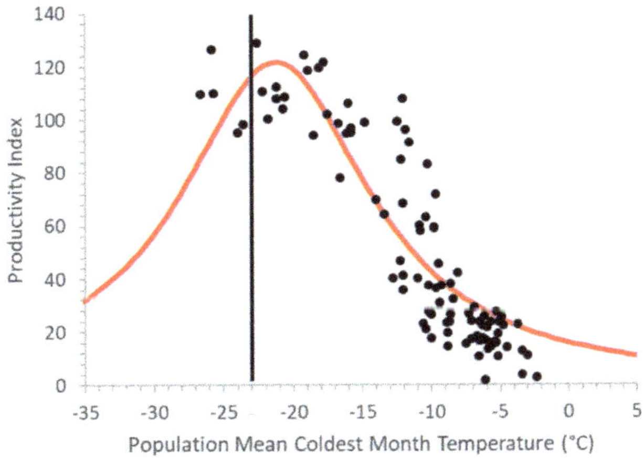

Figure 3. Genecology function for a provenance test site near the town of High Level, Alberta, showing that the most productive populations at the High Level site are those that originate from locations having a mean coldest month temperature (MCMT) approximately 3 °C warmer than the plantation. The vertical bar identifies the MCMT of High Level, Alberta (−23 °C).

Comparisons of the modelled trajectories of population productivity over time show that productivity is slightly greater for the local than the migrated population at establishment (i.e., when plantation MCMT = −18 °C); however, shortly after establishment (i.e., when plantation MCMT ≈ −15 °C), the relative productivity of the migrated population surpasses that of the local population, suggesting that assisted migration can help mitigate climate change impacts to productivity (Figure 4). Where forest productivity is low or long-lived plantations are desired (as in many conservation situations), the increasing advantage of migrated populations over time can be particularly valuable.

These analyses corroborate a growing body of literature suggesting that assisted migration can help forestall some of the negative impacts of climate change on forest productivity [47,59]. In lodgepole pine (*Pinus contorta*), where the results of long-term provenance trials have been widely examined, assisted migration is also expected to have a positive impact on productivity [46].

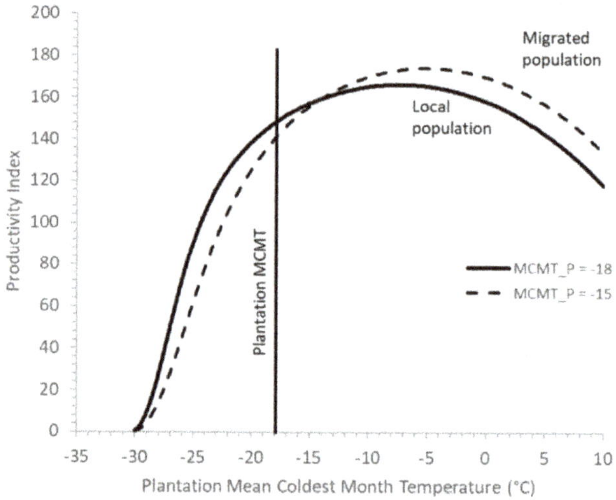

Figure 4. Modelled response function showing predicted relative productivity trajectory of "local" (solid line) and migrated (broken line) populations of *Picea glauca* × *engelmannii* hybrid complex (MCMT_P = −18 and −15 °C, respectively) when planted at the climate origin of the local population, represented by the vertical bar.

4.3. *México. Abies religiosa. The Species Performs Well When Moved 400 m Upward and Local Shrubs Are Used as Nurse Plants*

Upward transfer of seed sources by up to 400 m (equivalent to planting at a site 2 °C mean annual temperature colder than the seed source, given a lapse rate of 0.5 °C/100 m elevation [66]) does not appear to have any negative impacts on the planted seedlings if the seedlings are planted under the shade of local shrubs serving as nurse plants. When planted under nurse plants, survival was 83% averaged across sites and provenances; when planted outside the nurse plant crown, average survival was only 14%. Thus, at least for the case of this shade-tolerant *Abies religiosa* species, assisted migration must be protected assisted migration, by using nurse plants as a companion for the target migrated species [65]. This finding is consistent with that of St. Clair, Howe, and Kling [59] who found that Douglas-fir seed sources moved to sites 2 °C cooler than their origin still retained acceptable long-term survival and productivity.

5. Limitations

5.1. Canada. *Pinus albicaulis*

There are a number of limitations to implementing an operational assisted migration program for *Pinus albicaulis*. First, the major immediate challenge to this species is the pathogen *Cronartium ribicola*, which causes white pine blister rust. Genetic variation in rust resistance exists within this species, and there are numerous efforts underway to select for disease resistance and generate resistant seed for restoration. If this species is to be migrated, it makes sense to do this with rust resistant material as the disease will likely follow the species north. Current resources are probably best allocated to testing for disease resistance rather than assisted migration.

Secondly, *Pinus albicaulis* seeds are dispersed almost solely by the Clark's nutcracker (*Nucifraga columbiana*), a corvid that tears the non-dehiscent cones open and disperses the seeds into soil caches. If *Pinus albicaulis* populations are established farther north, well outside of the native range of this bird species, will the birds follow? If not, then migrated tree populations will not likely naturalize.

Finally, the assisted migration of this species is to a large extent an ethical issue and will require careful societal consideration of issues beyond biology. In BC, public support for assisted migration within species range is considerably higher than support for movements outside the current range [67]. In addition, opportunities to conserve this species and genetic diversity in climatic refugia within the existing species range should be acted upon [68]. Palmer and Larson [69] analyzed the ethics of *Pinus albicaulis* assisted migration and concluded that, on balance, there appear to be good reasons to justify moving this species. *Pinus albicaulis* enthusiasts may welcome this practice to increase the survival of this charismatic species. However, local people including indigenous communities who relate to subalpine and alpine ecosystems in recipient areas as they are today may resist the introduction of non-local species into those ecosystems, and such societal concerns might apply to other species. In fact, one indigenous community expressed opposition to this experiment at the time of establishment for this reason.

5.2. Canada. *Picea glauca × engelmannii* Hybrid Complex

The *Picea glauca × engelmannii* provenance trial contains one of the widest ranges of population and test site climates in North America, so one may expect a high likelihood of observing local adaptation at these sites. Nonetheless, transfer functions were nearly flat at almost half of the test sites at age six [56], suggesting (at those sites) that the species complex lacks local adaptation. Further, where local adaptation was observed, it varied widely, and was only weakly related to site climate. Sites that did not display strong local adaptation may have escaped extreme climate events or biotic agents or may have lacked local soil edaphic conditions needed to reveal local adaptation (i.e., needed to translate genetic differences among populations into phenotypic differences) [70]. Older tests or different test site locations may have resulted in stronger evidence of local adaptation or more test sites showing local adaptation. Therefore, when developing BC's climate-based seed transfer system, its designers chose to err on the side of caution by including data from only those sites having the steepest transfer functions where local adaptation was strongest [48].

How Far Is Too Far? Determining a Suitable Migration Distance

The climate distance populations are migrated in assisted migration initiatives (i.e., the migration distance) that can significantly impact the success of those efforts; migration distances that are too short may be ineffective, while those that are too long may result in early mortality due to cold injury. Comparisons of modelled response functions for populations from a range of climates can help identify optimum migration distances by identifying seed source climates expected to maximize productivity over the expected lifetime of the plantation ([47]; see also *Picea glauca × engelmannii* example, Figure 4).

Nonetheless, establishing plantations with seed sources whose 1940s climate matches a plantation's current climate would address any adaptation lag that has arisen since the onset of rapid anthropogenic climate warming in the 1950s. However, accounting solely for anthropogenic climate change by attempting to optimize adaptation at establishment could compromise adaptation in later decades, given continued climate change, especially for long-lived species. Therefore, the optimum migration distance should also consider climate change during the lifespan of a plantation by weighing the risk of maladaptation during seedling establishment with the risk of maladaptation towards the end of a plantation's lifespan [51]. A recently implemented assisted migration system in BC adopted the projected climate at one quarter of the plantation's expected lifespan as the adaptation target date [48].

The logic employed in this proposal is not a radical or risky departure from previous seed sourcing practices, as it attempts to re-establish pre-existing genotype–climate relationships and does not depend on projections of the late 21st century climate. Arguably, the migration distance calculation proposed here, being dependent on only the 1931–1960 climate mean of the seed source and the projected plantation climate of approximately 2040, is less prone to error than migration distances interpreted from adaptation lags shown in transfer functions (Figure 3) or by comparison of modelled response functions for populations from a range of climates (Figure 4), which can vary considerably with transfer range, model function, tree age, and different dependent and independent variables.

5.3. *México. Abies religiosa. Too Short Observational Period*

Due to the young ages of the field trials, interannual climatic variability may not have been fully represented, and population differentiation may not be fully expressed. Although the Las Palomas test site experienced an historic extreme winter storm resulting in the windfall of about 25,000 trees in a single night [65,71], during the observation period, there were no extreme heat waves or droughts that might have caused mortality. The short period of observation is directly linked to budget limitations.

6. Challenges

6.1. *Making Large Multisite Provenance Field Trials More Affordable*

Provenance tests were first used to identify superior seed sources for reforestation, and subsequently, to quantify safe seed transfer distance [72]. New analytical approaches have extended their application by providing estimates of impacts of seed transfer [73] and climate change [34,46,74–77] on the growth and survival of tree populations. However, few older multisite provenance trials contain a sufficiently wide climate range among populations or sites to allow accurate estimation of seed transfer or climate change impacts [3]. Examples of trials containing wide population or site sampling include *Quercus petraea* in the European Union [78]; *Pinus sylvestris* in Scandinavia [79]; the Assisted Migration Adaptation Trial in Canada and the USA [80]; and trials of *Pinus sylvestris* in the former Soviet Union [81–83].

Remarkably few older provenance trials contain test sites located outside the contemporary species distribution [78,84], limiting the ability to assess assisted migration outside of range. Further, costs of provenance trials can be prohibitive where numerous tree species are planted, or unjustifiable where few seedlings are planted. Nonetheless, collaboration among jurisdictions (provinces, states, and countries) can reduce costs to individual jurisdictions, while increasing the climate range of test populations and sites [84], arguably the most significant limitation of early provenance trials. Careful selection of populations (i.e., sampling widely and uniformly across climates) could reduce numbers of populations needed in future trials. Additionally, careful blocking, incomplete block alpha-design [85], and use of single-tree plots can reduce the number of seedlings needed for each population to have adequate statistical power to characterize patterns of adaptive variation. More strategic selection of populations and test sites using geographic information systems and climate models, sampling a wider climate range, and careful prioritization of species to be tested could reduce costs while increasing effectiveness [34]. Demonstration of similar climate clines for adaptation traits among a wide range of tree species mitigates the need for provenance trials for all species [6,86].

6.2. Replications over Time

As seedlings are perhaps the most vulnerable life stage, it would be ideal to replicate field experiments across several years, in order to capture effects of interannual variation on survival at early ages. Both longer periods of observation and among-year replications of experiments can be difficult to achieve due to the short-term nature of research funding, and an overall lack of funding, particularly for México.

6.3. The Dilemma of Species Prioritization in Megabiodiverse Countries

Given many tree species and limited resources, it is challenging to prioritize species for assisted migration research, either for conservation or for wood production objectives. This is particularly the case in México, a country with incredible biodiversity and very limited funding. Given the shortage of funding, and the need to test many species, how should priorities be set for assisted migration research? We suggest that priority should be granted first to species having a high risk of extirpation in the short term (e.g., Mexican spruces: *Picea chihuahuana*, *P. martinezii* and *P. mexicana*; see [50]). The second priority should go to species of the highest economic potential (e.g., *Pinus patula* and *Pinus pseudostrobus*). These experiments would provide the basis for adaptative management strategies needed to address climate change, particularly in those human communities that depend on sustainable forest management, for example, the indigenous communities of Nuevo San Juan Parangaricutiro and Cherán, of the Purhépecha ethnic group that manage *Pinus pseudostrobus*, at Michoacán state, in central-western México and the Ixtlán de Juárez, Oaxaca, Southern México, Zapoteco ethnic group that manage *Pinus patula*. The third high-priority group should be species that have the highest ecological and biological conservation values (e.g., *Abies religiosa* for establishing and conserving climate-resilient monarch butterfly overwintering sites).

6.4. When Poleward Migration Is Not an Option

The primary geographic direction of migration will vary among species and with local topography. In the case of *Abies religiosa*, the upper altitudinal limit of the natural distribution of the species approximately coincides with the summits of the Monarch Butterfly Biosphere Reserve (MBBR), between 3550 and 3600 m. It may not be possible to establish *A. religiosa* populations beyond the contemporary upper altitudinal limit inside the MBBR. An option would be to conduct assisted migration outside of range of *A. religiosa* at altitudes above 3600 m, on higher altitude mountains, such as extinct volcanoes along the Mexican Transvolcanic Belt: Nevado de Toluca, Popocatépetl, Iztachíhuatl, La Malinche, Cofre de Perote and Pico de Orizaba [87]. *A. religiosa* at present is also distributed on each of these mountains, but also with an upper elevation limit between 3550 and 3600 m. The challenge, then, would be to establish *A. religiosa* populations, for example, between 3800 and 4000 m on any of those mountains, as potential future sites for monarch butterfly overwintering sites. The task would not be easy, because at 3800 to 4000 m, the vegetation is largely alpine grassland (e.g., *Muhlenbergia macroura* with low density of *Pinus hartwegii*), where the extreme low temperatures make the environment quite harsh. Thus, the use of nurse plants to establish assisted migration tests at those altitudes would be needed. That of course assumes that monarch butterflies would be able to locate these afforested *A. religiosa* stands and have sufficient plasticity to overwinter at these sites.

6.5. Tackling Uncertainty by Promoting Genetic Diversification

While assisted migration has received significant attention as a climate change adaptation strategy [88,89], diversification of forestry practices (e.g., use of alternative silviculture systems, planting more tree species, prioritizing harvest in areas of high risk of disturbance) and genetic diversification in particular have also been proposed to buffer some of the uncertainty in climate change projections [1,31,90]. Crop diversification is commonly practiced by farmers to mitigate uncertainty associated with crop prices and variability in production due to pest outbreaks and extreme weather [91], and portfolio diversification is widely used by investors to stabilize returns from fluctuating markets [92]. To the extent that populations differ in adaptation, they will differ in their capacity to tolerate disturbances, including climate extremes. The need for genetic diversity in reforestation materials has been examined [93] and standards are in place in many jurisdictions [94]. Greater genetic diversity could be achieved in plantation forestry by planting multiple seedlots from different climates, or by increasing effective population size in orchard or natural stand seed collections [95]. Borrowing from economic theory, Crowe and Parker [96] proposed a robust climate change adaptation strategy in which the selection of genetic material (sets of genotypes) to be used in reforestation is optimally adapted to multiple, equally probable future climates. Importantly, genetic diversification and assisted migration are independent strategies, they can be used separately or jointly. Nonetheless, while provenance trials can provide strong support for assisted migration, and guidelines for assisted migration (i.e., migration distance) have been presented in this report and elsewhere [48,51], few trials or specific guidelines exist for genetic diversification as a climate change adaptation strategy (but see [31]).

7. Mitigating Risks

7.1. Introduction of Species That Might Become Invasive

Several concerns have been raised regarding assisted migration within range and, in particular, assisted migration outside of range [67]. The most frequent concern [38] pertains to assisted migration outside of range disrupting or displacing local ecological communities and that relocated populations could become invasive [97,98]. An assessment of the relationship between distance from source to introduction site and invasiveness by Mueller and Hellmann [99] concluded that while the proportion of species migrated within regions of continents that becomes invasive is relatively small, the impacts of those invasions can be large. Therefore, invasiveness needs to be addressed as part of assisted migration outside of range programs, through assessment of risks before translocations, and monitoring during and after implementation. Before implementing a program of assisted migration outside of range, it might be useful to consider if the species has weedy characteristics (e.g., high fecundity, early age of reproduction, wide soil type tolerance) or close relatives that have shown invasive behavior. Considering that assisted migration outside of range in forestry will mostly be aimed at woody species, Daehler [100] concluded through a taxonomic analysis that nitrogen-fixers and clonal trees show the highest risk of becoming natural area invaders, along with species that are primarily aquatic or semi-aquatic, and grasses. Therefore, special attention should be taken when considering assisted migration outside of range with these types of species.

During and after implementation of a program of operational assisted migration outside of range, population health and naturalization of the planted materials should be monitored regularly. Introduced species may show no invasive behavior for decades after planting, due to lags in reproduction and establishment, and then spread rapidly, a particular concern for invasion biology [101–103]. In the case of assisted migration outside of range, rapid spread may be considered a success since the species is being introduced intentionally or may be a concern due to ecological effects on the recipient ecosystem. How this is viewed will vary with individuals and their priorities for ecosystems.

7.2. Mismatch of Biotic Interactions

Concern exists that assisted migration within and outside of range may cause a mismatch in biotic interactions [31]. Empirical evidence on this issue is accumulating. Camarretta et al. [104] concluded in a study of species and provenance translocation into different community assemblages that performance was not affected by community context. Additionally, experimental evidence has suggested that migrating plants with the soil biota of the site of origin increases plant performance in the target site [105], and that assisted migration outside of range is possible along elevation gradients for some rare species [106]. These, and other studies, suggest that biotic interactions and the more complex set of interactions among biotic and abiotic factors will influence the outcome of assisted migration within or outside of range. More trials and close monitoring of performance are necessary to further clarify this issue for individual species as it will be difficult to generalize across species.

7.3. Outbreeding Depression

Genetic risks of assisted migration outside of range include outbreeding depression, disruption of local adaptation by non-climatic factors, and population replacement [40]. Outbreeding depression is only likely in wind-pollinated tree species when individuals are moved between long-diverged populations, e.g., subspecies or varieties. If outbreeding depression exists but is weak, natural selection will remove deleterious immigrant alleles over time, allowing mean fitness to recover and surpass that of the local population, according to population simulation modelling [40]. Genetic incompatibilities causing outbreeding depression are unlikely in widespread forest tree species such as those discussed here. Field trials and good knowledge of the biology and silvics of species can provide information about local adaptation to environmental factors such as soil type, pests, pathogens, or mutualists. If translocated individuals have higher fitness than the resident population, and if the population is naturally regenerating rather than being planted, over time the genetic composition of the population may gradually be replaced by the introduced genotypes. As this will increase the average fitness of the local population, this replacement is more likely a benefit than a risk.

Furthermore, to reduce ecological risks, care should be taken to avoid unintended introductions of species hitch-hiking with planted seedlings and their soil, as hitch-hikers, non-pathogenic in their native environment, may become pathogenic on naïve hosts in their new environments.

7.4. Balancing the Risks: The Core Dilemma

Assisted migration, in particular assisted migration outside of range, is perceived by some observers as a risky management measure, because it could cause: (a) negative interactions with other organisms of the recipient community (for example, when the translocated genotypes become invasive and displace other local plant species; [31]), and (b) outbreeding depression, disruption of local adaptation to non-climatic factors, and disruption of mutualist or parasitic relationships among species [31,40]. In our view, such risks exist to varying degrees, but need to be considered in relation to the risks of inaction, which can be dire, as numerous modelling studies have predicted [14,50,53,58]. What will happen to forest communities in the future if no interventions are made under different scenarios of greenhouse gas emissions, i.e., Representative Concentration Pathway (RCP) 8.5 (pessimistic) or RCP 4.5 (intermediate)? The fact is that atmospheric CO_2 concentrations are currently 414 ppm (July 2020 record; https://www.esrl.noaa.gov/gmd/ccgg/trends/), and climate records are being broken nearly every year [107]. Additionally, a more worrisome sign is that forest decline is already well underway in many forests of the world, linked to heat waves, drought stress, and climate-associated insect and disease outbreaks [19–21,45]. For *A. religiosa*, for example, there is already growing evidence of the lack of regeneration at lower elevations of the species' distribution [108,109]. Thus, we need to avoid inaction because that might make us simply witnesses of an upcoming ecological disaster.

8. Conclusions

Evidence from these field tests suggests that in general, tree populations grow well when moved to planting locations 2 to 3 °C mean annual temperature colder than their origin. Tree species can also establish and grow well when moved outside of their natural distributions to sites with climates within their modelled historic climatic niches.

In the case of a shade-tolerant species such as *Abies religiosa*, the use of nurse plants may be necessary to protect the target species by ameliorating extreme temperatures, either day/night fluctuations, the coldest temperatures of winter, or the warmest temperatures during the warm, dry season (March–May in México). With the use of nurse plants, an upward shift of populations by up to 400 m appears to have no negative impact on growth or survival.

However, as heat waves and unusual drought periods become more frequent, prolonged, and intense, it might be advisable to use nurse plants to increase the survival of the targeted seedlings even for shade-intolerant pine species. For example, British Columbia experienced two extreme droughts during the summer of 2017 and 2018, which caused significant mortality in young plantations [110]. Whereas shade from adjacent vegetation has long been considered undesirable for shade-intolerant tree species in British Columbia, frequent, extreme droughts may require reconsideration of that view in drought-prone regions or coarse, shallow soils.

Long-term, multi-site field tests located in disparate climates are needed to develop accurate climate transfer and response functions for priority species. Collaboration among neighbouring jurisdictions and careful field test design can help reduce costs.

For some species such as *Pinus albicaulis*, developing disease-resistant trees through selective breeding may be a higher priority than assisted migration, and have a greater impact on tree health and population size. However, once resistant material is available, it can be integrated into a restoration program that includes assisted migration.

Planning of field provenance trials requires a long-term budget and human resource commitment. Collecting and extracting seed, producing seedlings, and locating and preparing test sites can take 5 years; an additional 10–15 years of field testing, with frequent test site maintenance, may be needed before extreme climate events and pest disturbances reveal patterns of adaptation.

Assessment of invasion risk is warranted prior to, and regular monitoring after, implementation of an assisted migration test or program, particularly for tree species having weedy traits, if naturalization of the migrated species is not desired. Careful screening for hitch-hiking pests on seedlings prior to migration out of range is also recommended.

Author Contributions: C.S.-R. and S.N.A. conceived the paper. C.S.-R., G.O. and S.N.A. provided experimental data and analysis per species. R.L.-C. developed the mitigating risk section. C.S.-R., G.O. and S.N.A. led the writing. All authors have read and agreed to the published version of the manuscript.

Funding: Funding was provided by the Coordinación de la Investigación Científica of the Universidad Michoacana de San Nicolás de Hidalgo to C.S.-R.

Acknowledgments: This paper is an undertaking of the Forest Genetic Resources Working Group/North American Forest Commission/Food and Agricultural Organization of the United Nations. Thanks to Christine Chourmouzis for phenotyping the *Pinus albicaulis* experiment in 2018, and to Iain Reid for preliminary analyses of the resulting data. The *Picea glauca* × *engelmannii* provenance trial map was created by Amy Vallarino. Thanks to Barry Jaquish, co-investigator of the *Picea glauca* × *engelmannii* provenance trial.

Conflicts of Interest: The authors declare no conflict of interest.

Appendix A.

Methodological details for the analysis of *Picea glauca* × *engelmannii* hybrid complex (interior spruce) field tests.

Appendix A.1. Anchor Points for P. glauca × *engelmannii Analysis*

As the climate ranges of populations and test sites were mostly over-lapping, it became necessary to impute anchor points in order to develop peaked transfer functions at the warmest and coldest sites. Preliminary analyses identified mean coldest month temperature (°C) transfer distance (MCMT_tx, MCMT of the test site minus MCMT of the seed source) as the transfer distance climate variable most strongly related to PI10. Therefore, the mean MCMT_tx among the 2% of population X site transfers that experienced the greatest "coldward" transfer and the 1% of population X site ransfers that experienced the greatest "warmward" transfer was calculated. Estimated P110 of a "local" population (i.e., P110L, a population growing in a climate equal to its origin, that is, at MCMT_tx = 0) was obtained from the preliminary analyses, and used to calculate the relative productivity index, PI10R (PI10/PI10L), of each population X site combination, from which mean PI10R of transfers in each of the two groups (2% coldest and 1% warmest transfers) was calculated. P110L was multiplied by mean PI10R to obtain a single PI10 anchor point value at each site.

Appendix A.2. Universal Transfer Function and Ghost Test Site for P. glauca × engelmannii Analysis

A universal transfer function (UTF), which predicts the productivity of any population at any location, was developed following methods of O'Neill et al. (2008). Transfer functions were first fitted to each site by relating population PI10 to population MCMT_tx using a Cauchy function. Linear and quadratic relationships were then developed between the three coefficients and test site MCMT. Due to the lack of a test sites towards the upper thermal limit of interior spruce, it was impossible to develop a peaked function for the scale parameter A. Therefore, using an approach similar to that of Rehfeldt et al. (2017), a "ghost" point with A = 0 (all planted populations expected to die) was generated and used in fitting the scale parameter equation. The ghost point was created at MCMT = 6.8 °C by adding 10 °C to the warmest interior BC location at which interior spruce was found on the tree species layer of the provincial geospatial database.

Appendix A.3. Appendix References

O'Neill, G.A., Hamann, A., and Wang, T. 2008. Accounting for population variation improves estimates of the impact of climate change on species' growth and distribution. Journal of Applied Ecology 45: 1040–1049. http://onlinelibrary.wiley.com/doi/10.1111/j.1365-2664.2008.01472.x/full.

Rehfeldt GE, Leites LP, Joyce DG, Weiskittel AR. 2017. Role of population genetics in guiding ecological responses to climate. Global Change Biology 24:858–868.

References

1. Ledig, F.T.; Kitzmiller, J.H. Genetic strategies for reforestation in the face of global climate change. *For. Ecol. Manag.* **1992**, *50*, 153–169. [CrossRef]
2. Rehfeldt, G.E.; Wykoff, W.R.; Ying, C.C. Physiologic plasticity, evolution, and impacts of a changing climate on *Pinus contorta*. *Clim. Chang.* **2001**, *50*, 355–376. [CrossRef]
3. Aitken, S.N.; Yeaman, S.; Holliday, J.A.; Wang, T.; Curtis-McLane, S. Adaptation, migration or extirpation: Climate change outcomes for tree populations. *Evol. Appl.* **2008**, *1*, 95–111. [CrossRef] [PubMed]
4. Rehfeldt, G.E.; Jaquish, B.C.; López-Upton, J.; Sáenz-Romero, C.; St Clair, J.B.; Leites, L.P.; Joyce, D.G. Comparative genetic responses to climate for the varieties of *Pinus ponderosa* and *Pseudotsuga menziesii*: Realized climate niches. *For. Ecol. Manag.* **2014**, *324*, 126–137. [CrossRef]
5. Rehfeldt, G.E. Ecological genetics of *Pinus contorta* from the Rocky Mountains (USA): A synthesis. *Silv. Genet.* **1988**, *37*, 131–135.
6. Aitken, S.N.; Bemmels, J.B. Time to get moving: Assisted gene flow of forest trees. *Evol. Appl.* **2016**, *9*, 271–290. [CrossRef]
7. Rehfeldt, G.E.; Leites, L.P.; Joyce, D.G.; Weiskittel, A.R. Role of population genetics in guiding ecological responses to climate. *Glob. Chang. Biol.* **2017**, *24*, 858–868. [CrossRef]
8. Sáenz-Romero, C.; Guzmán-Reyna, R.R.; Rehfeldt, G.E. Altitudinal genetic variation among *Pinus oocarpa* populations in Michoacán, Mexico: Implications for seed zoning, conservation, tree breeding and global warming. *For. Ecol. Manag.* **2006**, *229*, 340–350. [CrossRef]
9. Alvarez-Maldini, C.; Acevedo, M.; Dumroese, R.K.; González, M.; Cartes, E. Intraspecific variation in drought response of three populations of *Cryptocarya alba* and *Persea lingue*, two native species from mediterranean central Chile. *Front. Plant Sci.* **2020**, *11*, 1–15. [CrossRef]
10. Peñuelas, J.; Ogaya, R.; Boada, M.S.; Jump, A. Migration, invasion and decline: Changes in recruitment and forest structure in a warming-linked shift of European beech forest in Catalonia (NE Spain). *Ecography* **2007**, *30*, 829–837. [CrossRef]

11. Iverson, L.R.; Prasad, A.M.; Matthews, S.N.; Peters, M. Estimating potential habitat for 134 eastern US tree species under six climate scenarios. *For. Ecol. Manag.* **2008**, *254*, 390–406. [CrossRef]
12. Rehfeldt, G.E.; Crookston, N.L.; Sáenz-Romero, C.; Campbell, E.M. North American vegetation model for land-use planning in a changing climate: A solution to large classification problems. *Ecol. Appl.* **2012**, *22*, 119–141. [CrossRef]
13. Rehfeldt, G.E.; Jaquish, B.C.; Sáenz-Romero, C.; Joyce, D.G.; Leites, L.P.; St Clair, J.B.; López-Upton, J. Comparative genetic responses to climate in the varieties of *Pinus ponderosa* and *Pseudotsuga menziesii*: Reforestation. *For. Ecol. Manag.* **2014**, *324*, 147–157. [CrossRef]
14. Gómez-Pineda, E.; Sáenz-Romero, C.; Ortega-Rodríguez, J.M.; Blanco-García, A.; Madrigal-Sánchez, X.; Lindig-Cisneros, R.; Lopez-Toledo, L.; Pedraza-Santos, M.E.; Rehfeldt, G.E. Suitable climatic habitat changes for Mexican conifers along altitudinal gradients under climatic change scenarios. *Ecol. Appl.* **2020**, *30*, e02041. [CrossRef] [PubMed]
15. Lenoir, J.; Gégout, J.-C.; Marquet, P.; De Ruffray, P.; Brisse, H. A significant upward shift in plant species optimum elevation during the 20th century. *Science* **2008**, *320*, 1768–1771. [CrossRef] [PubMed]
16. Alfaro-Ramírez, F.U.; Arredondo-Moreno, J.T.; Pérez-Suárez, M.; Endara-Agramont, Á.R. *Pinus hartwegii* Lindl. treeline ecotone: Structure and altitudinal limits at Nevado de Toluca, Mexico. *Rev. Chapingo. Ser. Cienc. For. Ambiente* **2017**, *23*, 261–273. [CrossRef]
17. Trant, A.; Higgs, E.; Starzomski, B.M. A century of high elevation ecosystem change in the Canadian Rocky Mountains. *Sci. Rep.* **2020**, *10*, 1–10. [CrossRef] [PubMed]
18. Alfaro, R.I.; Fady, B.; Vendramin, G.G.; Dawson, I.K.; Fleming, R.A.; Sáenz-Romero, C.; Lindig-Cisneros, R.A.; Murdock, T.; Vinceti, B.; Navarro, C.M. The role of forest genetic resources in responding to biotic and abiotic factors in the context of anthropogenic climate change. *For. Ecol. Manag.* **2014**, *333*, 76–87. [CrossRef]
19. Allen, C.D.; Macalady, A.K.; Chenchouni, H.; Bachelet, D.; McDowell, N.; Vennetier, M.; Kitzberger, T.; Rigling, A.; Breshears, D.D.; Hogg, E.T. A global overview of drought and heat-induced tree mortality reveals emerging climate change risks for forests. *For. Ecol. Manag.* **2010**, *259*, 660–684. [CrossRef]
20. Allen, C.; Breshears, D.; McDowell, N. On underestimation of global vulnerability to tree mortality and forest die-off from hotter drought in the Anthropocene. *Ecosphere* **2015**, *6*, 1–55. [CrossRef]
21. Hogg, E.H.; Michaelian, M.; Hook, T.I.; Undershultz, M.E. Recent climatic drying leads to age-independent growth reductions of white spruce stands in western Canada. *Glob. Chang. Biol.* **2017**, *23*, 5297–5308. [CrossRef] [PubMed]
22. Hampe, A.; Petit, R.J. Conserving biodiversity under climate change: The rear edge matters. *Ecol. Lett.* **2005**, *8*, 461–467. [CrossRef] [PubMed]
23. Jump, A.S.; Hunt, J.M.; Penuelas, J. Rapid climate change-related growth decline at the southern range edge of *Fagus sylvatica*. *Glob. Chang. Biol.* **2006**, *12*, 2163–2174. [CrossRef]
24. Mátyás, C. Forecasts needed for retreating forests. *Nature* **2010**, *464*, 1271. [CrossRef] [PubMed]
25. Mátyás, C.; Berki, I.; Czúcz, B.; Gálos, B.; Móricz, N.; Rasztovits, E. Future of beech in Southeast Europe from the perspective of evolutionary ecology. *Acta Silv. Lignaria Hung.* **2010**, *6*, 91–110.
26. Fréjaville, T.; Fady, B.; Kremer, A.; Ducousso, A.; Benito Garzón, M. Inferring phenotypic plasticity and population responses to climate across tree species ranges using forest inventory data. *Glob. Ecol. Biogeogr.* **2019**, *28*, 1259–1271. [CrossRef]
27. Gonzalez, A.; Ronce, O.; Ferriere, R.; Hochberg, M.E. Evolutionary rescue: An emerging focus at the intersection between ecology and evolution. *Philosoph. Trans. R. Soc. B* **2012**, *368*, 1–8. [CrossRef]
28. Hamrick, J.L.; Godt, M.J.W.; Sherman-Broyles, S.L. Factors influencing levels of genetic diversity in woody plant species. *New For.* **1992**, *6*, 95–124. [CrossRef]
29. St Clair, J.B.; Howe, G.T. Genetic maladaptation of coastal Douglas-fir seedlings to future climates. *Glob. Chang. Biol.* **2007**, *13*, 1441–1454. [CrossRef]
30. Michaelian, M.; Hogg, E.H.; Hall, R.J.; Arsenault, E. Massive mortality of aspen following severe drought along the southern edge of the Canadian boreal forest. *Glob. Chang. Biol.* **2011**, *17*, 2084–2094. [CrossRef]
31. Bucharova, A. Assisted migration within species range ignores biotic interactions and lacks evidence. *Restorat. Ecol.* **2017**, *25*, 14–18. [CrossRef]
32. Richardson, D.M.; Hellmann, J.J.; McLachlan, J.S.; Sax, D.F.; Schwartz, M.W.; Gonzalez, P.; Brennan, E.J.; Camacho, A.; Root, T.L.; Sala, O.E. Multidimensional evaluation of managed relocation. *Proc. Natl. Acad. Sci. USA* **2009**, *106*, 9721–9724. [CrossRef] [PubMed]
33. McLachlan, J.S.; Hellmann, J.J.; Schwartz, M.W. A framework for debate of assisted migration in an era of climate change. *Conserv. Biol.* **2007**, *21*, 297–302. [CrossRef] [PubMed]
34. Wang, T.; O'Neill, G.A.; Aitken, S.N. Integrating environmental and genetic effects to predict responses of tree populations to climate. *Ecol. Appl.* **2010**, *20*, 153–163. [CrossRef] [PubMed]
35. Gray, L.K.; Gylander, T.; Mbogga, M.S.; Chen, P.-y.; Hamann, A. Assisted migration to address climate change: Recommendations for aspen reforestation in western Canada. *Ecol. Appl.* **2011**, *21*, 1591–1603. [CrossRef] [PubMed]
36. Gray, L.K.; Hamann, A. Strategies for reforestation under uncertain future climates: Guidelines for Alberta, Canada. *PLoS ONE* **2011**, *6*, e22977. [CrossRef]
37. Leech, S.M.; Almuedo, P.L.; O'Neill, G. Assisted migration: Adapting forest management to a changing climate. *J. Ecosyst. Manag.* **2011**, *12*, 18–34. Available online: http://jem.forrex.org/index.php/jem/article/view/91/98 (accessed on 15 November 2020).

38. Hewitt, N.; Klenk, N.; Smith, A.L.; Bazely, D.R.; Yan, N.; Wood, S.; MacLellan, J.I.; Lipsig-Mumme, C.; Henriques, I. Taking stock of the assisted migration debate. *Biol. Conserv.* **2011**, *144*, 2560–2572. [CrossRef]
39. Pedlar, J.H.; McKenney, D.W.; Aubin, I.; Beardmore, T.; Beaulieu, J.; Iverson, L.; O'Neill, G.A.; Winder, R.S.; Ste-Marie, C. Placing forestry in the assisted migration debate. *BioScience* **2012**, *62*, 835–842. [CrossRef]
40. Aitken, S.N.; Whitlock, M.C. Assisted gene flow to facilitate local adaptation to climate change. *Annu. Rev. Ecol. Evol. Syst.* **2013**, *44*, 367–388. [CrossRef]
41. Lu, P.; Parker, W.H.; Cherry, M.; Colombo, S.; Parker, W.C.; Man, R.; Roubal, N. Survival and growth patterns of white spruce (*Picea glauca* [M oench] Voss) rangewide provenances and their implications for climate change adaptation. *Ecol. Evol.* **2014**, *4*, 2360–2374. [CrossRef] [PubMed]
42. McKenney, D.W.; Pedlar, J.H.; Lawrence, K.; Papadopol, P.; Campbell, K.; Hutchinson, M.F. Change and evolution in the plant hardiness zones of Canada. *BioScience* **2014**, *64*, 341–350. [CrossRef]
43. Castellanos-Acuña, D.; Lindig-Cisneros, R.; Sáenz-Romero, C. Altitudinal assisted migration of Mexican pines as an adaptation to climate change. *Ecosphere* **2015**, *6*, 1–16. [CrossRef]
44. Koralewski, T.E.; Wang, H.-H.; Grant, W.E.; Byram, T.D. Plants on the move: Assisted migration of forest trees in the face of climate change. *For. Ecol. Manag.* **2015**, *344*, 30–37. [CrossRef]
45. Sáenz-Romero, C.; Mendoza-Maya, E.; Gómez-Pineda, E.; Blanco-García, A.; Endara-Agramont, A.R.; Lindig-Cisneros, R.; López-Upton, J.; Trejo-Ramírez, O.; Wehenkel, C.; Cibrián-Tovar, D. Recent evidence of Mexican temperate forest decline and the need for ex situ conservation, assisted migration, and translocation of species ensembles as adaptive management to face projected climatic change impacts in a megadiverse country. *Can. J. For. Res.* **2020**, *50*, 843–854. [CrossRef]
46. Wang, T.; Hamann, A.; Yanchuk, A.; O'neill, G.; Aitken, S. Use of response functions in selecting lodgepole pine populations for future climates. *Glob. Chang. Biol.* **2006**, *12*, 2404–2416. [CrossRef]
47. Nigh, G. Mitigating the effects of climate change on lodgepole pine site height in British Columbia, Canada, with a transfer function. *For. Int. J. For. Res.* **2014**, *87*, 377–387. [CrossRef]
48. O'Neill, G.; Wang, T.; Ukrainetz, N.; Charleson, L.; McAuley, L.; Yanchuk, A.; Zedel, S. *A Proposed Climate-Based Seed Transfer System for British Columbia*; Technical Report 099; Ministry of Forests, Lands and Natural Resource Operations: Victoria, BC, Canada, 2017.
49. Peterson St-Laurent, G.; Hagerman, S.; Findlater, K.M.; Kozak, R. Public trust and knowledge in the context of emerging climate-adaptive forestry policies. *J. Environ. Manag.* **2019**, *242*, 474–486. [CrossRef]
50. Ledig, F.T.; Rehfeldt, G.E.; Sáenz-Romero, C.; Flores-López, C. Projections of suitable habitat for rare species under global warming scenarios. *Am. J. Bot.* **2010**, *97*, 970–987. [CrossRef]
51. Ukrainetz, N.K.; O'Neill, G.A.; Jaquish, B. Comparison of fixed and focal point seed transfer systems for reforestation and assisted migration: A case study for interior spruce in British Columbia. *Can. J. For. Res.* **2011**, *41*, 1452–1464. [CrossRef]
52. Warwell, M.V.; Rehfeldt, G.E.; Crookston, N.L. Modeling contemporary climate profiles of whitebark pine (Pinus albicaulis) and predicting responses to global warming. In *Proceedings of the Conference Whitebark Pine: A Pacific Coast Perspective*; Goheen, E., Ed.; R6-NR-FHP-2007-01; US Department of Agriculture, Forest Service: Ashland, OR, USA, 2007; pp. 139–142. Available online: https://www.fs.usda.gov/treesearch/pubs/33997 (accessed on 15 November 2020).
53. McLane, S.C.; Aitken, S.N. Whitebark pine (*Pinus albicaulis*) assisted migration potential: Testing establishment north of the species range. *Ecol. Appl.* **2012**, *22*. [CrossRef] [PubMed]
54. Seliger, B.J.; McGill, B.J.; Svenning, J.C.; Gill, J.L. Widespread underfilling of the potential ranges of North American trees. *J. Biogeogr.* **2020**. [CrossRef]
55. Statistics Canada. *Human Activity and the Environment: Forests in Canada 2017*; Catalogue No. 16 201 X; Statistics Canada: Ottawa, QC, Canada, 2018.
56. O'Neill, G.A.; Stoehr, M.; Jaquish, B. Quantifying safe seed transfer distance and impacts of tree breeding on adaptation. *For. Ecol. Manag.* **2014**, *328*, 122–130. [CrossRef]
57. Wang, T.; Hamann, A.; Spittlehouse, D.; Carroll, C. Locally downscaled and spatially customizable climate data for historical and future periods for North America. *PLoS ONE* **2016**, *11*, e0156720. [CrossRef] [PubMed]
58. O'Neill, G.A.; Hamann, A.; Wang, T. Accounting for population variation improves estimates of the impact of climate change on species' growth and distribution. *J. Appl. Ecol.* **2008**, *45*, 1040–1049. [CrossRef]
59. St. Clair, J.B.; Howe, G.T.; Kling, J.G. The 1912 Douglas-Fir heredity study: Long-term effects of climatic transfer distance on growth and survival. *J. For.* **2020**, *118*, 1–13. [CrossRef]
60. Rzedowski, J. *Vegetación de México*; Comisión Nacional para el Conocimiento y Uso de la Biodiversidad: Ciudad de México, México, 2006.
61. Sánchez-Velásquez, L.R.; Domínguez-Hernández, D.; Pineda-López, M.d.R.; Lara-González, R. Does Baccharis conferta shrub act as a nurse plant to the *Abies religiosa* seedling? *Open For. Sci. J.* **2011**, *4*, 67–70. [CrossRef]
62. Guzmán-Santiago, J.C.; Aguirre-Calderón, O.A.; Jiménez-Pérez, J.; Vargas-Larreta, B. Volume estimation of *Abies religiosa* (Kunth) Schltdl. & Cham. In different states Mexico. *Colomb. For.* **2020**, *23*, 99–113. [CrossRef]
63. Anderson, J.; Brower, L. Freeze-protection of overwintering monarch butterflies in Mexico: Critical role of the forest as a blanket and an umbrella. *Ecol. Entomol.* **1996**, *21*, 107–116. [CrossRef]

64. Montoya, E.; Guzmán-Plazola, R.A.; López-Mata, L. Fragmentation dynamics in an *Abies religiosa* forest of central Mexico. *Can. J. For. Res.* **2020**, *50*, 680–688. [CrossRef]
65. Carbajal-Navarro, A.; Navarro-Miranda, E.; Blanco, A.; Cruzado-Vargas, A.L.; Gómez-Pineda, E.; Zamora-Sánchez, C.; Pineda-García, F.; O'Neill, G.; Gómez-Romero, M.; Lindig-Cisneros, R. Ecological restoration of *Abies religiosa* forests using nurse plants and assisted migration in the Monarch Butterfly Biosphere Reserve, Mexico. *Front. Ecol. Evol.* **2019**, *7*, 421. [CrossRef]
66. Sáenz-Romero, C.; Rehfeldt, G.E.; Crookston, N.L.; Duval, P.; St-Amant, R.; Beaulieu, J.; Richardson, B.A. Spline models of contemporary, 2030, 2060 and 2090 climates for Mexico and their use in understanding climate-change impacts on the vegetation. *Clim. Chang.* **2010**, *102*, 595–623. [CrossRef]
67. Peterson, S.-L.G.; Hagerman, S.; Kozak, R. What risks matter? Public views about assisted migration and other climate-adaptive reforestation strategies. *Clim. Chang.* **2018**, *151*, 573–587. [CrossRef]
68. Barrows, C.W.; Ramirez, A.R.; Sweet, L.C.; Morelli, T.L.; Millar, C.I.; Frakes, N.; Rodgers, J.; Mahalovich, M.F. Validating climate-change refugia: Empirical bottom-up approaches to support management actions. *Front. Ecol. Environ.* **2020**, *18*, 298–306. [CrossRef]
69. Palmer, C.; Larson, B.M. Should we move the whitebark pine? Assisted migration, ethics and global environmental change. *Environ. Values* **2014**, *23*, 641–662. [CrossRef]
70. Pickles, B.J.; Twieg, B.D.; O'Neill, G.A.; Mohn, W.W.; Simard, S.W. Local adaptation in migrated interior Douglas-fir seedlings is mediated by ectomycorrhizas and other soil factors. *New Phytol.* **2015**, *207*, 858–871. [CrossRef]
71. Brower, L.P.; Williams, E.H.; Jaramillo-López, P.; Kust, D.R.; Slayback, D.A.; Ramírez, M.I. Butterfly mortality and salvage logging from the march 2016 storm in the Monarch Butterfly Biosphere Reserve in Mexico. *Am. Entomol.* **2017**, *63*, 151–164. [CrossRef]
72. Campbell, R.K. Use of phenology for examining provenance transfers in reforestation of Douglas-fir. *J. Appl. Ecol.* **1974**. [CrossRef]
73. Raymond, C.; Lindgren, D. Genetic flexibility-a model for determining the range of suitable environments for a seed source. *Silv. Genet.* **1990**, *39*, 112–120.
74. Mátyás, C. Modeling climate change effects with provenance test data. *Tree Physiol.* **1994**, *14*, 797–804. [CrossRef]
75. Carter, K. Provenance tests as indicators of growth response to climate change in 10 north temperate tree species. *Can. J. For. Res.* **1996**, *26*, 1089–1095. [CrossRef]
76. Leites, L.P.; Robinson, A.P.; Rehfeldt, G.E.; Marshall, J.D.; Crookston, N.L. Height-growth response to climatic changes differs among populations of Douglas-fir: A novel analysis of historic data. *Ecol. Appl.* **2012**, *22*, 154–165. [CrossRef] [PubMed]
77. Breed, M.F.; Stead, M.G.; Ottewell, K.M.; Gardner, M.G.; Lowe, A.J. Which provenance and where? Seed sourcing strategies for revegetation in a changing environment. *Conserv. Genet.* **2013**, *14*, 1–10. [CrossRef]
78. Sáenz-Romero, C.; Lamy, J.B.; Ducousso, A.; Musch, B.; Ehrenmann, F.; Delzon, S.; Cavers, S.; Chałupka, W.; Dağdaş, S.; Hansen, J.K. Adaptive and plastic responses of *Quercus petraea* populations to climate across Europe. *Glob. Chang. Biol.* **2017**, *23*, 2831–2847. [CrossRef] [PubMed]
79. Berlin, M.; Persson, T.; Jansson, G.; Haapanen, M.; Ruotsalainen, S.; Bärring, L.; Andersson Gull, B. Scots pine transfer effect models for growth and survival in Sweden and Finland. *Silv. Fenn.* **2016**, *50*, 1562. [CrossRef]
80. O'Neill, G.; Carlson, M.; Berger, V.; Ukrainetz, N. *Assisted Migration Adaptation Trial: Workplan*; BC Ministry of Forests, Lands and Natural Resource Operations: Vernon, BC, Canada, 2013.
81. Shutyaev, A.; Giertych, M. Height growth variation in a comprehensive Eurasian provenance experiment of (*Pinus sylvestris* L.). *Silv. Genet.* **1998**, *46*, 332–348.
82. Shutyaev, A.; Giertych, M. Genetic subdivisions of the range of Scots pine (*Pinus sylvestris* L.) based on a transcontinental provenance experiment. *Silv. Genet.* **2000**, *49*, 137–151.
83. Sáenz-Romero, C.; Kremer, A.; Nagy, L.; Újvári-Jármay, É.; Ducousso, A.; Kóczán-Horváth, A.; Hansen, J.K.; Mátyás, C. Common garden comparisons confirm inherited differences in sensitivity to climate change between forest tree species. *PeerJ* **2019**, *7*, e6213. [CrossRef]
84. Marris, E. Planting the forest of the future: While conservation biologists debate whether to move organisms threatened by the warming climate, one forester in British Columbia is already doing it. *Nature* **2009**, *459*, 906–908. [CrossRef]
85. Fu, Y.-B.; Clarke, G.P.Y.; Namkoong, G.; Yanchuk, A.D. Incomplete block designs for genetic testing: Statistical efficiencies of estimating family means. *Can. J. For. Res.* **1998**, *28*, 977–986. [CrossRef]
86. Yeaman, S.; Hodgins, K.A.; Lotterhos, K.E.; Suren, H.; Nadeau, S.; Degner, J.C.; Nurkowski, K.A.; Smets, P.; Wang, T.; Gray, L.K. Convergent local adaptation to climate in distantly related conifers. *Science* **2016**, *353*, 1431–1433. [CrossRef] [PubMed]
87. Sáenz-Romero, C.; Rehfeldt, G.E.; Duval, P.; Lindig-Cisneros, R.A. *Abies religiosa* habitat prediction in climatic change scenarios and implications for monarch butterfly conservation in Mexico. *For. Ecol. Manag.* **2012**, *275*, 98–106. [CrossRef]
88. Etterson, J.R.; Cornett, M.W.; White, M.A.; Kavajecz, L.C. Assisted migration across fixed seed zones detects adaptation lags in two major North American tree species. *Ecol. Appl.* **2020**, e02092. [CrossRef] [PubMed]
89. Prasad, A.; Pedlar, J.; Peters, M.; McKenney, D.; Iverson, L.; Matthews, S.; Adams, B. Combining US and Canadian forest inventories to assess habitat suitability and migration potential of 25 tree species under climate change. *Divers. Distrib.* **2020**, *26*, 1142–1159. [CrossRef]
90. Millar, C.I.; Stephenson, N.L.; Stephens, S.L. Climate change and forests of the future: Managing in the face of uncertainty. *Ecol. Appl.* **2007**, *17*, 2145–2151. [CrossRef]

91. Lin, B.B. Resilience in agriculture through crop diversification: Adaptive management for environmental change. *BioScience* **2011**, *61*, 183–193. [CrossRef]
92. Markowitz, H.M. Portafolio selection. *Finance* **1952**, *7*, 77–91.
93. Yanchuk, A.D. A quantitative framework for breeding and conservation of forest tree genetic resources in British Columbia. *Can. J. For. Res.* **2001**, *31*, 566–576. [CrossRef]
94. Ministry of Forests, Lands, Natural Resource Operations and Rural Development. *Chief Forester Standards for Seed Use 2020*; Ministry of Forests, Lands, Natural Resource Operations: Victoria, BC, Canada, 2020.
95. Sgrò, C.M.; Lowe, A.J.; Hoffmann, A.A. Building evolutionary resilience for conserving biodiversity under climate change. *Evol. Appl.* **2011**, *4*, 326–337. [CrossRef]
96. Crowe, K.A.; Parker, W.H. Using portfolio theory to guide reforestation and restoration under climate change scenarios. *Clim. Chang.* **2008**, *89*, 355–370. [CrossRef]
97. Ricciardi, A.; Simberloff, D. Assisted colonization is not a viable conservation strategy. *Trends Ecol. Evol.* **2009**, *24*, 248–253. [CrossRef] [PubMed]
98. Seabrook, L.; Mcalpine, C.A.; Bowen, M.E. Restore, repair or reinvent: Options for sustainable landscapes in a changing climate. *Landsc. Urban Plan.* **2011**, *100*, 407–410. [CrossRef]
99. Mueller, J.M.; Hellmann, J.J. An assessment of invasion risk from assisted migration. *Conserv. Biol.* **2008**, *22*, 562–567. [CrossRef] [PubMed]
100. Daehler, C.C. The taxonomic distribution of invasive angiosperm plants: Ecological insights and comparison to agricultural weeds. *Biol. Conserv.* **1998**, *84*, 167–180. [CrossRef]
101. Kowarik, I. Time lags in biological invasions with regard to the success and failure of alien species. In *Plant Invasions: General Aspects and Special Problems*; Pyšek, P., Rejmánek, K.P.M., Wade, M., Eds.; SPB Academic Publishing: Amsterdam, The Netherlands, 1995; pp. 15–38.
102. Binggeli, P. Time−Lag between Introduction, Establishment and Rapid Spread of Introduced Environmental Weeds. In Proceedings of the III International Weed Science Congress, MS no. 8. International Weed Science Society, Oxford, UK, 6–11 June 2001.
103. Crooks, J.A. Lag times and exotic species: The ecology and management of biological invasions in slow-motion1. *Ecoscience* **2005**, *12*, 316–329. [CrossRef]
104. Camarretta, N.; Harrison, P.A.; Bailey, T.; Davidson, N.; Lucieer, A.; Hunt, M.; Potts, B.M. Stability of species and provenance performance when translocated into different community assemblages. *Restor. Ecol.* **2020**, *28*, 447–458. [CrossRef]
105. Remke, M.J.; Hoang, T.; Kolb, T.; Gehring, C.; Johnson, N.C.; Bowker, M.A. Familiar soil conditions help *Pinus ponderosa* seedlings cope with warming and drying climate. *Restor. Ecol.* **2020**, *28*, S344–S354. [CrossRef]
106. García-Hernández, M.d.l.Á.; Toledo-Aceves, T. Is there potential in elevational assisted migration for the endangered *Magnolia vovidesii*? *J. Nat. Conserv.* **2020**, *53*, 125782. [CrossRef]
107. World Meteorological Organization. *WMO Provisional Statement on the State of the Global Climate in 2019*; WMO: Geneva, Switzerland, 2020; p. 34.
108. Guzmán-Aguilar, G.; Carbajal-Navarro, A.; Sáenz-Romero, C.; Herrerías-Diego, Y.; López-Toledo, L.; Blanco-García, A. Abies religiosa Seedling limitations for passive restoration practices at the monarch butterfly biosphere reserve in Mexico. *Front. Ecol.* **2020**, *13*. [CrossRef]
109. Romahn-Hernández, L.F.; Rodríguez-Trejo, D.A.; Villanueva-Morales, A.; Monterroso-Rivas, A.I.; Pérez-Hernández, M.J. Altitudinal range: Forest vigor factor and determinant in the natural regeneration of the sacred fir. *Entreciencias* **2020**, *8*, 1–17. [CrossRef]
110. MacLauchlan, M.; Buxton, K. *Overview of Forest Health Conditions in Southern British Columbia*; Ministry of Forests, Lands and Natural Resource Operations and Rural Development: Kamloops, BC, Canada, 2018.

forests

MDPI

Article

Species, Climatypes, Climate Change, and Forest Health: A Conversion of Science to Practice for Inland Northwest (USA) Forests

Gerald E. Rehfeldt [1,*], Marcus V. Warwell [2] and Robert A. Monserud [3]

1 Independent Researcher, 2424 D Street, Moscow, ID 83843, USA
2 U.S. Department of Agriculture, Forest Service, Rocky Mountain Research Station, 1221 South Main Street, Moscow, ID 83843, USA; marcus.v.warwell@usda.gov
3 Independent Researcher, 8210 SW Woods Creek Court, Portland, OR 97219, USA; robert.monserud@comcast.net
* Correspondence: jrehfeldt@gmail.com

Received: 6 September 2020; Accepted: 20 November 2020; Published: 24 November 2020

Abstract: *Research Highlights*: This paper integrates disparate research results pertaining to climate change impacts to 12 co-occurring forest tree species and their climatypes such that management options for the ecosystem as a whole become discernible. *Background and Objectives*: The ecosystem under analysis is the Thuja-Tsuga forest ecosystem, occupying ca. 121,500 km^2 in a largely mountainous setting in the interior northwest, USA. Our goal is to present land management options tied directly to climate-change in a straightforward framework for both the current and future generations. *Materials and Methods*: By merging synecological and genecological concepts in a climatic framework, we simplify complex interactions in a manner that relates directly to climate change impacts. Species and climatype distributions are redefined in terms of mean annual temperature and elevation of forested landscapes. *Results*: For each 2 °C increase in temperature, plant associations should shift upwards ca. 400 m, provided precipitation remains at or near contemporary levels, which, for this ecosystem, vary between 300 mm and 1450 mm. Management guidelines are developed for (a) selecting climatypes of the species suited to the climate at the leading edge of the migration front, (b) anticipating decline at the trailing edge, and (c) converting climatypes in areas where species should persist. *Conclusions*: Our results can provide robust strategies for adapting forest management to the effects of climate change, but their effectiveness is dependent on the implementation of global warming mitigation actions.

Keywords: climate-change impacts; ecosystem responses to climate; species distributions; climatype distributions; adaptive management

1. Introduction

Warming climates increase atmospheric energy which, in turn, increases weather variability such that extreme climatic events are reaching new thresholds [1]. Direct effects to the world's terrestrial and aquatic ecosystems have been pronounced and widespread. Species are coping by altering physiology, morphology, phenology, or genetics, leading toward an erosion of the fitness of organisms to their environments [2]. For forest trees, climate change has increased the incidence of insects and disease [3–6], advanced spring phenologies [7,8], and lengthened growing seasons [9]. As trees become less attuned physiologically to their environment, forest health deteriorates and mortality accrues. Restoring a semblance of balance between ecological distributions and the climate requires contraction at the trailing edge of the migration front [10–17] and immigration at the leading edge [16–18].

Vegetation change, however, is not necessarily an orderly process. Time lags exist between the forest renewal processes encompassing physiological decline, mortality, disturbance, migration, and colonization. These lags, known collectively as the migration lag [19], are exacerbated when climate change is rapid [20], thereby increasing the time required for a semblance of equilibrium to be approached between the vegetation and the new climate.

Climate change impacts to forests are complicated additionally by the genetic structures of tree species. Most broadly distributed species encounter more climate variation across their range than can be accommodated by the adaptability of individual trees [21,22]. In such species, adaptation to heterogeneous environments has been achieved through environmental selection operating on genetic variation to produce clines in fitness-related traits that parallel the environmental gradients [23]. In many temperate and boreal species, the clines reflect a trade-off between growth potential and cold hardiness [23–25]. This trade-off develops as selection favors adaptation to either specific environments or adaptability across a range of environments (specialization vs. flexibility) to produce clines that can be either steep when specialization is favored or gentle when adaptability is favored [26–29]. Although clines are continuous, populations occurring along a cline can be grouped into climatypes [30,31], the climatic ecotypes of Turesson [32], that are composed of adaptively similar populations. Climatypes, therefore, are merely classifications of continuous genetic variation. Species with steep clines (e.g., *Pseudotsuga menziesii* (Mirb.) Franco) will have many climatypes. Those with gentle clines (e.g., *Thuja plicata* Donn ex D. Donn) will have a few climatypes and those with flat clines (e.g., *Pinus monticola* Douglas ex D. Donn) will have none.

When clines and climatypes are prominent features of the adaptive genetic system, impacts from a change in climate can reverbate throughout a species distribution as well as at the trailing and leading edges. While species with flat clines will be affected by climate change primarily at the distribution margins, those with steep clines will lose fitness throughout their distribution [27]. Under the natural system, the maintenance of population fitness and forest health during change invokes the evolutionary process. Since evolutionary changes accumulate over generations, another lag, the adaptation lag [33], is superimposed on migration lags to further delay adjustment of the vegetation to a new climate. Many analyses (e.g., [34,35]) have shown, in fact, that the rate the climate is changing is far faster than natural systems are able to overcome the migration and adaptation lags. Fossil records [36] indicate that such conditions result in vegetation of diminished richness and diversity.

Land managers face the task of integrating these biological complexities with (a) rates of climate change that are rapid compared to the lifespan of trees [20], and (b) the uncertainty surrounding climate-change projections [37]. Adaptive planning, therefore, may seem opaque at best and unsolvable at worst [38,39]. While tools such as niche models [40], species distribution models [41], or models of species importance [42] can provide guidelines for management of individual species, with few exceptions (e.g., [43,44]) management of ecosystems as a whole has received less emphasis.

Our goal is to integrate disparate research results pertaining to climate-change impacts to species and their climatypes such that options for land managers can be viewed in a straightforward framework for both the current and future generations. The foundation for our analysis is the ecological classifications of Rexford Daubenmire [45,46] for the Thuja-Tsuga forest ecosystem of the inland northwestern USA and adjacent Canada (Figure 1). We deal with only the USA portion, ca. 121,500 km^2 where (a) the Daubenmire system is basic to contemporary forest planning, and (b) genecological analyses have been completed for most of the region's endemic conifers. We combine the genecology and synecology with their climatic descriptors to address our goals.

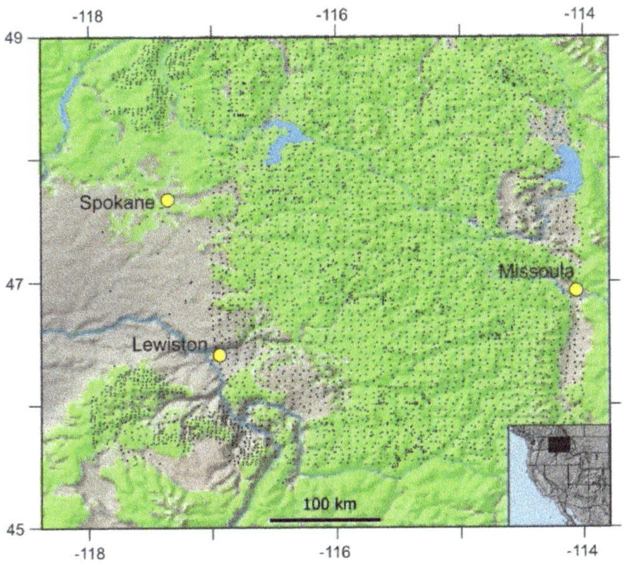

Figure 1. Forested lands of the Thuja-Tsuga ecosystem (green shading) of the inland Northwest USA with dots locating forest inventory's permanent sample plots.

2. Methods

The portion of Thuja-Tsuga ecosystem with which we deal, roughly −118 to −114 longitude by 45.5 to 49 latitude (Figure 1), is a mountainous home to 12 conifer species: *Pinus ponderosa* Laws., *Pseudotsuga menziesii*, *Larix occidentalis* Nutt., *Pinus contorta* Douglas ex Louden, *Abies grandis* (Douglas ex D. Donn) Lindl, *Pinus monticola*, *Thuja plicata*, *Tsuga heterophylla* (Raf.) Sarg., *Picea engelmannii* Parry ex Engelm., *Abies lasiocarpa* (Hook.) Nutt., *Tsuga mertensiana* (Bong.) Carrier, *Pinus albicaulis* Engelm. Forest inventory in this region is served by ca. 5100 ground plots (Figure 1) at elevations ranging from 300 to 1000 m. Modeled estimates of climate [47] for these ground plots show mean annual temperature of forested lands to vary between −2.8 °C and 12.5 °C and mean annual precipitation to range from 331 mm to 1455 mm. Climate estimates used in our analyses are the modeled climate estimates from the ground plots of Figure 1 that are derived from 1961–1990 normals and were used in the analyses of Rehfeldt et al. [47]. We consider these normals to be the reference period from which global warming and its effects are measured.

The ecosystem's conifers were arranged by Daubenmire [46] in an altitudinal ordination (Figure 2, top), which he visualized as paralleling a climatic gradient from warm and dry at low elevation to cool and moist at high. Subsequent analyses [47] designed to quantify the underlying climatic bases of Daubenmire's ordinaton found the lower altitudinal distribution limits to be most closely related to a summer dryness index and the upper limits to winter cold, measured by negative degree-days (Figure 2, bottom). These analyses found no other climate variables that could convert the two-dimensional ordination of Figure 2 into three-dimensional climate space. While Figure 2 (bottom) clearly demonstrates the climatic basis of Daubenmire's ordination, topoclimatic and edaphic effects along with successional trends that would separate, for instance, the occurrence *P. menziesii* and *T. heterophylla* are not represented.

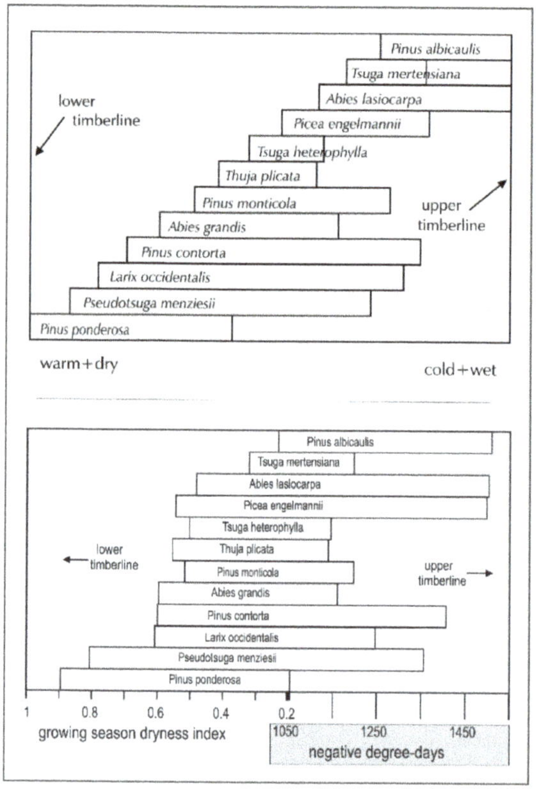

Figure 2. (**top**) Conceptual ordination of 12 species along an environmental gradient from warm-dry to cool-moist (left to right) as re-drawn from Daubenmire [46]. (**bottom**) Climatic quantification [47] of the conceptual ordination.

While Figure 2 relates ecological distributions directly to climate, the analytical results tend to be opaque from the practical viewpoint, that is, they lack an intuitive transparency tying Figure 2 directly to topographic landscapes. The ordination also lacks a direct correspondence to the variables used to report global temperature increases, that is, mean temperature.

To convert Figure 2 from its conceptual and climatic bases into a form more intuitive and, therefore, better suited for management, we used the relationships in Figure 3 to replace the bivariate climate gradient with a singled variable, mean annual temperature, which is closely related to both the summer dryness index (Figure 3, top left) and negative degree-days (Figure 3, top right), and incorporate an additional scale for elevation, which is also closely related to mean annual temperature (Figure 3, bottom). The conversions were made visually. Regression models were not fit to the scatter plots of Figure 3 largely because fitting a nonlinear function to these distributions seemed superfluous.

The addition of an approximate elevation to Daubenmire's ordination allowed species distributions to be subdivided into climatypes according to the research results of Table 1. The addition of mean annual temperature then allowed the ordination to be re-cast for an increase in temperature. We illustrate the effects of global warming with +2 °C and +4 °C temperature increases and assume no change in precipitation, that is, the reference period variation in precipitation does not change.

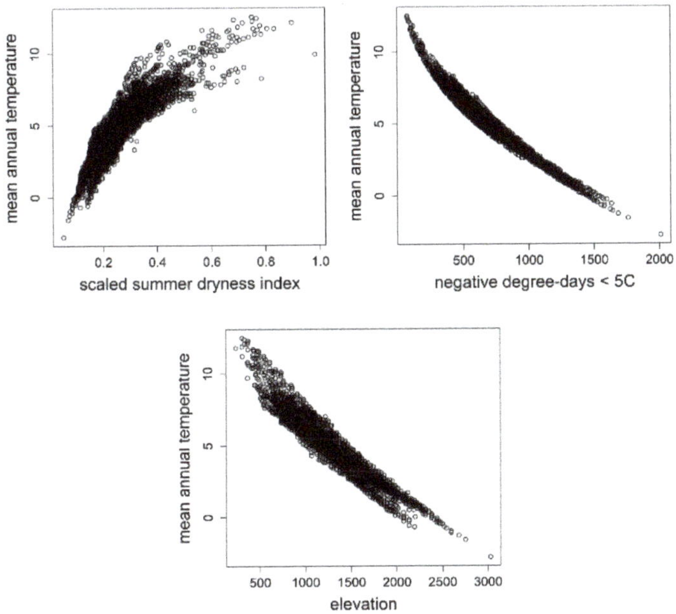

Figure 3. (**top left**) Scatter plot relating the summer dryness index to mean annual temperature. (**top right**) Scatter plot relating negative degree-days to mean annual temperature. (**bottom**) Scatter plot relating elevation to mean annual temperature. Climate data are modeled climate estimates for the ca. 5100 ground plots of Figure 1 that were used in the analyses of Rehfeldt et al. [47].

Table 1. Altitudinal breadth of climatypes for eight of the tree species in the Thuja-Tsuga ecosystem and their literature references. Genetic variation has not been studied in *A. grandis, T. heterophylla, T. mertensiana*, and *A. lasiocarpa*.

Species	Climatype Breadth (m Elevation)	Reference
Pseudotsuga menziesii	Nonlinear: ca. 240 at low elevation, 350 at mid elevation, 1 climatype for elevation > 2000	[48]
Pinus contorta	300	[49]
Pinus ponderosa	400	[50]
Picea engelmannii	420	[51]
Larix occidentalis	450	[52]
Pinus albicaulis	450	[53,54]
Thuja plicata	600	[55]
Pinus monticola	no detectable climatypes	[56]

Note: low elevation, ≈1000 m, mid elevation, ≈1500 m, high elevation, >2000 m.

3. Results

Figure 4 shows Daubenmire's ordination (Figure 2, top) in relation to mean annual temperature and elevation. Species distributions are plotted linearly for mean annual temperature, which forces the scale for elevation to be nonlinear (see Figure 3). This nonlinear relationship has its greatest impact for low elevations. Having a scale showing elevation provides a link to topographic maps and, therefore, forested landscapes.

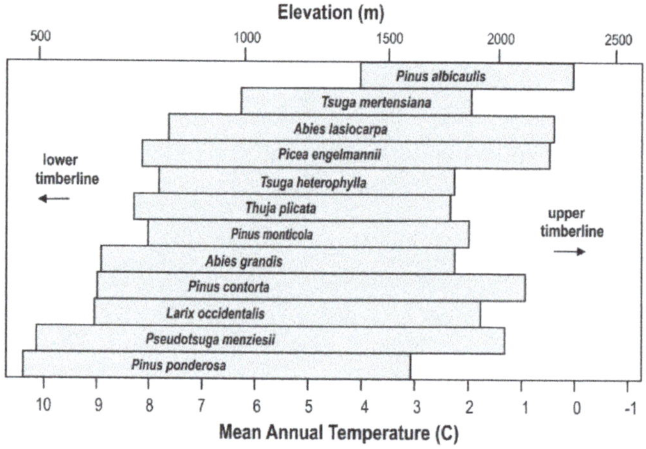

Figure 4. Ordination of 12 conifer species of Figure 2 according to mean annual temperature and elevation for USA's inland northwest.

The scale for elevation in Figure 4 also provides a link to the genecology studies that have been conducted for 8 of the 12 species. Since most of these studies were completed before climate data became readily available, their results related genetic differentiation to elevation, widely assumed to be a surrogate for the climate. This genecological research produced the data in Table 1 that we use to segment species distributions into climatypes (Figure 5, top). The ordination in Figure 5 has been drawn using a linear scale for elevation, thereby forcing the scale for mean annual temperature to be nonlinear. Elevation is truncated at 2600 m because the land base >2600 m is miniscule in this region (see Figure 3, bottom graph). In this figure, the limits of distribution of climatypes in the top panel were drawn in reference to mean annual temperatures during the 1961–1990 reference period. The climatypes, therefore, reflect the clines in genetic variability that were produced by climatic selection in the generations leading up to and including the reference period. Differences among species in the number of climatypes and their climatic breadth reflects the ecological amplitude of the species as well as the steepness of the genetic cline.

With species distributions now depicted as functions of temperature at both the leading and trailing edges (Figure 5, top), plotting the impact of a change in temperature becomes straightforward. We chose +2 °C (Figure 5, middle) and +4 °C (Figure 5, bottom) to illustrate responses. The middle panel of this figure shows that a warming of +2 °C would shift the climate currently occupied by these species and their climatypes upwards by ca. 400 m at the trailing edge and ca. 350 m at the leading edge. A warming of +4 °C would shift the altitudinal distribution of the contemporary habitat an additional 400 m at the trailing edge, but immigration at the leading edge would most likely be limited by both the elevation of the ridgeline and substrate, that is, rock. Figure 5 is based on temperature alone and the assumption, therefore, that precipitation will remains similar to that of the reference period (see discussion).

All panels in Figure 5 depict a quasi-equilibrium state at which this group of 12 species and their climatypes inhabit climates in which their fitness is suitable for optimal growth and productivity. For the middle and bottom panels, this would occur when the species and their climatypes somehow occur in the new location of the temperature regime they inhabited in the reference period. While there are no means for projecting a timeline for this to happen, the process by which the top panel in Figure 5 is converted to the middle and bottom panels is known.

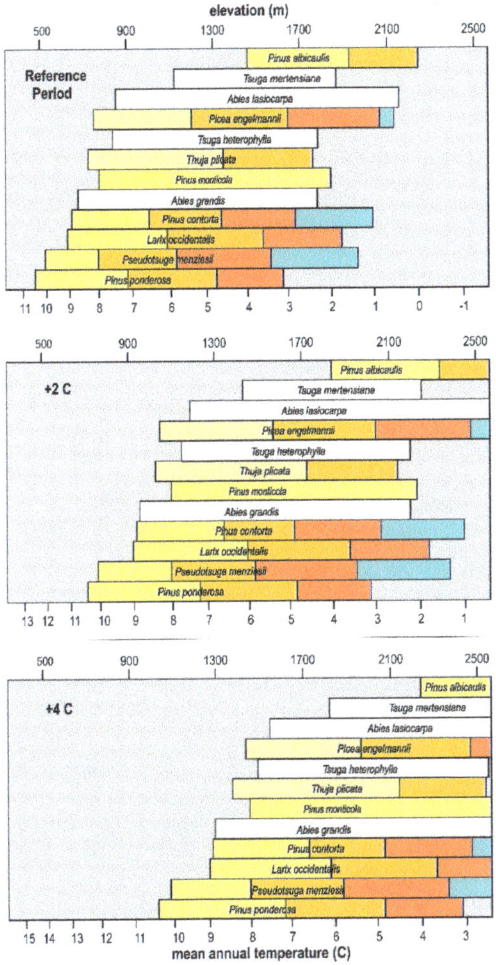

Figure 5. Ordination of Figure 4 with species distributions segmented into climatypes (colors) and plotted with a linear elevation scale for mean temperatures of the 1961–1990 reference period (**top**), and for an increase in temperature of 2 °C (**middle**) and 4 °C (**bottom**). The number and breath of climatypes for each species corresponds to Table 1. White bars code species whose genetic structures have not been studied.

Provenance tests in forestry are climate-change experiments (see [57]) in which populations or their wind-pollinated descendants are moved from their home climate to the climate of a planting site. In a re-analysis of *P. contorta* provenances, temperature-driven response functions were produced for 10 climatypes to describe growth in relation to site temperature and temperature change [58]. In Figure 6, the functions for four climatypes have been re-cast to show the innately controlled plastic responses of the climatypes. The top graph shows growth in relation to a temperature change. The location of a climatype in the reference period is denoted by a dot at a climate change of zero. The bottom panel shows growth responses in relation to mean annual temperature of forest sites. In re-casting this figure, the x-axis was converted from negative degree-days to mean annual temperature in the same way that Figure 4 was produced using Figure 3.

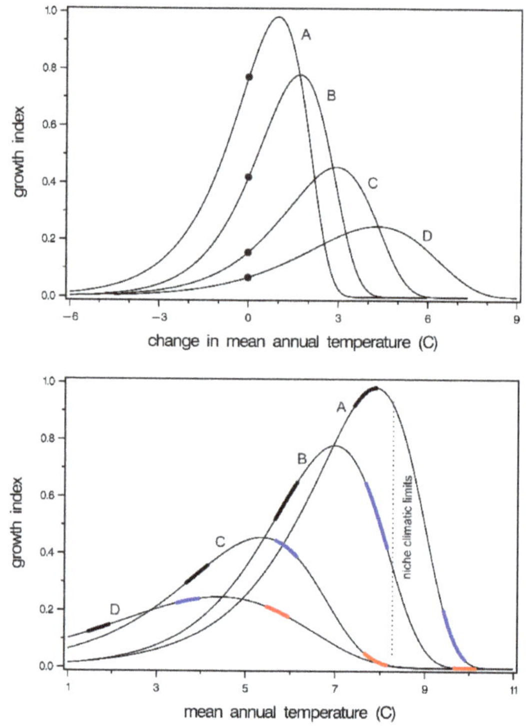

Figure 6. (**top**) Response of four climatyes (lines) of *Pinus contorta* to change in temperature. Dots show the growth expected at the home provenance having, therefore, a climate change of zero. Lines labeled A to D depict climatypes from lowest to highest elevation. (**bottom**) Response of the same four climatypes to mean annual temperature. Thickened portion of response functions shows growth of climatypes at the temperature of their home (black), +2 °C (blue), and +4 °C (red). Niche climate limits are based on ground plot data. Both panels are redrawn from Rehfeldt et al. [55].

The analyses on which Figure 6 is derived illustrated two principles: (1) climatypes occupy climates colder than where they grow best, presumably by competitive exclusion by faster growing trees, and (2) the temperature difference between the site where they occur and where they grow best increases as the climate where they occur becomes colder. Thus, the top panel of Figure 6 shows that, at a climate change of zero, the climate is colder than where the climatypes grow the best. Both panels show that the discrepancy increases as the inhabited climate becomes colder. The bottom graph shows further that the populations within a climatype occur where they can be competitively exclusive while still being competitively excluded from where they grow the best. The growth potential-temperature cline, incidentally, would be represented by a line connecting the apex of each curve in the bottom graph.

The panels together illustrate clearly that climatypes will respond differently to change. Climatypes from cold climates will respond favorable to warming climates for a much longer period than climatypes from warm climates. This differential response is clearly illustrated in the bottom graph where the thickened portion of the response functions colored black shows growth in the reference period climate, that is, the temperature of their home provenance. The portion colored blue shows their growth as if their home temperature had increased by 2 °C, and the portion colored red shows their growth at +4 °C. With warming of +2 °C, two climatypes would be in decline. At +4 °C, all climatypes would be beyond their ecological optimum and, therefore, be in decline. The climatic limits of the species as indicated by the populations of Figure 1 is shown by the hashed line for reference.

The graphs of Figure 6 pertain to *P. contorta*, but the same principles are evident in *Pinus banksiana* [59], *Pinus sylvestris* [35], and three species of Eurasian *Larix* [60]. It is also well known for many additional species that populations inhabit climates cooler than where they grow the best [61–63]. It is likely, therefore, the adaptive system illustrated in Figure 6 is a widespread feature of species exhibiting clines controlled by the trade-off between maximizing growth to survive competition and requirements for cold hardiness [22,34]. The resolution of this compromise will produce species-specific response curves (e.g., Figure 6) with the number of climatypes and the breadth of the functions being dependent on the steepness of the cline.

4. Discussion

Our analysis concerns the Tsuga-Tsuga ecosystem, a part of the temperate coniferous forests of western North America. The ecosystem is home to 12 conifers that occur in an altitudinal sequence, with species distributions limited by the winter cold at upper elevations and by a balance between temperature and precipitation at lower elevations. We assume that precipitation during global warming will remain unchanged from the 1961–1990 reference period and illustrate climate-change impacts at +2 °C and +4 °C. Resulting impacts would be direct and immediate at high elevation thermic limits but would depend also on precipitation anomalies at the low elevations. Because our projections are specific to the ecological associations of and climatic relationships in the Tsuga-Tsuga ecosystem, they pertain solely to this ecosystem.

We describe potential impacts on the current generation of forests and project them for future generations as a response to temperature increase. For each 2 °C warming, climatic niche of 12 species are projected to shift upwards in elevation by about 400 m. Yet, we treat the species assemblage as reacting to the same climate variables when, in fact, plant associations did not respond as a unit to the Pleistocene climate changes [36,64]. It is well established that species respond individualistically to changes in climate. Our position is that if the reference period and future climates are analogous, then species and climatypes adapted to reference period climates will be adapted to the same climate in the future.

When future climates have no reference period analogs, the outcome of competitive interactions is expected to change, potentially altering the climatic limits of species distributions [64,65]. Climate analyses suggest, moreover, that no-analog climates should become more frequent in USA's inland northwest as global warming ensues (Figure 8 in Ref. [65], Figure 16 in Ref. [66]). In fact, analyses of North American biomes suggest that no-analog climates late in the current century may result in the Thuja-Tsuga ecosystem becoming of increasing prominence in Canada, while USA's interior northwest becomes suited to vegetation occurring today in coastal biomes (Figure 8 in Ref. [67]). As no-analog climates increase in frequency, projections lose relevance. Consequently, our results pertain to forests of the current generation and to those of the near future but become increasingly less germane for more distant generations. We limit our assessment of climate change impacts to +2 °C and +4 °C for the same reason.

Plant distributions are controlled by precipitation and the interaction of precipitation with temperature as well as by temperature alone (Figure 2). We over-simplify climatic limits of distribution by ordering species along a temperature gradient, thereby forcing projections for the future to rely solely on change in temperature. The relevance of Figure 5, therefore, depends on the confidence one has in temperature increasing while precipitation remains as it was during the reference period. With regard to temperature, general circulation models tend to be consistent and robust [68] because energy and heat are reasonably well understood by atmospheric physicists. Yet air masses, clouds, and precipitation are more poorly understood. For our window, a compilation of 42 models projects future precipitation to increase between 0% and 10%, but the increase is small compared to the variability among predictions. In fact, only 80% of the models agree that precipitation will increase while the remainder predict a decrease [68]. For the tree species of the Thuja-Tsuga ecosystem, small increases in precipitation might act to reduce impacts somewhat at the trailing edge, which tends to be under the control of a dryness index, but should, at the same time, have little effect at the leading edge, which is a thermal limit controlled by winter cold (see Figure 2). Large changes in precipitation, despite their low probability, could hasten development of climates without reference period analogs [65,69], thereby leading to a realignment of plant associations [65,69].

Climatypes are designed by humans classifying continuous genetic variation. They are constructed by grouping populations along a cline such that the groups contain populations that are similar in terms of adaptation. In forest trees, the most prominent adaptive clines parallel temperature gradients. Since temperature gradients tend to be continuous, so too are the clines. Because continuous systems are difficult to convert into managerial guidelines, discrete units, that is, climatypes, are defined by classifying the continuous genetic variation. Yet, climatypes do not exist as discrete ecological entities. The climatypes of Figure 4 are one set of innumerable classifications, all of which could be valid. While the breadth of the climatype is important, its placement along the gradient is not.

An upward shift of the Thuja-Tsuga communities is directly relevant to management. Obviously, both lower and upper timberlines will rise. This means that the forest area will decrease because species occupy smaller and smaller areas as they move upwards. Upward migration is also limited by the altitude of ridge lines and the substrate typical of high elevations. In addition, migration in our ecosystem will affect diversity, as the species-rich forests of the mid-elevations invade the subalpine zone. As warming temperatures relax limits governing distributions [70], upward migration of species has enriched European forest communities [17,71–73]. Yet, as an increase in diversity at upper elevations increases management options, a concomitant loss of diversity and fewer options occur at lower elevations.

Even though species may react to climate change individualistically in the long term, Figures 2–5 illustrate clearly that the process by which vegetation adjusts to change forces an interdependence on species' responses. Each species has a leading edge and a trailing edge. Therefore, in species-rich assemblages, the trailing edge of one species is often the leading edge of another. Consequently, rates of immigration of one species often are limited by rates of demise and mortality in other species. Rates are dependent also by the occurrence of disturbance to prepare sites for colonization. Because lags in these ecological processes are inevitable, the conclusion reached by many [74] is that maintaining forest health, modifying composition, and maintaining productivity while providing amenities will require intensive management.

4.1. Managing the Current Generation

The two panels of Figure 6 illustrate together the fate of existing populations. Because climatypes occupy climates colder than where they grow the best, most climatypes initially gain in growth as the climate begins to warm. As warming continues, growth of those climatypes near the species' trailing edge soon reaches the ecological optimum and then rapidly declines. Climatypes in cool climates near the species' leading edge can absorb warming temperatures for a prolonged period before growth declines. Yet, with warming of +4 °C, all climatypes would be beyond their ecological optimum, with some projected to be in climates beyond the current climatic limits of the species. Decline, however, may be abrupt in cases associated with extreme weather, insect outbreaks, and disease, or decline may be prolonged when arising from physiological deterioration.

As decline in growth ensues, trees become less fit for their environment. Forest health deteriorates as productivity declines. Poorly adapted plants, moreover, are under physiological stress, thereby becoming vulnerable to insects and diseases [2]. An appropriate stratagem for managers to maintain healthy, productive forests would be to anticipate the period of decline, recognize that decline will occur at different times for disparate climatypes, harvest, and convert to a suitable species and/or climatype.

4.2. Considerations for Future Generations

The greatest challenge to the maintenance of healthy and productive forests in the next and subsequent generations is in having species and climatypes growing at the new location of the climate to which that climatype is genetically fit. The goal, therefore would be to maintain a semblance of equilibrium between plant distributions and climate. While widespread planting programs, including assisted migration when appropriate, would address the problem directly, the immense scope of a reforestation program required for making the conversions implied in Figure 5 leads toward the conclusion that the ecological consequences of poorly adapted forests cannot be averted everywhere by means of artificial regeneration. Natural reproduction, therefore, likely will be a common default approach for forest renewal.

As a default, natural regeneration carries liabilities except for those lands inhabited today by climatypes also suited for the future. The uncertainties surrounding seed production, seed dispersal, and seedling establishment make timely immigration problematic, especially when the distance from seed source to reforestation site is large [19]. Likewise, conversion of one climatype to another via natural reproduction and natural selection could require several generations, with each generation needing to survive long enough to reproduce [34,35]. These contingencies promote migration and adaptation lags that will act to impede timely immigration.

Regardless of the forest renewal strategy, however, it seems clear that management objectives concentrating on forest health and productivity are likely to supplant objectives focusing on conservation, recreation, watersheds, grazing, or wildlife, which commonly drive current forest renewal plans [75].

5. Conclusions

Like innumerable similar analyses, our synthesis projects climate-change impacts to the Thuja-Tsuga ecosystem to be extraordinarily large. For each +2 °C change in temperature, plant distributions are projected to rise ca. 400 m. Changes of this magnitude would result in widespread disruption of the balance between plant distributions and climate. We provide a template for maintaining forest health and productivity during these times of change. While our modeled results provide a foundation for adaptive planning, models and their predicted effects are guidelines. Models cannot replace the intimate knowledge of professionals dealing with species—environmental interactions. Model output needs to be integrated with edaphic, microsite, and topographical effects; seral stages; and the unique silvical characteristics of the species. Models can guide managers toward robust strategies that will be useful for dealing with a changing climate, but their effectiveness is dependent on the implementation of global warming mitigation actions. Efficacy of guidelines such as ours will be voided by continual warming, particularly if rapid, as (1) management actions become negated before reaching fruition, (2) evolution of novel climates accelerates to alter competitive relationships among species, and (3) human survival itself takes on an overriding priority.

Author Contributions: Conceptualization, G.E.R., M.V.W., and R.A.M. Methodology, G.E.R., M.V.W., and R.A.M. Writing, G.E.R. and M.V.W. Visualization, G.E.R., M.V.W., and R.A.M. Funding acquisition, M.V.W. All authors have read and agreed to the published version of the manuscript.

Funding: Partial funding was provided by the Special Technology Development Program of the USDA Forest Service R1-2020-01 and R1-2020-02.

Acknowledgments: We appreciate critical reviews provided by Ned Klopfenstein, James Worrall, Mary Williams, and Andrew Bower.

Conflicts of Interest: The authors declare no conflict of interest.

References

1. Schwartz, J. The Siberian Heat Wave? Yes, Climate Change Was a Big Factor. *New York Times*, 15 July 2020.
2. Scheffers, B.R.; De Meester, L.; Bridge, T.C.L.; Hoffmann, A.A.; Pandolfi, J.M.; Corlett, R.T.; Butchart, S.H.M.; Pearce-kelly, P.; Kovacs, K.M.; Dudgeon, D.; et al. The broad footprint of climate change from genes to biomes to people. *Science* **2016**, *354*, aaf7671. [CrossRef] [PubMed]

3. Sáenz-Romero, C.; Mendoza-Maya, E.; Gómez-Pineda, E.; Blanco-García, A.; Endara-Agramont, A.R.; Lindig-Cisneros, R.; López-Upton, J.; Trejo-Ramírez, O.; Wehenkel, C.; Cibrián-Tovar, D.; et al. Recent evidence of Mexican temperate forest decline and the need for ex situ conservation, assisted migration, and translocation of species ensembles as adaptive management to face projected climatic change impacts in a megadiverse country. *Can. J. For. Res.* **2020**, *50*, 843–854. [CrossRef]
4. Bebber, D.P.; Ramotowski, M.A.T.; Gurr, S.J. Crop pests and pathogens move polewards in a warming world. *Nat. Clim. Chang.* **2013**, *3*, 985–988. [CrossRef]
5. Anderegg, W.R.L.; Kane, J.M.; Anderegg, L.D.L. Consequences of widespread tree mortality triggered by drought and temperature stress. *Nat. Clim. Chang.* **2013**, *3*, 30–36. [CrossRef]
6. Weed, A.S.; Ayres, M.P.; Hicke, J.A. Consequences of climate change for biotic disturbances in North American forests. *Ecol. Monogr.* **2013**, *83*, 441–470. [CrossRef]
7. Parmesan, C. Ecological and evolutionary responses to recent climate change. *Annu. Rev. Ecol. Evol. Syst.* **2006**, *37*, 637–669. [CrossRef]
8. Root, T.L.; Price, J.T.; Hall, K.R.; Schneider, S.H.; Rosenzweig, C.; Pounds, J.A. Fingerprints of global warming on wild animals and plants. *Nature* **2003**, *421*, 57–60. [CrossRef]
9. Reyes-Fox, M.; Steltzer, H.; Trlica, M.J.; McMaster, G.S.; Andales, A.A.; LeCain, D.R.; Morgan, J.A. Elevated CO2 further lengthens growing season under warming conditions. *Nature* **2014**, *510*, 259–262. [CrossRef]
10. Allen, C.D.; Macalady, A.K.; Chenchouni, H.; Bachelet, D.; McDowell, N.; Vennetier, M.; Kitzberger, T.; Rigling, A.; Breshears, D.D.; Hogg, E.H.; et al. A global overview of drought and heat-induced tree mortality reveals emerging climate change risks for forests. *For. Ecol. Manag.* **2010**, *259*, 660–684. [CrossRef]
11. Breshears, D.D.; Cobb, N.S.; Rich, P.M.; Price, K.P.; Allen, C.D.; Balice, R.G.; Romme, W.H.; Kastens, J.H.; Floyd, M.L.; Belnap, J.; et al. Regional vegetation die-off in response to global-change-type drought. *Proc. Natl. Acad. Sci. USA* **2005**, *102*, 15144–15148. [CrossRef]
12. Gitlin, A.R.; Sthultz, C.M.; Bowker, M.A.; Stumpf, S.; Paxton, K.L.; Kennedy, K.; Muñoz, A.; Bailey, J.K.; Whitham, T.G. Mortality gradients within and among dominant plant populations as barometers of ecosystem change during extreme drought. *Conserv. Biol.* **2006**, *20*, 1477–1486. [CrossRef] [PubMed]
13. Jump, A.S.; Hunt, J.M.; Peñuelas, J. Rapid climate change-related growth decline at the southern range edge of *Fagus sylvatica*. *Glob. Chang. Biol.* **2006**, *12*, 2163–2174. [CrossRef]
14. Michaelian, M.; Hogg, E.H.; Hall, R.J.; Arsenault, E. Massive mortality of aspen following severe drought along the southern edge of the Canadian boreal forest. *Glob. Chang. Biol.* **2011**, *17*, 2084–2094. [CrossRef]
15. Shaw, J.D.; Steed, B.E.; DeBlander, L.T. Forest Inventory and Analysis (FIA) annual inventory answers the question: What is happening to pinyon-juniper woodlands? *J. For.* **2005**, *103*, 280–285.
16. Woodall, C.W.; Oswalt, C.M.; Westfall, J.A.; Perry, C.H.; Nelson, M.D.; Finley, A.O. An indicator of tree migration in forests of the eastern United States. *For. Ecol. Manag.* **2009**, *257*, 1434–1444. [CrossRef]
17. Lamprecht, A.; Semenchuk, P.R.; Steinbauer, K.; Winkler, M.; Pauli, H. Climate change leads to accelerated transformation of high-elevation vegetation in the central Alps. *New Phytol.* **2018**, *220*, 447–459. [CrossRef]
18. Tchebakova, N.M.; Parfenova, E.I.; Soja, A.J. Potential climate-induced vegetation change in Siberia in the 21st century. In *Environmental Change in Siberia*; Baltzer, H., Ed.; Springer: Dordrecht, The Netherlands, 2010; pp. 67–82.
19. Davis, M.B. Lags in vegetation response to greenhouse warming. *Clim. Chang.* **1989**, *15*, 75–82. [CrossRef]
20. Davis, M.B.; Shaw, R.G. Range Shifts and Adaptive Responses to Quaternary Climate Change. *Science* **2001**, *292*, 673–679. [CrossRef]
21. Aitken, S.N.; Yeaman, S.; Holliday, J.A.; Wang, T.; Curtis-McLane, S. Adaptation, migration or extirpation: Climate change outcomes for tree populations. *Evol. Appl.* **2008**, *1*, 95–111. [CrossRef]
22. Rehfeldt, G.E.; Wykoff, W.R.; Ying, C.C. Physiologic Plasticity, Evolution, and Impacts of a Changing Climate on *Pinus contorta*. *Clim. Chang.* **2001**, *50*, 355–376. [CrossRef]
23. Morgenstern, E.K. *Geographic Variation in Forest Trees*; UBC Press: Vancouver, BC, Canada, 1996.
24. Grime, J.P. *Plant Strategies and Vegetation Processes*; John Wiley and Sons: Hoboken, NJ, USA, 1979.
25. Loehle, C. Height Growth Rate Tradeoffs Determine Northern and Southern Range Limits for Trees. *J. Biogeogr.* **1998**, *25*, 735–742. [CrossRef]
26. Rehfeldt, G.E. Microevolution of conifers in the northern Rocky Mountains: A view from common gardens. In Proceedings of the Eighth North American Forest Biology Workshop, Logan, UT, USA, 30 July–1 August 1984; pp. 32–46.

27. Rehfeldt, G.E. Evolutionary genetics, the biological species and the ecology of the interior cedar-hemlock forests. In Proceedings of the Symposium: Interior Cedar-Hemlock-White Pine Forests: Ecology and Management, Spokane, WA, USA, 2–4 March 1993; Baumgartner, D.M., Lotan, J.E., Tonn, J.R., Eds.; Department of Natural Resources, Washington State University: Pullman, WA, USA, 1994; pp. 91–100.
28. Leites, L.P.; Rehfeldt, G.E.; Steiner, K.C. Adaptation to climate in five eastern North America broadleaf deciduous species: Growth clines and evidence of the growth-cold tolerance trade-off. *Perspect. Plant Ecol. Evol. Syst.* **2019**, *37*, 64–72. [CrossRef]
29. Rehfeldt, G.E.; Leites, L.P.; St. Clair, J.B.; Jaquish, B.C.; Sáenz-Romero, C.; López-Upton, J.; Joyce, D.G. Comparative genetic responses to climate in the varieties of *Pinus ponderosa* and *Pseudotsuga menziesii*: Clines in growth potential. *For. Ecol. Manag.* **2014**, *324*, 138–146. [CrossRef]
30. Tchebakova, N.M.; Rehfeldt G.E.; Parfenova, E.I. Impacts of climate change on the distribution of *Larix* spp. and *Pinus sylvestris* and their climatypes in Siberia. *Mitig. Adapt. Strateg. Glob. Chang.* **2006**, *11*, 861–882. [CrossRef]
31. Rehfeldt, G.E.; Tchebakova, N.M.; Parfenova, E.I.; Pandalai, S.G. Genetic responses to climate and climate-change in conifers of the temperate and boreal forests. *Recent Res. Dev. Genet Breed* **2004**, *1*, 113–130.
32. Turesson, G. The genotypical response of the plant species to the Habitat. *Hereditas* **1922**, *3*, 211–350. [CrossRef]
33. Mátyás, C. Adaptation lag: A general feature of natural populations. (Invited lecture). In Proceedings of the WFGA-IUFRO Symposium, Olympia, WA, USA, 20–25 August 1990; No. 2.226, 10p.
34. Rehfeldt, G.E.; Ying, C.C.; Spittlehouse, D.L.; Hamilton, D.A. Genetic responses to climate in *Pinus contorta*: Niche breadth, climate change, and reforestation. *Ecol. Monogr.* **1999**, *69*, 375–407. [CrossRef]
35. Rehfeldt, G.E.; Tchebakova, N.M.; Parfenova, Y.I.; Wykoff, W.R.; Kuzmina, N.A.; Milyutin, L.I. Intraspecific responses to climate in *Pinus sylvestris*. *Glob. Chang. Biol.* **2002**, *8*, 912–929. [CrossRef]
36. Betancourt, J.L. Late Quaternary Biogeography of the Colorado Plateau. In *Packrat Middens, the Last 40,000 Years of Biotic Change*; Betancourt, J.L., Van Devender, T.R., Martin, P.S., Eds.; Univ. Arizona Press: Tucson, AZ, USA, 1990; pp. 259–302.
37. Millar, C.I.; Stephenson, N.L.; Stephens, S.L. Climate change and forests of the future: Managing in the face of uncertainty. *Ecol. Appl.* **2007**, *17*, 2145–2151. [CrossRef]
38. Keenan, R.J. Climate change impacts and adaptation in forest management: A review. *Ann. For. Sci.* **2015**, *72*, 145–167. [CrossRef]
39. Yousefpour, R.; Jacobsen, J.B.; Thorsen, B.J.; Meilby, H.; Hanewinkel, M.; Oehler, K. A review of decision-making approaches to handle uncertainty and risk in adaptive forest management under climate change. *Ann. For. Sci.* **2012**, *69*, 1–15. [CrossRef]
40. Wang, T.; Campbell, E.M.; O'Neill, G.A.; Aitken, S.N. Projecting future distributions of ecosystem climate niches: Uncertainties and management applications. *For. Ecol. Manag.* **2012**, *279*, 128–140. [CrossRef]
41. Elith, J.; Leathwick, J.R. Species distribution models: Ecological explanation and prediction across space and time. *Annu. Rev. Ecol. Evol. Syst.* **2009**, *40*, 677–697. [CrossRef]
42. Iverson, L.; Prasad, A.; Matthews, S. Modeling potential climate change impacts on the trees of the northeastern United States. *Mitig. Adapt. Strateg. Glob. Chang.* **2008**, *13*, 487–516. [CrossRef]
43. Gómez-Pineda, E.; Sáenz-Romero, C.; Ortega-Rodríguez, J.M.; Blanco-García, A.; Madrigal-Sánchez, X.; Lindig-Cisneros, R.; Lopez-Toledo, L.; Pedraza-Santos, M.E.; Rehfeldt, G.E. Suitable climatic habitat changes for Mexican conifers along altitudinal gradients under climatic change scenarios. *Ecol. Appl.* **2020**, *30*, e02041. [CrossRef]
44. Aubry, C.; Devine, W.; Shoal, R.; Bower, A.; Miller, J.; Maggiulli, N. *Climate Change and Forest Biodiversity: A Vulnerability Assessment and Action Plan for National Forests in Western Washington*; US Department of Agriculture, Forest Service, Pacific Northwest Region: Olympia, WA, USA, 2011.
45. Daubenmire, R.; Daubenmire, J.B. *Forest Vegetation of Eastern Washington and Northern Idaho*; Technical Bulletin 6; Washington State University: Pullman, WA, USA, 1968; p. 104.
46. Daubenmire, R. Vegetation: Identification of typal communities. *Science* **1966**, *151*, 291–298. [CrossRef]
47. Rehfeldt, G.E.; Ferguson, D.E.; Crookston, N.L. Quantifying the abundance of co-occurring conifers along inland northwest (USA) climate gradients. *Ecology* **2008**, *89*, 2127–2139. [CrossRef]
48. Rehfeldt, G.E. Ecological adaptations in Douglas-Fir (*Pseudotsuga menziesii* var. *glauca*): A Synthesis. *For. Ecol. Manag.* **1989**, *28*, 203–215. [CrossRef]

49. Rehfeldt, G.E. Ecological Genetics of *Pinus contorta*: A synthesis. *Silvae Genet.* **1988**, *37*, 131–135.
50. Rehfeldt, G.E. A model of genetic variation for *Pinus ponderosa* in the Inland Northwest (USA): Applications in gene resource management. *Can. J. For. Res.* **1991**, *21*, 1491–1500. [CrossRef]
51. Rehfeldt, G.E. Adaptation of *Picea engelmannii* populations to the heterogeneous environments of the Intermountain West. *Can. J. Bot.* **1994**, *72*, 1197–1208. [CrossRef]
52. Rehfeldt, G.E. Genetic variation, climate models and the ecological genetics of *Larix occidentalis*. *For. Ecol. Manag.* **1995**, *78*, 21–37. [CrossRef]
53. Warwell, M.V.; Shaw, R.G. Climate-related genetic variation in a threatened tree species, *Pinus albicaulis*. *Am. J. Bot.* **2017**, *104*, 1205–1218. [CrossRef] [PubMed]
54. Bower, A.D.; Aitken, S.N. Ecological genetics and seed transfer guidelines for *Pinus albicaulis* (Pinaceae). *Am. J. Bot.* **2008**, *95*, 66–76. [CrossRef]
55. Rehfeldt, G.E. Genetic structure of western red cedar populations in the Interior West. *Can. J. For. Res.* **1994**, *24*, 670–680. [CrossRef]
56. Rehfeldt, G.E. Ecotypic differentiation in populations of *Pinus monticola* in north Idaho—Myth or Reality? *Am. Nat.* **1979**, *114*, 627–636. [CrossRef]
57. Mátyás, C. Modeling climate change effects with provenance test data. *Tree Physiol.* **1994**, *14*, 797–804. [CrossRef]
58. Rehfeldt, G.E.; Leites, L.P.; Joyce, D.G.; Weiskittel, A.R. Role of population genetics in guiding ecological responses to climate. *Glob. Chang. Biol.* **2018**, *24*, 858–868. [CrossRef]
59. Mátýas, C.; Yeatman, C.W. Effect of geographical transfer on growth and survival of jack pine (*Pinus banksiana* Lamb.) populations. *Silvae Genet.* **1992**, *41*, 370–376.
60. Rehfeldt, G.E.; Tchebakova, N.M.; Milyutin, L.I.; Parfenova, E.I.; Wykoff, W.R.; Kouzmina, N.A. Assessing population responses to climate in *Pinus sylvestris* and *Larix* spp. of Eurasia with climate-Transfer models. *Eurasian J. For. Res.* **2003**, *6*, 83–98.
61. Namkoong, G. Nonoptimality of local races. In Proceedings of the 10th Southern Forest Tree Improvement Conference, College Station, TX, USA, 17–19 June 1969; pp. 149–153.
62. Schmidtling, R.C. Use of provenance tests to predict response to climatic change: Loblolly pine and Norway spruce. *Tree Physiol.* **1994**, *14*, 805–817. [CrossRef] [PubMed]
63. Carter, K.K. Provenance tests as Indicators of growth response to climate change in 10 north temperate tree species. *Can. J. For. Res.* **1996**, *26*, 1089–1095. [CrossRef]
64. Jackson, S.T.; Overpeck, J.T. Responses of plant populations and communities to environmental changes of the late Quaternary. *Paleobiology* **2000**, *26*, 194–220. [CrossRef]
65. Williams, J.W.; Jackson, S.T. Novel climates, no-analog communities, and ecological surprises. *Paleoecology* **2007**, *5*, 475–482. [CrossRef]
66. Rehfeldt, G.E.; Crookston, N.L.; Warwell, M.V.; Evans, J.S. Empirical analyses of plant-climate relationships for the western United States. *Int. J. Plant Sci.* **2006**, *167*, 1123–1150. [CrossRef]
67. Rehfeldt, G.E.; Crookston, N.L.; Saenz-Romero, C.; Campbell, E.M. North American vegetation model for land-use planning in a changing climate: A solution to large classification problems. *Ecol. Appl.* **2012**, *22*, 119–141. [CrossRef]
68. Kirtman, B.; Power, S.B.; Adedoyin, A.J.; Boer, G.J.; Bojariu, R.; Camilloni, I.; Doblas-Reyes, F.; Fiore, A.M.; Kimoto, M.; Meehl, G.; et al. Near-term climate change: Projections and predictability. In *Climate Change 2013: The Physical Science Basis. The Working Group I Contribution to the Fifth Assessment Report of the Intergovernmental Panel on Climate Change*; Cambridge University Press: Cambridge, UK; Cambridge, MA, USA, 2013; Volume 9781107057, pp. 953–1028.
69. Jackson, S.T.; Betancourt, J.L.; Booth, R.K.; Gray, S.T. Ecology and the ratchet of events: Climate variability, niche dimensions, and species distributions. *Proc. Natl. Acad. Sci. USA* **2009**, *106*, 19685–19692. [CrossRef]
70. Körner, C. *Alpine Treelines. Functional Ecology of the Global High Elevation Tree Limits*, 1st ed.; Springer: Basel, Switzerland, 2012; ISBN 978-3-0348-0395-3.
71. Matteodo, M.; Wipf, S.; Stöckli, V.; Rixen, C.; Vittoz, P. Elevation gradient of successful plant traits for colonizing alpine summits under climate change. *Environ. Res. Lett.* **2013**, *8*, 024043. [CrossRef]
72. Stöckli, V.; Wipf, S.; Nilsson, C.; Rixen, C. Using historical plant surveys to track biodiversity on mountain summits. *Plant Ecol. Divers.* **2012**, *4*, 415–425. [CrossRef]
73. Grabherr, G.; Gottfried, M.; Pauli, H. Climate effects on mountain plants. *Nature* **1994**, *369*, 448. [CrossRef]

74. Rehfeldt, G.E.; Jaquish, B.C.; Sáenz-Romero, C.; Joyce, D.G.; Leites, L.P.; St. Clair, J.B.; López-Upton, J. Comparative genetic responses to climate in the varieties of *Pinus ponderosa* and *Pseudotsuga menziesii*: Reforestation. *For. Ecol. Manag.* **2014**, *324*, 147–157. [CrossRef]
75. Keiter, R.; Boling, T.; Milkman, L. Legal perspectives on ecosystem management: Legitimizing a new federal land management policy. In *Ecological Stewardship, A Common Reference for Ecosystem Management, Volume III*; Sexton, W.T., Malk, A.J., Szaro, R.C., Johnson, N.C., Eds.; Elsevier Science Ltd.: Kidlington, UK, 1999; pp. 9–41.

Publisher's Note: MDPI stays neutral with regard to jurisdictional claims in published maps and institutional affiliations.

MDPI
St. Alban-Anlage 66
4052 Basel
Switzerland
Tel. +41 61 683 77 34
Fax +41 61 302 89 18
www.mdpi.com

Forests Editorial Office
E-mail: forests@mdpi.com
www.mdpi.com/journal/forests

www.ingramcontent.com/pod-product-compliance
Lightning Source LLC
Chambersburg PA
CBHW061503210926
51850CB00008B/1810
9783036522814